Sinking Chicago

In the series ***Urban Life, Landscape, and Policy***, edited by David Stradling, Larry Bennett, and Davarian Baldwin. Founding editor, Zane L. Miller.

ALSO IN THIS SERIES:

Pamela Wilcox, Francis T. Cullen, and Ben Feldmeyer, *Communities and Crime: An Enduring American Challenge*

J. Mark Souther, *Believing in Cleveland: Managing Decline in "The Best Location in the Nation"*

Nathanael Lauster, *The Death and Life of the Single-Family House: Lessons from Vancouver on Building a Livable City*

Aaron Cowan, *A Nice Place to Visit: Tourism and Urban Revitalization in the Postwar Rustbelt*

Carolyn Gallaher, *The Politics of Staying Put: Condo Conversion and Tenant Right-to-Buy in Washington, DC*

Evrick Brown and Timothy Shortell, eds., *Walking in Cities: Quotidian Mobility as Urban Theory, Method, and Practice*

Michael T. Maly and Heather Dalmage, *Vanishing Eden: White Construction of Memory, Meaning, and Identity in a Racially Changing City*

Harold L. Platt, *Building the Urban Environment: Visions of the Organic City in the United States, Europe, and Latin America*

Kristin M. Szylvian, *The Mutual Housing Experiment: New Deal Communities for the Urban Middle Class*

Kathryn Wilson, *Ethnic Renewal in Philadelphia's Chinatown: Space, Place, and Struggle*

Robert Gioielli, *Environmental Activism and the Urban Crisis: Baltimore, St. Louis, Chicago*

Robert B. Fairbanks, *The War on Slums in the Southwest: Public Housing and Slum Clearance in Texas, Arizona, and New Mexico, 1936–1965*

Carlton Wade Basmajian, *Atlanta Unbound: Enabling Sprawl through Policy and Planning*

Scott Larson, *"Building Like Moses with Jacobs in Mind": Contemporary Planning in New York City*

Gary Rivlin, *Fire on the Prairie: Harold Washington, Chicago Politics, and the Roots of the Obama Presidency*

William Issel, *Church and State in the City: Catholics and Politics in Twentieth-Century San Francisco*

Jerome Hodos, *Second Cities: Globalization and Local Politics in Manchester and Philadelphia*

Julia L. Foulkes, *To the City: Urban Photographs of the New Deal*

William Issel, *For Both Cross and Flag: Catholic Action, Anti-Catholicism, and National Security Politics in World War II San Francisco*

Lisa Hoffman, *Patriotic Professionalism in Urban China: Fostering Talent*

John D. Fairfield, *The Public and Its Possibilities: Triumphs and Tragedies in the American City*

Andrew Hurley, *Beyond Preservation: Using Public History to Revitalize Inner Cities*

HAROLD L. PLATT

Sinking Chicago

CLIMATE CHANGE AND THE REMAKING OF A FLOOD-PRONE ENVIRONMENT

TEMPLE UNIVERSITY PRESS
Philadelphia • Rome • Tokyo

TEMPLE UNIVERSITY PRESS
Philadelphia, Pennsylvania 19122
www.temple.edu/tempress

Published 2018

Library of Congress Cataloging-in-Publication Data

Names: Platt, Harold L., author.
Title: Sinking Chicago : climate change and the remaking of a flood-prone environment / Harold L. Platt.
Description: Philadelphia : Temple University Press, 2018. | Series: Urban life, landscape, and policy | Includes bibliographical references and index.
Identifiers: LCCN 2017050476 (print) | LCCN 2017059175 (ebook) | ISBN 9781439915509 (E-book) | ISBN 9781439915486 (cloth) | ISBN 9781439915493 (pbk.)
Subjects: LCSH: Climatic changes—Illinois—Chicago. | Floodplains—Illinois—Chicago. | Flood control—Illinois—Chicago. | Water quality—Illinois—Chicago. | Chicago (Ill.)—Environmental conditions.
Classification: LCC QC903.2.U6 (ebook) | LCC QC903.2.U6 P53 2018 (print) | DDC 363.6/10977311—dc23
LC record available at https://lccn.loc.gov/2017050476

Contents

List of Illustrations

PART I

The Dry Years

Introduction

Cities, Sprawl, and Climate Change

The *Entrepôt* City

Chicago is a city built on water. Its *raison d'état* as a place of Euro-American settlement was a river, which linked the Great Lakes to the Mississippi River. Among the earliest explorers of the water routes into the interior of the mid-continent were Father Jacques Marquette and Louis Jolliet. In 1673, Native Americans guided their canoes along the western shore of Lake Michigan to the Green Bay peninsula, the Fox and Wisconsin Rivers, and down the Mississippi to within 450 miles/724 kilometers (km) of the Gulf Coast. Turning back and paddling upstream, they took a shortcut into the Illinois and Des Plaines Rivers and then crossed a marshy portage into the Chicago River. Returning in the winter of 1674, Marquette became too sick to continue a second trip farther downstream. He kept a journal while encamped about 6 miles/9.7 km from the lake, near the site of the portage where the south branch of the Chicago River petered out. He spent four months there, waiting for his health to return and the water highway's ice to melt.

The following spring, the Jesuit priest recorded the first account of a flood. "On the 29th [of March]," he writes, "the waters rose so high that we had barely time to decamp as fast as possible, putting our goods in the trees, and trying to sleep on a hillock. The water gained on us nearly all-night, but there was a slight freeze, and the water fell a little, while we were near our packages." Two days later, he resumed his mission but could no longer find the portage—later called Mud Lake—to the Des Plaines River. On the contrary, he observes in his journal, "the very high lands alone are not flooded. At the place where we are, the water has risen more than 12 feet [3.7 meters]. This is where we began our portage eighteen months ago." After a few more days' delay to let the "dangerous rapids" subside, he was able to proceed across the temporary lake without having to carry his heavily loaded canoe overland.[1]

Looking closer at Marquette's account reveals the fundamental relationship between Chicagoans and their flood-prone environment. The explorer was taken by surprise, because the water rose without the usual forewarning of a rainstorm. Instead, a spring thaw was enough to cause a surge of surface runoff. Marquette did not need a geologist to inform him that the source of the "dangerous rapids" had to be the Des Plaines River. The terrain was so flat that runoff could flow in any direction. He could see that the water was flowing in the opposite direction from its normal course from east to west. What he could not see was the impermeable layer of clay just below the surface soils. The melting snow quickly saturated this ground, forming innumerable tiny rivulets that fed small brooks and streams, leading into slightly larger creeks that poured into the river, which overflowed its banks and turned the portage into a lake. Encompassing the entire Chicago area, this topography is what I call a prairie wetland (see Figure I.1).[2]

The priest's journey also sheds light on humans' dependence on the waterways as an essential economic asset. Before the coming of the railroad 175 years later, moving people and goods on water was cheaper, faster, and more efficient than any alternative means of travel on land. Although Marquette's primary mission was saving the souls of the Indians, a related goal was establishing trade with them. During his layover in Chicago, he exchanged tobacco for beaver and ox skins, corn, and pumpkins. As he learned, even floods could be beneficial when they submerged overland routes and eased transportation across them. At the same time, they could result in lost property and lives as a consequence of human failure. In his case, building a cabin on low-lying ground near the river put his material possessions at risk. In other words, floods are natural; flood *damages* are caused by humans.[3]

Another piece of vital information about the site of Chicago can be extracted from Father Marquette's journal: it was a land of plenty. He describes not only the Indians' willingness to trade their surplus of food supplies but also an abundance of wildlife to supplement his diet, including deer, partridges, and pigeons. He was at the center of what William Cronon has aptly called *Nature's Metropolis*. In this milestone study of environmental history, he reunites the urban and the rural in symbiotic relationship. Cronon shows how the city and the surrounding hinterland grew in tandem, each shaping the development of the other. To harvest nature's bounty, Chicago remained a frontier outpost until 1825, when the completion of the Erie Canal opened a water route between New York City and the Great Lakes. Then, the Euro-Americans poured in, pushing out the Native Americans. A new landscape of farms, city homes, grain warehouses, factories, and shipping docks was erected on this prairie wetland. Chicago became an *entrepôt*, the place where the water stopped and the Great West began.[4]

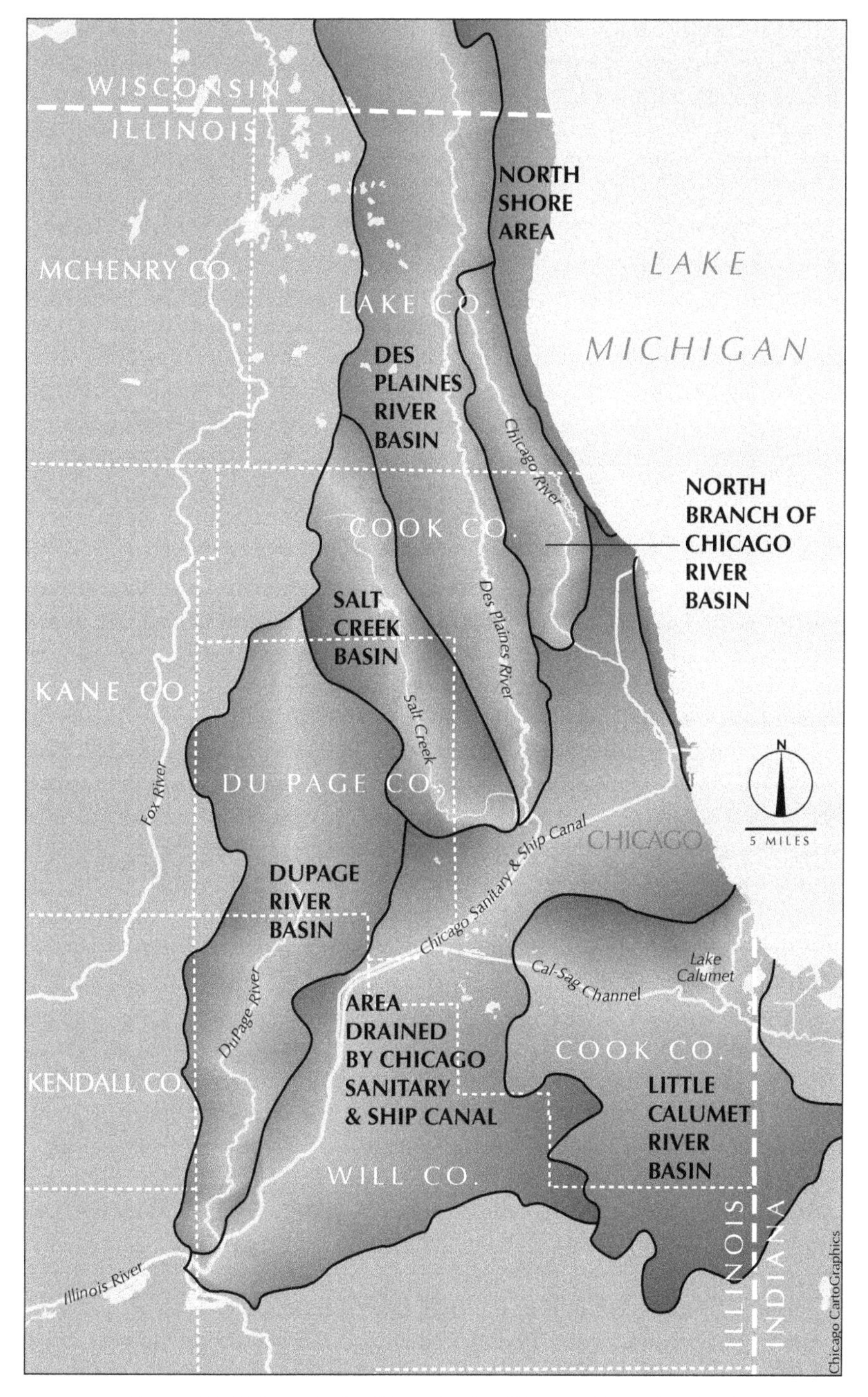

Figure I.1. Chicago's Six Watersheds (Chicago CartoGraphics)

In contrast to Cronon's book, the aim of this environmental history of Chicago is to reunite the water and the land. This book is the first study of the effects of long-term climate change on an American city. *Sinking Chicago* is based on 130 years of weather records, which show a pattern of dry years from 1885 to 1945, followed by a still ongoing wet period of more-frequent and more-severe rainstorms.[5] Chicago is flood-prone because it was built on flat land with poor drainage and an impervious layer of clay just below the surface. This book examines how residents, politicians, and policy makers have grappled with the dual problem of flood control and water quality. In building a city on top of a prairie wetland, people forgot the environmental connection between land use and water management. As they paved over the land, they turned the rain into ever-more-damaging runoff surges of storm water that sank their city and contaminated their drinking supplies with raw sewage.

The Exploding City

The Great Chicago Fire of 1871 is the place to begin this story about the ways in which climate change taught a city how to adapt to living in a flood-prone environment. Costing at least three hundred lives, this disaster was the price paid for losing sight of the links between water and land. Without water, a small accidental fire ended up burning down an entire city built of wood. Starting on the near West Side, flickering embers soon blew across the river, torching the main water-pumping station in the central business district. The conflagration consumed eighteen thousand buildings and made a hundred thousand people, or one-third of the population, homeless. Ironically, a new water tower and pumping station on the North Side were among the few surviving structures. Despite the expensive construction of this public utility, the firemen were left defenseless against the spreading flames. They were completely overwhelmed by another regular weather pattern of the midwestern prairie that makes Chicago known as the "Windy City." From spring to winter, currents of hot air from the southwest prevail, sometimes with gusts of more than 30 miles/48 km per hour. On that October 8, the winds were especially strong, following a three-month drought that left the built environment tinder dry. The fire ultimately burned itself out two days later, as it reached the outer edges of the city.[6]

This human-made disaster exposed the shortcomings of building an urban environment under a regime in which "property ruled."[7] A political culture of privatism adhered to the Jeffersonian ideal that the best government was the least government. In contemporary terms, Chicago's civic leaders were advocates of public policies that delegated most city-building functions to the private sector. Paving streets and sidewalks; supplying water; laying roads to the suburbs; providing streetlights, gas service, and public transit; and even protecting against fires were all consigned to it. Although the waterworks was the one significant exception, it, too, was run like a private business by a city council devoted to maximizing the profits of the municipal corporation. Their siphoning of most of the utility's revenue to pay for other things kept its expansion lagging far behind the explosive growth of the city. Since the coming of the railroad in the late 1840s, Chicago had become the fastest growing urban center in the Western world, roughly doubling in population every decade to the end of the nineteenth century.[8]

This surge of newcomers caused another human-made disaster: a cholera epidemic from contaminated supplies of drinking water. Chicagoans were forced to take collective charge of the quality of drinking supplies and the removal of disease-breeding effluents. Beginning in the 1850s, the separation between land use and water management became increasingly institutionalized in private associations of real-estate brokers and government departments of public health and public works. As urban growth paved over a flood-prone environment on a metropolitan scale during the Gilded Age, the ability of individuals to protect their property against the effects of climate change diminished, while the power of government increased at each level of the federal system. Landowners and other residents looked to their policy makers to solve Chicago's dual problem of water quality and flood control.[9]

The low-lying peninsula of marshland formed by the lake and the river had become the geographical center of the city. The main street, appropriately named Market Street, paralleled the Main Branch of the river, just behind the docks and warehouses lining it and stretching along its South and North Branches. As the aspiring metropolis swelled with newcomers, the river became its greatest economic asset and a physical barrier to growth. For ships, it was a safe haven from the dangerous storms that swept across the lake. But for people, the river blocked movement, because every bridge thrown across it impeded the crucial flow of traffic in and out of this harbor. For the remainder of the nineteenth century, Chicago expanded mostly to the south, while development of the North and the West Sides was stunted in comparison.[10]

The river at flood stage also played a dual role of natural advantage and human-made disaster. The spring thaw recorded by Marquette that submerged the portage and created a wave sweeping down the South Branch had the effect of flushing the sedimentation on the riverbed into the lake. Local sanitary officials considered these "freshets" a blessing that saved the city the cost of dredging the river's bottom. When this annual event coincided with heavy rains, however, an ice-chunk-laden wave could barrel down the river with destructive force. On March 12, 1849, just such an extreme weather event tore through the downtown area, smashing all the bridges and piling the ships in the harbor on the riverbanks like a stack of cordwood.[11]

Thus, Chicago was forced to confront its first civic crisis: water management. Following the flood, a cholera epidemic from contaminated drinking supplies turned Chicago into the unhealthiest city in the country. Its citizens had to figure out a way to bring good water in and move the bad out. The problems of water supply and wastewater removal are inseparable. People recognized that they would have to follow the lead of other urban centers, such as Philadelphia and Boston. In 1850, they decided to construct a waterworks that pumped fresh water from the lake through a distribution network of mains and pipes. At the same time, bringing more water into a flood-prone environment compounded all the problems of draining the city's muddy, vermin-infested streets and alleys. As the growth of Chicago was on the verge of turning an imagined metropolis into a real one, the city's future depended on establishing a new relationship with water.[12]

In the wake of the great fire, questions of land-use regulation paralleled public debate on restoring the water supply and the fire hydrants. Massive protests erupted when the city council attempted to enact a building code requiring a "fire-proof" city of bricks, which would have significantly raised the cost of housing compared to wood construction. Chicago entered a general state of class warfare that lasted a quarter of a century. Punctuating the relentless struggle between capital and labor were major pitched battles, including the first national (railroad) strike of 1877, the eight-hour-work-day movement leading to the Haymarket riots of 1886, and the Pullman strike of 1894, when the army occupied the city. The extent to which the environmental conditions of the working class contributed to these events cannot be considered separately from other causes of discontent.[13]

In part, human capacities to build decent housing and public utilities were simply overwhelmed in this exploding city. Yet, a political culture of privatism and a regime of corrupt ward bosses helped widen the gap of environmental inequality between the rich and the poor. The betrayal of the working class on behalf of the slumlords contributed to horrendous rates of infant mortality as well as disproportionate levels of sickness and death among the masses of the city's inhabitants. By the time Jane Addams and Ellen Gates Starr moved into Hull-House in one of the notorious river wards in 1889, for example, reformers estimated that the city had forty thousand illegal outhouses. In plain sight and smell of the health inspectors, they could exist only with the tacit approval of city hall. By then, new waves of water-borne diseases had again reached epidemic levels. This public-health crisis would force the political establishment to undertake another round of heroic public-works projects to regain control of nature.[14]

The Metropolitan City

Chicago always had suburbs, but the Great Fire of 1871 gave them an instant infusion of inhabitants and money. While many of these refuges became residents of shantytowns on burnt-over land, others were absorbed into the city's spokelike strings of railroad suburbs. A brief look at three different outlying communities linked with a river watershed serves to introduce the history of the metropolitan idea in Chicago.[15]

About 12 miles/19 km north of the city center, a narrow ridge of high land between the North Branch of the Chicago River and the lakefront was a privileged patch of dry, tree-shaded land. One of the first suburbs on this "North Shore" was Evanston, a religious community formed in the 1850s around its school, Northwestern University. An equal distance west of downtown on the Des Plaines River—a thirty-minute train ride—was another high point, Riverside. Just after the Civil War, speculators hired New York City's famous landscape architects Frederick Law Olmsted and Calvert Vaux to turn it into a residential enclave for the well-to-do.[16] To the southeast the same distance along the lakefront was the working-class settlement of Ainsworth, sited at the mouth of the Grand Calumet River and the bottleneck of rail lines coming around the bottom of the lake. Already a thriving, second *entrepôt*, it would be renamed South Chicago after the fire, when it became a favorite site of heavy industry looking to relocate a safe distance from the river congestion and pollution problems in the city's center.

From frontier days, the settlement of the Chicago area can be characterized as villages arising on islands of high land surrounded by a treacherous "sea of mud." What is called "urban sprawl" has never been a one-way flow from the core to the periphery: it is more complicated, because people have always been moving into and within the center at the same time. Consider that each household deserting the city for the suburbs leaves behind a vacant dwelling for another to occupy. Since the coming of the railroad and the telegraph in the late 1840s, Chicago has been going through a simultaneous process of centralization and decentralization. Its earliest suburban settlements served as outposts of future development. "Many were near area rivers and lakes," historian Ann Durkin Keating observes. "Camp meetings, golf courses, picnic groves, beer gardens, amusement parks, cemeteries, and religious and public institutions served as the centers of these settlements."[17]

Before the laying of iron rails on raised embankments, overland travel by horse-drawn, wheeled vehicles across Chicago's prairie wetland impeded the growth of residential, commuter suburbs. The situation of wagons getting inescapably stuck was so common that it earned a special term, "being 'slewed' [sloughed] on the prairie," and gave birth to a local hero, the "mud pilot." These young men learned that the best way to navigate across such wet, featureless landscapes was to strike out across them where no one had previously laid any tracks. "Their beaten paths are deceptive," a veteran of these times remembered, "for once broken through, this sea of mud will prove bottomless. Your wagon will go down to the axles, and you will soon see the way strewn with such wreaks."[18]

In searching for a protected place for their settlement, the decisive factor for the Methodist founders of Evanston was a suburban location with railroad access to the city. In selecting a 379-acre/153-hectare (ha) farm on the North Shore's lakefront in 1853, they purchased a soggy, albeit beautiful site on which to build an oasis of religious values and moral order. The key to their creation of a self-image as a sanctified village was the "4-mile/6 km limit," which banned the sale of alcohol within this island community. The sudden influx of refugees from the great conflagration only strengthened the resolve of its leaders to enforce prohibition and to maintain an identity separate from the city. During the next two decades, Evanston became the national headquarters of the Woman's Christian Temperance Union and the neighborhood center of twenty-two churches.[19]

Already the richest suburb, with three thousand residents listed in the 1870 census, Evanston underwent a 50 percent growth spurt following the Chicago fire. One of the children born there because of his parents' being burned out of their home on the North Side was Maurice Webster. His mother told him that she and his father had been able to load up a wagon and get the family to her brother's house. They lived there for several months while a new house and barn were being built nearby. The "strangers," the local newspaper boasted, "are learning, now, how much cleaner and more quiet and pleasant is our village than the city, and how much better it is for their children." Apparently, the Websters, with six boys to raise, were convinced.[20]

However, the vast majority of Chicagoans were reluctant to abandon their urban amenities for the country life. Besides all the modern conveniences of indoor plumbing, affluent city dwellers could enjoy artificial gas lighting in their homes and an array of street improvements, including pavement, lighting, drainage, and, perhaps, public transit services. In contrast, suburbanites depended on backyard wells and privies, kerosene lamps, jerry-built "plank roads," and, perhaps, a staff of servants to do the endless chores of house cleaning. Only a few prosperous railroad suburbs, including Evanston, could afford urban infrastructure, such as a waterworks or a sewer system; most had to suffer annual damaging floods and perennial seas of mud. To overcome the shortcomings of life in the country, real-estate developers after the Civil War promoted a plan to create an exclusive residential nature park with all the urban amenities included.[21]

The speculators selected a patch of exceptionally high land called Riverside Park that had recently been crossed by a railroad heading west from the center. Now that the rigors of overland travel had been replaced by a half-hour ride in comfort, the destination point of the mud pilots was ripe for redevelopment from a seasonal racetrack and picnic grounds into a pastoral retreat for the city's best men. Reflecting the counterintuitive nature of Chicago's topography, this heavily wooded 1,900-acre/769 ha piece of land surrounds the Des Plaines River as it snakes back and forth across the prairie. To the southeast lay Mud Lake (present-day Lyons and Stickney), but here the river stayed in its banks most of the time. Gaining control of the farmland in exchange for improvement company stock, the company directors hired the designers of New York's Central Park, Olmsted and Vaux, to draw a master plan of homes nestled in a picturesque setting of the river and the woods. Using credit secured by the land, they built an impressive waterworks tower, an arcade building, a first-class hotel near the train station, a coal-gas manufacturing plant, lighted streets and sidewalks paved with concrete and stones, and designer houses with a $5,000-minimum price tag.[22]

Initial enthusiasm for the success of Riverside was overwhelming. Like the planners, more and more city dwellers believed that "'the most attractive, the most refined and the most soundly wholesome forms of domestic life' were to be found in residential suburbs."[23] Beginning in 1868, its developers sponsored weekend excursions for elite members of society to showcase their property to these potential buyers. Journalists along for the (free) ride aboard luxury Pullman club cars outdid themselves describing the modern technology and the natural beauty of this residential suburb-in-the-making. On one such junket in May 1869, the guests were greeted by Olmsted and Vaux, who became their tour guides. Making the hard sell, Olmsted "claimed that Riverside presented advantages over any park ever laid out in the United States. The land was both fertile and high; the Des Plaines River was unsurpassed for its general beauty; the forest was abundant; and there was no good reason why Riverside Park should not be made the most attractive place on the continent."[24]

In fact, good reasons shattered the suburban vision of its designers and developers. Bad timing undermined the financial house of cards holding up the improvement company. The great fire drained off its construction workers, who suffered no loss of commuter fares or work time in choosing to accept jobs close at hand. In 1873, a national depression halted home sales, while the company's unpaid bills for its infrastructure projects piled up. In a desperate attempt to stay financially afloat, its directors issued more and more stocks and bonds until the unbearable load of this paper debt collapsed. In July 1874, investors sued them for fraud, building a four-year case in equity court that set a new record with 12,000 pages of documents, including the 103-page decision of the trial judge. Although the bursting of this land bubble did not diminish the natural beauty of a riverine landscape, it had the effect of lowering the suburb's rate of growth, land values, and social status.[25]

If the timing of the Chicago fire helped cause the crash of Riverside, it sparked a sustained boom in Ainsworth/South Chicago. The city's second port was well prepared to meet the needs of industry and trade, the result of a public-private partnership. Before the fire destroyed a large proportion of the piers, lumberyards, and factories on the Chicago River, the U.S. Army

Corps of Engineers had begun dredging the entrance and mouth of the Grand Calumet River as an alternative safe harbor at the bottom of Lake Michigan. A private company had installed docks, shipyards, and railroad sidings and was selling riverfront sites for storage facilities, including grain, lumber, coal, and iron ore. These warehouses, in turn, fed heavy industries, such as steel mills, machine shops, oil refineries, fertilizer plants, and breweries, that generated lots of toxic wastewater.[26]

The growth of the population of South Chicago is a good case study of the simultaneous process of decentralization and centralization in a metropolitan region. When industry moved to the suburbs, the construction of housing for its workers followed close behind. These people had to walk to work, because they could not afford to take streetcars or commuter trains. After the fire of 1871, workers followed their jobs from the core to the periphery. At the same time, newcomers were moving to Chicago in expectation of finding jobs in its expanding metropolitan economy. Thousands of these immigrants and transplanted residents alike would land in a South Chicago neighborhood called "Bush." Choked by smoke and crammed with wood-frame two-flat apartment buildings and cottages, it was totally lacking in urban infrastructure and utility services. "Bush," local historians Dominic A. Pacyga and Ellen Skerrett say, "was a typical Victorian industrial slum." In spite of a suburban location, it was not unlike equally disease-ridden counterparts within the city limits along the Chicago River and in Packingtown, the back-of-the-[stock]yards residential district. Bush was one of several ethnic/racial/religious working-class neighborhoods of South Chicago and other mill-gate communities being erected around the southeastern side of the lake.[27]

While manufacturing districts in the periphery, like South Chicago, were being built into environments that were more industrial than residential, plenty of wild places remained in the Chicago area. Just to the west of this industrial suburb arising on the lakefront was some of the best duck hunting in the entire country. Within earshot of the factory whistles of what was known as the Calumet District, the gunshots of hunters rang out thousands of times during the birds' semiannual migrations. Dating to before the Civil War, elite society had established men's clubs in this vast expanse of shallow lakes, swampy marshes, and interconnected rivers. One sportsman in an 1884 issue of *American Field* magazine asks rhetorically, "Who has not been to Calumet hunting?" He recounts "how the blood fairly races through one's veins as just at daybreak the first flocks are seen skimming along the river! How your gun flies to your shoulder, and the bang! bang! followed by the swash, swash of from 2 to 6 of the little beauties." Anyone could "bag" at least fifty birds without effort, he testifies, and a full day's shootout could bring down five or six times as many.[28]

Five years later, a reporter for *Forest and Stream* would take note of his outing at the Grand Calumet Heights [Shooting] Club. Emerson Hough's description of the marshland along the river explains why the migrating water birds sought out this environment in which to rest and feed. "The Grand Calumet," he writes,

> is a necessarily slow, deliberate, tortuous and torturingly crooked stream. . . . It just strolls off among the sandhills and pine barrens toward the foot of the lake . . . and then chang[es] its mind and takes a while over in the opposite direction. Its general appearance is that of a long crooked valley of rice and cane, running between low wooded banks and stretching out from half a mile to three-quarters of a mile [0.8 to 1.2 km] or more in width. Somewhere in this winding marsh, hidden by what a poet would call the lush and dank sedges of the marsh, creeps the deliberate Grand, ten to fifteen feet deep in much of its channel, a lake creek rather than a river, and a darling for ducks.

Naturalists would pay similar homages well into the twentieth century to patches of beauty within the five other rivers' watersheds, the lake's sand dunes, and the prairie's wetlands of the metropolitan region.[29]

In the postfire era of the Gilded Age, the mutual interdependence of urban, suburban, and rural areas became more tightly woven by the railroad and the telegraph. In 1874, real-estate speculator Everett C. Chamberlin published a four-hundred-plus-page volume titled *Chicago and Its Suburbs*. This boosterish travel guide includes a map that marks the birth of a metropolitan idea of Chicago on a regional scale. As Elaine Lewinnek underscores, "Maps helped people imagine their burgeoning metropolis by making implicit projections about urban growth." Drawn by Rufus Blanchard, half of the map has a series of concentric circles superimposed over a territory radiating from downtown to embrace Cook and DuPage Counties as well as pieces of three other collar counties. Each of the twenty-two rings is 1 mile/1.6 km farther from the center, reaching nearly twice as far as Evanston, Riverside, and South Chicago (see Figure I.2).[30]

The map also highlights the twelve railroad lines fanning in all directions that had "transformed Chicagoland," according to suburban historian Ann Durkin Keating. She attributes the suburban origins of thirty-one farm centers, seventy industrial towns, thirty-five bedroom communities, and thirty-two recreational and institutional retreats as being linked to these iron rails. Like pearls on a string, the railroad suburbs and farming communities in between grew in tandem with the expanding, outer edges of the core. Together, this centralizing and decentralizing process of sprawl began to bring into focus a coherent image of the region as a single economy and society.[31]

A second wellspring of a metropolitan idea of Chicago came from the downside of this success: damaging floods over ever-more-widespread areas.

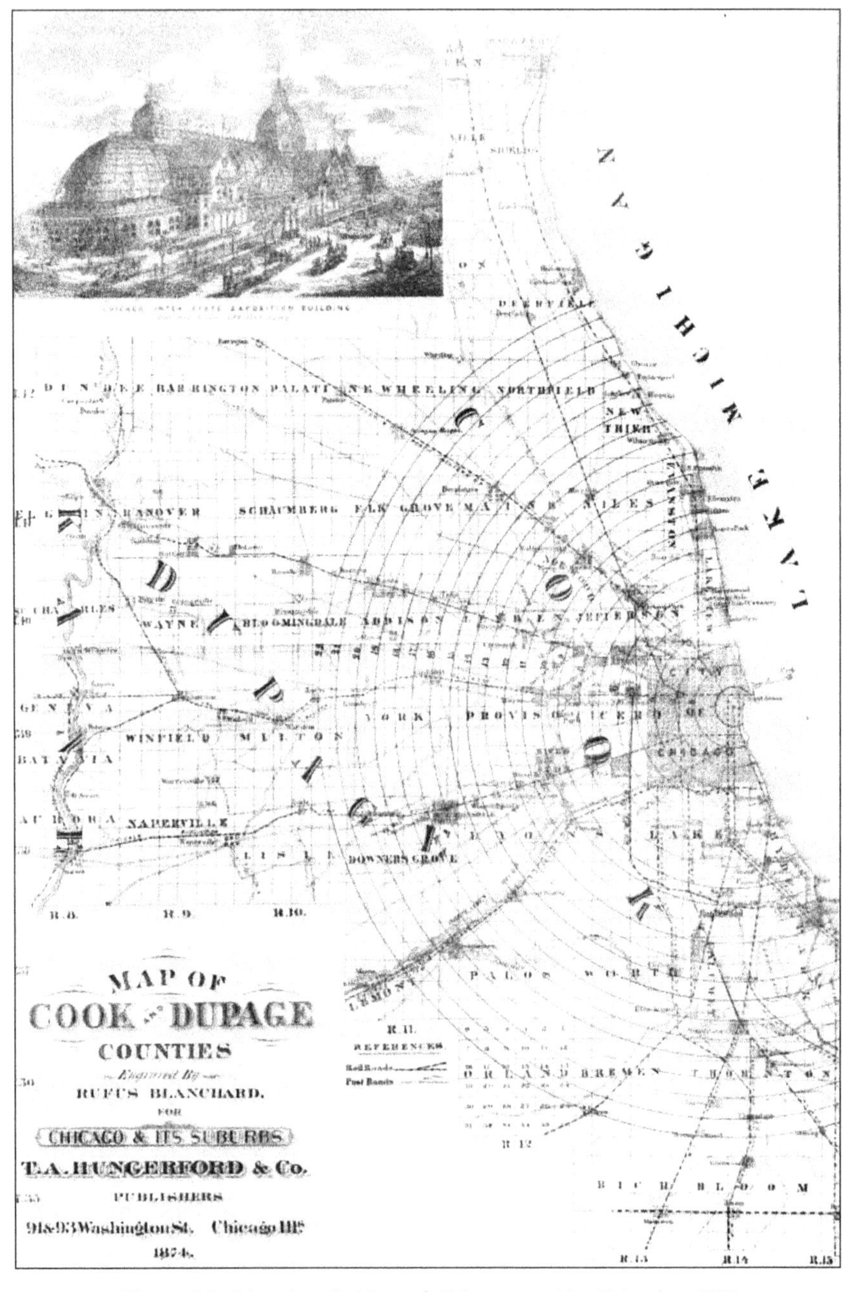

Figure I.2. Blanchard's Map of Chicago and Its Suburbs, 1874

In addition to the region's crazy quilt of overlapping governmental jurisdictions, rainwater has no respect for political boundaries. Outward extensions of the center combined with suburbanization to pave over and drain the land in a piecemeal fashion by public and private actors. Engineering a prairie wetland was producing greater volumes of surface runoff, which was being diverted from one property and watershed to another.

The most troublesome of these drainage projects was the Ogden-Wentworth Ditch, which regularly caused a repeat of Marquette's flood. Others paid the costs of draining their land near Mud Lake when its damlike embankment caused the Des Plaines River to overflow backward into the South Branch of the Chicago River. Bridgeport and other riverfront neighborhoods in the city suffered extensive inundations in February 1871, November 1872, May 1874, March 1875, April 1877, July 1878, June 1879, February and April 1881, February and November 1883, March 1884, and June 1885. Five years before the greatest rainfall yet of August 1885, the burgeoning suburbs on the banks of the 643-square-mile/1,665 square km watershed of the Des Plaines above Riverside began reporting flood damages. During this wet period, punctuated by 3-inch-plus/7 centimeter (cm) plus downpours, thousands of residents in the 450-square-mile/1,165 square km catchment of the Calumet District also witnessed their neighborhoods' seas of mud become lakes of sewage with increasing predictability.[32]

The Second City

By the time of the great flood of 1885, the explosive growth of Chicago and the interrelated, environmental problems of flood control and pollution control had brought into focus an idea of the city as a regional-scale metropolis. Adding a half million people during the 1880s, Chicago became the second-biggest city in the nation. Its built-up areas sprawled into and merged with its bordering communities, such as Edgewater, Jefferson, Cicero, Englewood, and Hyde Park. Extensions of the street railways and telegraph networks also knit the social and spatial fabrics of the urban and the suburban more tightly together.

Nonetheless, the gap between them in terms of modern infrastructure and utility services threatened to widen in coming years. An array of promising new technologies was coming online in the city center, including electric light and power, rapid (electric) transit, telephones, and an advanced generation of gas lighting and heating equipment. But with a few wealthy exceptions, such as Evanston, most outlying communities remained literally and figuratively stuck in a sea of mud. The public blamed the politicians for the chronic failure of government to manage storm runoff surges and safe drinking supplies.[33]

The real flooding on August 2–3, 1885, and the imagined threat of an epidemic outbreak in its wake led Chicago to undertake systemic reforms of its governmental institutions to regain control of nature. The city's best men were stirred into action against city hall by fear that the gigantic freshet of heavily contaminated water from the historic, 6.2 inches (in.)/15.7 cm of rain in a twenty-four-hour period was reaching the 2-mile/3.2 km water supply intake crib. Like the game-changing electric technologies, germ theories of disease were establishing new standards for measuring the public health of the urban environment. Armed with petri dishes and microscopes, sanitary experts could now identify previously invisible bacterium in the drinking water as well as test the efficacy of various ways to filter and purify it. Over the next four years of civic debate and political contestation, the reformers gained approval for an annexation plan that would burst the city limits from 40 to 190 square miles/64 to 306 square km. Adding 220,000 people during this period, Chicago recorded more than a million inhabitants in the 1890 census. In addition, its civic elites created an independent agency, the Sanitary District of Chicago (SDC), which would soon embrace the 385 square miles/971 square km of the expanded city and its surrounding suburbs.[34]

In some respects, the reformers followed the plan laid out thirty years earlier by the city's first great public-works engineer, Ellis S. Chesbrough, albeit on a much larger, metropolitan scale. The SDC would dig a deep channel to permanently reverse the flow of the Chicago River, install interceptor sewers along the lakefront, and dilute the sewage emptying into the river with fresh water from Lake Michigan. But in other respects, they were influenced by many other novel ideas in addition to germ theories of disease. They embraced modern theories about urban society and healthy bodies, homeownership and gender roles, family life and child rearing, and organized sports and outdoor recreation.

This mix of old plans and new ideas was the result of what engineers call a "path dependency," or a technical system that severely limits alternative options in the future. Thirty years earlier, the city had put the manager of its waterworks in charge of investigating options for solving the drainage problem. At the beginning of his report, Chesbrough states in bold type that "the main object of the sewers is to improve and preserve the health of the city." But by the end, the bottom line of lowest cost became the decisive factor in choosing among the alternative plans. The engineer had done his homework, paying particular attention to London's sophisticated system of wastewater removal and treatment. In the biggest city in the Western world, large interceptor pipes under the embankments of the Thames River carried the effluents to "sewage farms," where the manure became fertilizer. This option and two others were rejected in favor of combined sewer and drain pipes that would empty into the river and the lake. Its advantage of being the cheapest to build and maintain overweighed objections "that it would endanger the health of the city, especially during the warm, dry portions of the year."[35]

This plan became the single-most-important decision in the city's history of its relationship with water and land for four reasons. The first was the path dependency created by building a combined versus separate system of sewerage and drainage, because pollution control and flood control became inseparable. Chesbrough acknowledged that the experts recommended a two-pipe solution, but with few such systems in operation, it could be dismissed as theoretical. To keep costs down, the pipes were designed to handle a maximum of 1 in./2.5 cm of rainfall in a twenty-four-hour period. For a city build on a prairie wetland, this would result in chronic flood damages from storm runoff contaminated with raw sewage.[36]

Second, the decision set the precedent of the subservience of technical expertise to political expediency. One does not have to read between the lines to gain a sense that Chesbrough felt ill at ease with the report's convoluted conclusions. His anxiety is expressed in several sidebars that list possible add-ons to the plan in the future, including building interceptor sewers parallel to the river and deepening the Illinois and Michigan Canal or installing pumps at its Bridgeport entrance to reverse the flow of the river away from the lake. The following year, his lack of confidence in the choice of the city's commercial-civic elite led him to travel to London and other cities in Europe to see for himself how their systems worked.

Upon returning, he wrote a supplementary report to reemphasize the goal of protecting the health of the city. "Public sentiment in favor of sanitary reform has been so thoroughly aroused," he claims, "that a problem so important as this will not be suffered to rest till a satisfactory solution is obtained. The feeling is becoming very general, that wherever practicable, sewage should not be allowed to pollute water courses of any kind." Pleading for a second look, he reiterates his recommendation for interceptor sewers and pins his hopes on the enlargement of the canal to cleanse the river with a steady flow of lake water. Some half measures along these lines were adopted in the coming, postfire years, but Chesbrough and his successors understood that their jobs depended on the subordination of their advice to the demands of politics. Engineers like Chesbrough became semantic experts in dismissing their own technical recommendations as "impracticable."[37]

Another enduring legacy of the city's sanitary strategy was an obsession with heroic, big technology projects to engineer the environment. Although the plan required a minimum of public expenditure, it extracted a tremendous cost of money and inconvenience from the city's residents. It depended on raising the grade several feet, high enough to let gravity empty the sewers. Following this bold scheme to lift the city out of the mud, Chesbrough proposed an even-more-audacious plan to protect the water supply from the human and animal wastes pouring into the lake. Underwritten by the huge profits being generated by the waterworks, the city built a tunnel under the lakebed through the clay to an intake crib 2 miles/3.2 km out from shore.

Begun in 1863, during the Civil War, the project was completed four years later to national fanfare and civic celebration at the new water tower and pumping station on the North Side.[38]

Finally, the decision sealed the fate of the river as an open sewer and the lake as the ultimate sink of the city's effluents. Earning the apocryphal euphemism of a "working river," the waterway served not only as a shipping lane but also as the dumping ground for industries lining its banks. Perhaps the best known of these sources of pollution was the Union Stockyards, which had opened for business on Christmas Day, 1865. They were located on the South Branch to take advantage of a free way to dispose of the tons of organic wastes produced daily in the slaughterhouses. The stench arising from the river became permanent; this degradation of the environment created a social geography of class that divided the city into the lakefront and the riverfront districts. The rebuilding of the city after the Great Fire of 1871 had reinforced this spatial separation between those who could afford to live at a distance from the river and those forced by poverty to make their homes next to it.

In contrast to the first approach to water management, what became a self-proclaimed "Progressive Era" took pride in distinguishing itself from the past by its reliance on scientific experts in the formation of public policy. In the case of flood control, the establishment of a weather bureau in 1872 provided reformers with a consistent set of precipitation and temperature records. Moreover, other professionals had been mapping the underground geography and the surface topography of Chicago's prairie wetlands with much greater precision. Together with the medical practitioners of the new public health, they calculated the design specifications of a heroic public-works project to engineer the environment.

Starting with the crisis of the great flood of 1885, this book follows the policy debate over remaking the city's flood-prone environment during an extended, 125-year period of climate change and geographic deconcentration. Its two subperiods are defined by an equally important combination of dramatic swings in rainfall patterns and evolutionary shifts in society's attitudes toward the environment. From roughly 1885 to 1945, Chicago went through a dry period of rainfall compared to a wet period since then of more-frequent and more-severe storms. This change in its climate is documented in more than 140 years of daily weather records.[39] The causes of change are not at issue here, only the impact on a city built in a flood-prone environment.

The weather bureau's database can be turned into two charts that illustrate long-term patterns of rainfall on a regional area undergoing explosive urban sprawl. Figure I.3 simply plots Chicago's annual precipitation from 1871 to 2014 on a line representing the average amount of 34.5 in./87.6 cm. It clearly shows an above-the-line spike in the decade before the great flood of

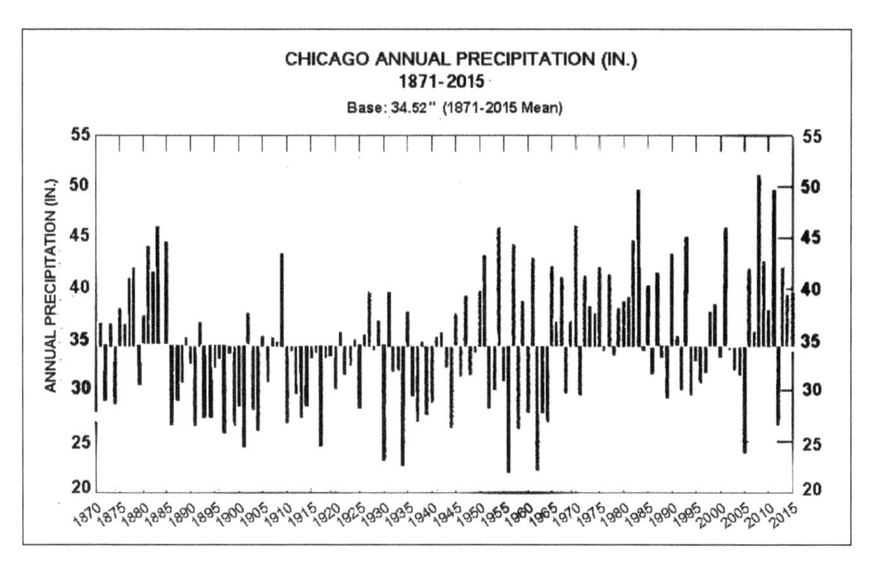

Figure I.3. Annual Precipitation, 1871–2015 (Climate Stations, available at https://www.climatestations.com/wp-content/uploads/2016/01/CHIPRCP.gif)

1885, followed by a sustained, sixty-year period of below-average rainfall. Then, since 1945, Chicago has experienced an equally long period of above-average amounts of rainfall. It has also endured years of record-breaking heat and drought, contributing to this period's pattern of more-frequent and more-extreme weather events.

Yearly totals, however, do not indicate the frequency or severity of downpours that can cause flooding. In the case of Chicago, a 1 in./2.5 cm. rainfall in a twenty-four-hour period can be taken as the benchmark for potential flood damages. On a 1-square-mile/2.6 square km area, it will drop 17,500,000 gallons (gal.)/66,245,000 liters (L) of water. The city's combined sewer system reaches capacity, backflows into basements, and overflows into the rivers and the lake. Historically, a deluge of this amount was considered to be a proverbial "Act of God," or a "hundred-year rainfall," because it would have been enough to sink the city. A true "hundred-year rainfall" is an event with a 1 percent chance of happening in any given year. Figure I.4 marks out a 50 percent increase since the mid-century in the number of rainfalls of 1 in./2.5 cm or more. From 1900 to 1950, the average number of times the metropolitan region's flood-control system was overwhelmed remained flat, at six times per year, but since then, the number has increased to nine. From 1958 to 2007, the number of "heavy rainfall events," or the top 1 percent, increased by a third.[40]

Adding to the risk of damaging floods, more-imperious surfaces have resulted in proportionate increases in storm-surge runoff. Since the 1960s, as little as one-half to two-thirds of an inch of rain has been sufficient to defeat the city's flood-control system.

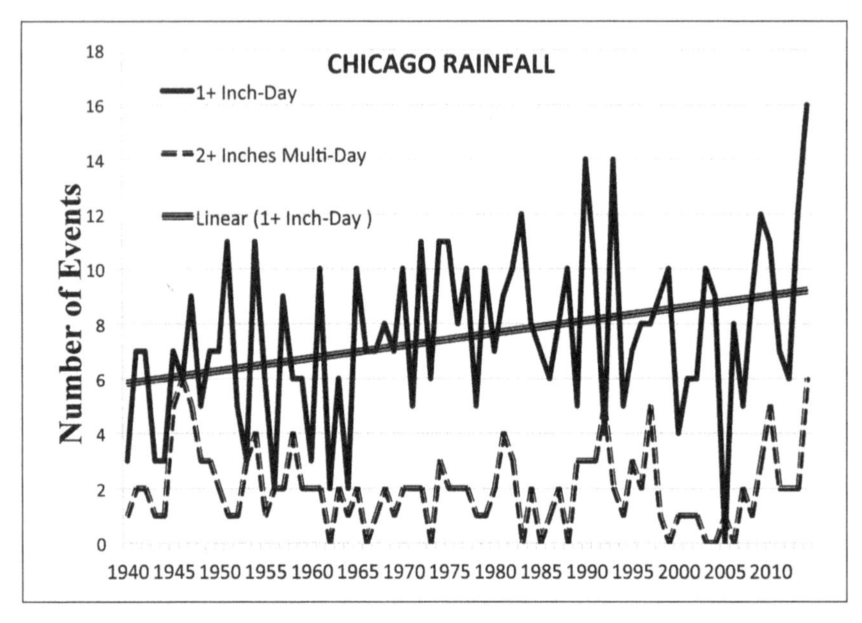

Figure I.4. One-Inch-Plus Rainfalls, 1940–2010 (Data from U.S. National Oceanic and Atmospheric Administration, National Climatic Data Center, *Record of Climatological Observations, 2014*, available at http://www.ncdc.noaa.gov/)

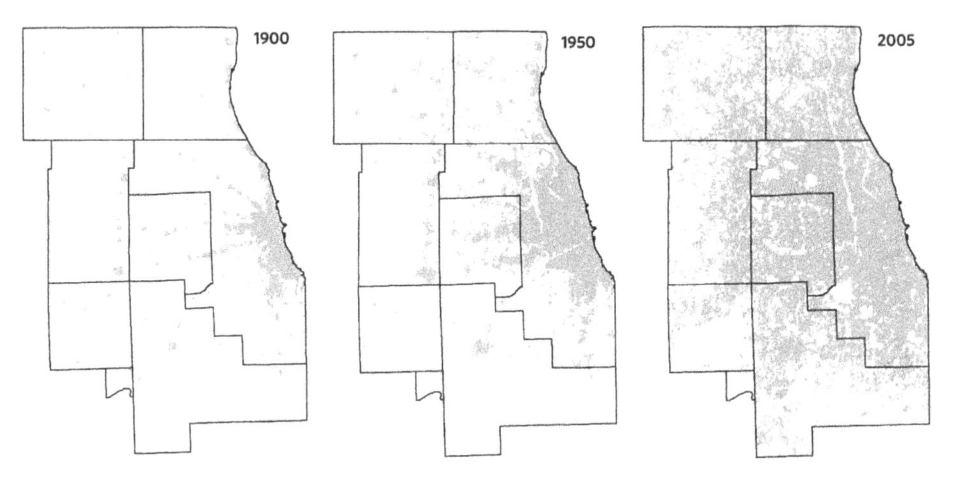

Figure I.5. Regional Sprawl, 1900, 1950, 2005 (City of Chicago and Chicago Metropolitan Agency for Planning, *2040—Comprehensive Regional Plan* [Chicago: Chicago Metropolitan Agency for Planning, October 2010], fig. 10, available at http://www.cmap.illinois.gov/documents/10180/17842/long_plan_FINAL_100610_web.pdf/1e1ff482-7013-4f5f-90d5-90d395087a53)

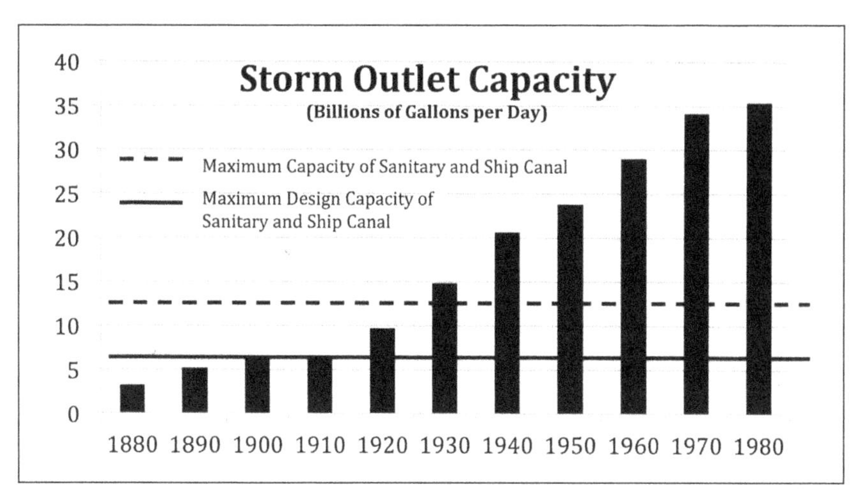

Figure I.6. Storm Outlet Capacity, 1880–1980 (Data from Frank E. Dalton, Victor Koelzer, and William J. Bauer, "The Chicago Area Deep Tunnel Project: A Use of the Underground Storage Resource," *Journal of the Water Pollution Control Federation 41* [April 1969], fig. 9)

Two additional illustrations show how the rate of sprawl—paving over a prairie wetland—has outpaced Chicagoans' efforts to gain control over nature, especially during the wet period of climate change. Figure I.5 maps the built-up area of the metropolitan region at half-century intervals, beginning at 1900.

Figure I.6 traces the growth of the sewer system's storage capacity, crossing a line representing the maximum amount of storm-surge overflow the Sanitary and Ship Canal can release at the Lockport Dam before forcing the sanitary authorities to release polluted storm water into the lake. As early as 1910, the river-in-reverse had already reached its design specifications, and it was at three times the maximum amount by the end of World War II.

An explosive quarter century of suburban homebuilding enlarged the underground network of sewer pipes to five times the limit of the drainage canal to get rid of storm surges. As the property and the human costs from flood damages mounted, policy makers responded in the mid-1970s with a big technology, the Tunnel and Reservoir Plan (TARP). Combined with climate change, the sanitary authorities have needed to discharge a continuing series of record-breaking amounts of effluents into the city's ultimate sink, Lake Michigan, closing its beaches and facilitating the invasion of alien species back and forth between the Great Lakes and Mississippi River basins.

A final illustration, Figure I.7, graphically presents the frequency and amount of sewage-laden wastewater that has been released into Lake Michi-

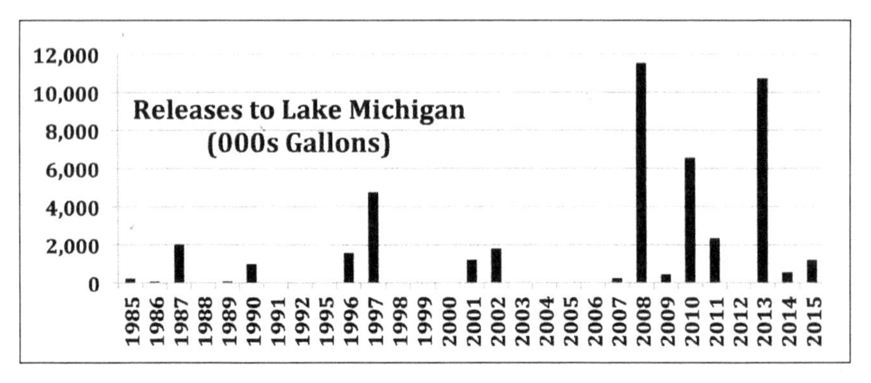

Figure I.7. Storm-Water Releases into Lake Michigan, 1985–2015 (Data from City of Chicago, Metropolitan Water Reclamation District, "Reversals to Lake Michigan [1985–Present]," 2015, available at https://www.mwrd.org/irj/go/km/docs/documents/MWRD/internet/protecting_the_environment/Combined_Sewer_Overflows/pdfs/Reversals.pdf)

gan since the 1985 opening of this never-ending, multi-billion-dollar public-works project.

Despite its heroic imagery as the technological sublime, TARP was built at one-half of the scale and capacity needed to prevent flood damages. Its original designers badly underestimated the amount of storm-surge runoff in the built-up areas.[41] A 2007 study found that 42 percent of Cook County's 925 square miles/2,400 square km of land had been paved over on average, which would amount to 6.8 billion gal./25.7 billion L of runoff, without counting the amount from the other 58 percent of the land. Another recent study calculated that 820 square miles/207.2 square km of the metropolitan region had impervious surfaces, or a runoff of more than 14 billion gal./53 billion L from 1 in./2.5 cm of rainfall. With new, record-breaking storms in 2013 and 2015, the wet period of climate change and the outward thrust of the region's settlement patterns continue, promising more record-breaking flood damages.[42]

The two subperiods of this book correspond not only to climate change but also to a shift from natural resource conservation to ecosystem restoration in popular thought about the environment. In Part I, Chapter 1 traces how Chicago's architects became its nature conservation experts during the Progressive Age. As Daniel Burnham, Dwight Perkins, and Jens Jensen transformed themselves into city planners, they took a leading role in bringing lakefront beaches, neighborhood parks, and riverine forest preserves to the masses of city dwellers. At same time, they engaged in the larger civic debate over how to build the proper mix of the city efficient, the city beautiful, and the city humane. Chapters 2 and 3 take up the post-Progressive, interwar years, when the rainfall continued to be below average. From the 1910s to

1930, Chicagoans engaged in a love affair with homeownership and outdoor living. Subdivisions were built over the city's remaining open spaces, while a ring of county forest preserves opened around this bungalow belt to protect the six watersheds. The New Deal supported a vast expansion of recreational facilities in public open spaces.

Part II examines the growing conflict between the grassroots, politicians, and policy makers over the remaking of Chicago's paved-over prairie wetland from the perspectives of modern ecological thought. By 1945, scientists had accumulated a half century of intense investigations of how engineering the environment changed the aquatic ecology of inland waterways from studying Chicago's six rivers and their downstream impacts on the Illinois River. Chapter 4 examines a twenty-year period when the continuing growth of enthusiasm for the outdoors would turn city dwellers and suburbanites alike from nature conservationists into proto-environmentalists. The postwar period of an unprecedented baby boom would underwrite a building boom of single-family homes in the suburbs. Gaining an ecological understanding of the interconnectedness of all living things from Rachel Carson's book *Silent Spring*, they organized grassroots movements to save the lake from pollution and the rivers from being turned by the sanitary district into sewage and drainage canals.

From 1965 to 1985, city engineers built an alternative big technology system, the Deep Tunnel. Chapter 5 covers this period of infrastructure construction, suburban white flight, and increasing flood damages from more-frequent and more-severe storm-surge runoffs. The failures of local government to prevent flooded basements and polluted waters gave birth to a modern environmental movement that linked the ecology of the lake and the rivers. Record-breaking rainfalls forced the engineers to divert billions of gallons of contaminated storm water into Lake Michigan, closing the beaches and threatening epidemic outbreaks. Adding to this old problem of using the lake as Chicago's ultimate sink were new ones caused by the movement of disruptive, invasive species between the St. Lawrence River/Great Lakes and the Illinois/Mississippi Rivers systems. As the crises of widespread flood and environmental damages grew worse, the grassroots mobilized to oppose the plans of city hall. They formed the Friends of the River and the Open Land Project to save nature in the metropolitan area.

Chapter 6 and the Conclusion take a look at the successes and failures of TARP during a period of mounting political demands for higher standards of environmental quality. From the opening of the deep tunnel in 1986 until the retirement of Mayor Richard M. Daley in 2011, Chicago became the country's last big-city holdout against coming into compliance with the landmark Clean Water Act of 1972. Nonetheless, the return of aquatic and wildlife to the Chicago River produced a self-reinforcing process of restoring urban nature and growing political support for its continued reclamation. The improvement of the waterway added to the renaissance of an expanding city center, transforming the "working river's" image from the city's back alley to its centerpiece, a scenic boulevard of light, air, and flowing water.

The goal of *Sinking Chicago* is to shed light on how people respond over an extended period to climate change. Chicago's flood-prone environment furnishes a good case study of a community's attempts to grapple with long-term swings in rainfall. Like those of other American cities, its citizens, politicians, and policy makers lost sight of the connections between land use and water management. Driven by urban rivalries for economic hegemony and their faith in progress, they attempted to conquer nature rather than adapt to its forces, which are beyond human control. Chicagoans became exceptionally obsessed with heroic, big-technology solutions to its dual problem of water quality and flood control. But in the post–World War II period of explosive suburban growth, this approach failed to keep up with the size of storm-water runoff surges caused by sprawl during an increasingly wet period of more-frequent and more-severe rainfalls. Searching for alternative solutions, environmental reform activists have been finding ways to adjust to climate change by reconnecting land and water in holistic ways.

1

The Triumph of Metropolitanism, 1885–1910

The Costs and Benefits of City Building

On February 16, 1883, the *Chicago Daily Tribune* sang the praises of the city's "natural advantages." As if graced by God, "Chicago enjoys an immunity which has come to be regarded as certain." Forgetting the disastrous freshet of 1849, the cholera epidemic, and all the subsequent public health emergencies, the editorial called the Great Chicago Fire of 1871 the city's one and only catastrophe. Unlike other places, "when floods come in the springtime . . . Chicago goes free. . . . The location of this city assures exemption from plague and disaster so long as the people take ordinary sanitary precautions." Like a garden in heaven, Chicago was shielded against heat waves and cold snaps, ensuring that "there is no need for special protection against flood, storm, and pestilence." "All this is said," the editor confessed disingenuously, "without any purpose to boast, and certainly with no intention of gloating over the trails which other communities are now undergoing." Instead, he simply wanted to remind Chicagoans to count their blessings for living with "the broad expanse of the lake on one side and the vast prairie on the other."[1]

Without question, the editorial writer knew better. Besides habitually damaging rainstorms, his paper had printed a story titled "The Threatened Flood" just four days earlier. It reported the failure of the city's commissioner of public works, DeWitt Clinton Cregier, to gain the cooperation of the lumbermen now in possession of the Mud Lake area. He sent a tugboat equipped as an icebreaker from the South Branch of the Chicago River to open a channel for melting ice and heavy rainfall, which had been forecast. He had not forgotten the lessons of previous catastrophes. However, Cregier ordered the captain to turn back when he learned that the property owners were unwilling to pay for the work. Nonetheless, the commissioner was keeping this boat and another makeshift icebreaker at the mouth of the river on standby alert against the threat of a destructive freshet sweeping through the central business district (CBD).[2]

If the newspaperman entertained any thoughts that he was living in a Garden of Eden, these illusions were shattered the very next day by widespread flooding. The *Tribune*'s headline noted that "a considerable portion of the city of Chicago [was] under water." As predicted, thawing ice and a heavy rainfall—the equivalent of a "regular April thunderstorm"—inundated the Southwest Side of the city. Bridges were swept away, telegraph wires downed, train services halted, and basements swamped with sewage. Moreover, the freshet stirred up the rotting sedimentation from the riverbed, causing more than an ordinary stink. "From Chicago's sweet-scented river yesterday," the newspaper carped sardonically, "there arose a stench as foul as to turn the stomach of those who were compelled to cross its bridges—a nauseating odor was wafted by the breeze over the entire city." Without the feared ice jam at the mouth of the river, this poisonous brew gushed into the lake.[3]

The price paid in property losses and human suffering from the flood reminded Chicagoans one more time of the natural disadvantages of building a city on a prairie wetland. Before the Progressive Age of urban reform, they considered the lake and the rivers to be unruly, albeit profitable, resources. They sought to control nature to maximize its economic benefits at a minimum of costs. As an *entrepôt* and manufacturing center, Chicago remained a city built on water. Its economy depended on cheap water-borne shipping, abundant supplies of pure water, and unregulated disposal of liquid wastes. The annual rainfall and seasonal freshets usually kept the Chicago River and the Illinois and Michigan Canal open to navigation, reducing freight tolls as well as the cost of dredging their bottoms. In addition, commercial fishing and mussels harvesting contributed to the urban economy. While the lake supplied a virtually free and unlimited supply of fresh water, the rivers served an equally beneficial purpose as sewage channels.

Before the mid-1890s, in contrast, Chicagoans did not apply a calculus of environmental costs and benefits in the formation of public policy. The combined sewer and drainpipe system was the single-most-important public-works decision that had discounted the costs of polluting the river and using the lake as the city's ultimate sink. At most, incremental advances in medicine and epidemiology reinforced traditional, sanitary efforts to clean the city and get rid of its stenches. Yet the extension of the combined system in response to urban sprawl continued to use the lake and the river as the outlets for increasing amounts of untreated organic wastes, toxic chemicals, and polluted runoff surges.

The 1893 Chicago World's Fair marked a crossroads in the city's attitude toward having a lake on one side and a prairie on the other. The timing of this historic event in American culture and technology coincided with the rise of a new, urban version of the Arcadian myth, spawning a multidimensional, mass movement to get "back to nature."[4] It spread the belief that the social and spiritual benefits of saving patches of open space outweighed the economic costs. Private groups of all kinds formed country clubs in the suburbs. Elite reformers and neighborhood associations mobilized civic efforts to make the city beautiful and to conserve its natural resources from overexploitation.

Modern, comprehensive planning was invented to weigh the environmental costs against the economic benefits of urban growth. Social settlement workers, school teachers, and club women called for more urban nature: neighborhood gardens, playgrounds, parks, and beaches as well as rural retreats to save the children from the debilitating influences of life on the city streets. As planning historian Robin F. Bachin explains, "Parks became a central feature of plans for reform of the urban landscape precisely because they were spaces of leisure."[5] Culminating in the justly famous *Plan of Chicago* of 1909 by Daniel Burnham and Edward Bennett, the Progressives would build nature into their visionary designs of the metropolis of the future.[6]

In addition, the city gave birth to related popular movements for outdoor sports and recreation. With the emergence of a more-affluent, secular society came more time to consume leisurely pursuits and commercial entertainments. One response was "muscular Christianity."[7] Fearing the overfeminization of the church and the moral evils of the city, white male Protestants forged a new identity of masculinity that sought to balance "mind, body, and spirit." Gaining many followers, it generated Young Men's Christian Association (YMCA) gyms, athletic team contests, the Boy Scouts, and summer camps in rural areas. Girls, too, were encouraged to participate in outdoor, albeit noncompetitive, sports and nature clubs, such as the Girl Scouts and the Campfire Girls. Entire families were attracted to the spectacle of professional teams playing in increasingly grand stadiums, especially baseball parks. Skyrocketing in popularity during the 1890s, moreover, were amateur outdoor sports, including golf, tennis, and swimming. Cycling became a mass craze to get back to nature, setting the stage for the ultimate "freedom machine," the automobile. In the meantime, the coming online of cable and electric streetcars helped make homeownership on the "crabgrass frontier" more accessible, appealing, and affordable.[8]

Chicago's reformers harnessed the burgeoning enthusiasm for the outdoors into political support for the funding of open space and organized play. For the first time, the public put a value on the environmental benefits of nature that outweighed their economic costs. The taxpayers approved bond issues to expand the city's parks. They also voted to establish an "outer belt" of forest preserves to save the suburban watersheds from the fate of the Chicago River. But these plans to save nature from the city for future generations came into conflict with the sanitary engineers' plans to turn all the rivers into economic resources as shipping, sewage, and drainage channels. From the announcement of the 1909 *Plan of Chicago* until the end of World War I a decade later, civic debate raised fundamental questions about who should decide the future course of the city's sprawling development and who should pay for building its infrastructure improvements and preserving its natural amenities.[9]

This chapter examines the introduction of environmental values in the formation of public policy. Following the disaster of 1885, the answer to the "drainage question" of flood control was decided on the narrow basis of economic costs and benefits at the expense of the environment. However, the sanitary and ship channel plan failed to stop the sewage of a quarter million people living along the lakefront from contaminating the water supply. Fearful epidemics of typhoid fever, small pox, and other contagious diseases kept the larger question of the link between water quality and public health at the top of municipal reform agendas. Coinciding with the adoption of new, germ theories of disease in the Health Department at the time of the 1893 World's Fair was the rise of the city beautiful, neighborhood playgrounds, and outdoor recreation movements.

Over the next twenty years, these reform crusades and many similar cultural impulses to get back to nature jelled into a much larger idea that contemporaries called "conservation." The age-old ethos of utilitarianism was expanded to include human minds and bodies as valuable natural resources that needed to be saved. At the heart of the Progressives' new, all-embracing concept was the conviction that the masses of people living in urban-industrial centers were alienated from the nonhuman world. Reformers considered spiritual renewal to be just as important as making the most efficient use of the material supplies required by modern civilization, such as wood, fresh water, iron, and coal. Conservation meant restoring the bonds between nature and society in a proper relationship of mutual benefit. "The conservation movement," historian Benjamin Heber Johnson states, "created lasting legacies in politics, the structure of the American state, the landscapes of cities, and environmental thought."[10]

As a case study, Chicago's efforts to remake a flood-prone environment support his contention that the conservation idea had long-term impacts on the formation of public policy. Beginning with the World's Fair and persisting into the post–World War II period, reformers strove to create an urban nature that would heal city residents' minds, spirits, and bodies. Chicago became a leader in drawing this grand plan of moral restoration and urban renewal. Jane Addams, Daniel Burnham, and Jens Jensen—to name just a few—helped mix environmental, human, and aesthetic values into a new, all-embracing concept of conservation. Pioneers in social work, city planning, and landscape architecture, respectively, their successful efforts to transform Chicago's landscape became national and international models of reform. But, as Johnson demonstrates in tracing the creation of the national parks and forests, the conservationists' production of urban space also came at a price that was paid by ethnic/racial minorities and the working class.[11] Chicago's Janus-faced social geography of class became indelibly imprinted on the built environment and the public imagination. In the twentieth century, the lakefront became the city's "front yard" for the well-heeled, and the riverfront became its "back alley," where the poor people lived. After the ship canal opened in 1900, the goal of the Sanitary District of Chicago (SDC) to expand its jurisdiction over all the drainage watersheds of the built-up area came into conflict with elite conservationists, who believed that the economic costs of saving nature were outweighed by its social benefits. They, in turn, faced opposition from local grassroots organizations, which argued that only the upper classes would be able to afford access to a distant "outer belt" of forest preserves. Moreover, the federal government began to take a stand in the way Chicago managed its water. The U.S. Army Corps of Engineers opposed the SDC's plan to drain so much water from Lake Michigan as to cause serious damage to navigation on all the Great Lakes. The ensuing struggles for control of the waters of the lakes and the rivers would reflect the triumph of metropolitanism in the environmental politics of Chicago.

Engineering a Flood-Prone Environment, 1885–1893

The city's greatest storm of more than 6 inches/15 centimeters of rain triggered a political crisis that lasted for eight years until May 1893, the official opening of the World's Fair, also called the World's Columbian Exposition to celebrate the four hundred years since Columbus's landing. For the first four years, fear of an imagined epidemic of cholera was the prime engine driving reformers to find an answer to the drainage question. The Haymarket bombing of 1886, the ensuing police crackdown, and the court trail of the anarchists added to the intensity of the city's state of emergency. From 1889 to 1892, deaths from a real outbreak of typhoid fever threatened to cancel the city's grand spectacle. Armed with the authority of science, public-health experts pointed to the contamination of Lake Michigan by the sewer systems of the city and the suburbs bordering the lakefront. Desperate to bring the epidemic under control, city hall upgraded its Health Department to modern standards and appointed a blue-ribbon Pure Water Commission.[12]

The battle for a sanitary strategy on a metropolitan scale and a campaign for annexation of the city's collar townships took place concurrently for closely related reasons. Science and technology promised to uplift the quality of daily life to unprecedented levels of comfort and convenience. But with a few exceptions, such as Evanston and Riverside, most suburbs could not afford to install the new generation of networked technologies, including gas mains, electric lines, telephone wires, rapid-transit tracks, and asphalt road pavements. Like building a flood-control system for Chicago's hydraulic geography, only Chicago's municipal government had the institutional and financial resources to administer these large-scale infrastructure projects and politically charged public-utility franchises. The desire of the outlying communities to live in the most up-to-date environments accounts for their pro-annexation and drainage district support.[13]

For the same reasons, some suburbs opposed inclusion in the new, bigger Chicago and in the sanitary district. Evanston, for example, took pride in keeping current with technological innovations and utility services, including a municipal waterworks and sewerage system. Families moving there knew they could build houses and plug them into all the modern conveniences of domestic life. Evanston's spokesmen protested against the inevitable rise in their property taxes to pay for other people's benefit. The "antis" also rallied around the flag of the community's identity as a sanctified village. They called for the defense of local self-government against the danger of being swallowed up and overrun by the monster city of booze, sin, and corruption.[14]

The question of where to draw the borders of a flood-control district on a metropolitan scale raised the larger related issue: the size of the city itself. Its 40 square miles/104 square kilometers (km) were so densely packed with more than eight hundred thousand people that its architects invented the skyscraper. In the downtown area of the cable-car "Loop," the only room left was the air above the ground. At the same time, land development was spreading outward in all directions. Crabgrass frontiers, such as Lake and Lake View Townships, were quickly becoming paved over, losing their suburban character. Together, these forces of centralization and decentralization blurred the dividing lines between urban and suburban landscapes as well as between residential, commercial, and industrial zones.[15]

Another reason for the overlapping politics of the geographic boundaries of the city and the sanitary district was the dream to make Chicago the nation's biggest urban center. To become number two, the city's boosters orchestrated a steady drumbeat of support in favor of expanding its borders fourfold to encompass 190 square miles/492 square km and a quarter million additional inhabitants. In 1887, the first test vote on annexation brought mixed results. The referendum was approved in Chicago, Jefferson Township to the northwest, Cicero on the western border, and Hyde Park Township on the southern lakefront, including the Calumet District. But the "antis" prevailed in the townships of Lake on the southwestern edge and Lake View on the North Side. Disappointed, but not deterred, the boosters renewed their campaign with an even greater determination to expand the city to metropolitan proportions in all directions.[16]

In contrast to the popular vote on the size of the city, policy makers put the decision of what to do about water supplies and flood control in the hands of experts. The panels of engineers who responded to the great flood of 1885 did not deviate from the path laid out by the sewerage board in the mid-1850s. They used the same calculus of economic costs and benefits to come to the same conclusions. In the name of public health, they materialized the boosters' dream of a heroic, big technology project. The engineers proposed to construct a gigantic canal to link ocean shipping between the Great Lakes and the Gulf of Mexico. "[It] will be by far," the *Chicago Tribune* exuded, "the largest municipal improvement the city has ever undertaken."[17]

Following through on one of Ellis S. Chesbrough's recommendations, the experts laid out a plan to dig a supersize, deep-cut channel through the clay and bedrock to permanently reverse the flow of the Chicago River. A steady current of fresh water from Lake Michigan would help keep ships afloat down to the gulf. Moreover, the city's drinking supplies would be protected from contamination by the river. Its organic wastes would be washed away and rendered harmless by dilution at no cost, except to the lake and riverine environments. The sponsors of legislation to establish the SDC predicted that the economic benefits of this "navigational waterway" would soon make the independent agency a self-supporting profit center.

In a similar way, the planners dismissed the only one of Chesbrough's options that considered environmental values, leaving aside for the moment its major gains for public health. In 1887, city hall brought in a nationally respected engineer, Rudolph Hering, to mediate the raging civic debate over the drainage question. Like his predecessor, Hering advised the city to follow the London plan of interceptor sewers to keep all liquid wastes out of the lake and the rivers. In place of sewage farms as the ultimate sink of the city's effluents, he reported, modern methods were available to recycle them in industrial-scale treatment plants.

Nonetheless, a majority of the city's policy makers on the Drainage and Water Supply Commission ruled the most advanced approaches of science, medicine, and technology impracticable. They used the same kinds of financial manipulation as the original sewage board had employed thirty years earlier. They maximized all the future costs of the London model against a minimum-size drainage channel to arrive at a bottom line of a two-to-one price differential. Although claiming to represent the taxpayers, the policy makers secretly intended to construct a much larger ship canal at double the projected price tag. The 28-mile/45 km waterway eventually cost $55 million, an amount equal to the estimated cost of building the superior, albeit rejected, option.[18]

After making the decision in favor of a big technology project to reengineer Chicago's flood-prone environment, the public sphere of civic debate shifted gears to engage in a politics of metropolitan geography. Reformers fought over the boundary lines of the proposed drainage district. The postfire history of damaging floods leading up to the deluge of October 2–3, 1885, left no doubt that the city and its surrounding suburbs shared overlapping watersheds of flat, low-lying marshland. The four-year battle of municipal reform came to a decisive climax in a compressed period of five months. Between the end of May and the middle of October 1889, the state legislature enacted a sanitary district law, the voters approved a second annexation referendum, and a panel of three county judges decided where to draw the new agency's borderlines (see Figure 1.1). Under the terms of the enabling act, the SDC was charged with constructing a "navigable waterway." It also had to be large enough to conduct a current with a sufficient rate of flow of water drained out of Lake Michigan as to render the city's sewage harmless before it reached river towns downstream.

During this period of transition from chemical to microbiological methods of analyzing water quality, the experts still generally agreed that diluting organic effluents with oxygen-rich, fresh water was the safest and the cheapest way to treat the city's sewage. During the long-running debate leading up to the passage of the reform act, engineers, doctors, and scientists had come to an agreement on a minimum flow rate of 5,000 cubic feet/142 cubic meters per second, or 3.2 billion gallons/12.1 billion liters per day for the million inhabitants of the built-up areas. The law allowed the SDC to divert up to twice as much water from the lake in anticipation of rapid growth of the population, commerce, and industry in the near future.[19]

While laying out detailed specifications for the design of the Sanitary and Ship Canal, the legislators studiously avoided going on the record on the hotly contested issue of the new agency's boundaries. They foisted this important decision on the municipal reform leader and circuit court judge Richard Prendergast and two of his colleagues on the bench, who were instructed to hear all sides on the issue before drawing their map. The governor signed the bill into law just a month before a second referendum vote on the annexation of the holdout collar townships. During June, contending civic, taxpayer, and community groups linked the politics of the two geographies.[20]

The decision of which outlying suburbs to include and exclude also had deeper implications for the balance of power between the two major parties. Consider that South Chicago would give most of the votes of its twenty thousand working-class residents to the Democrats, while Evanston could be expected to deliver most of its thirteen thousand middle-class homeowners

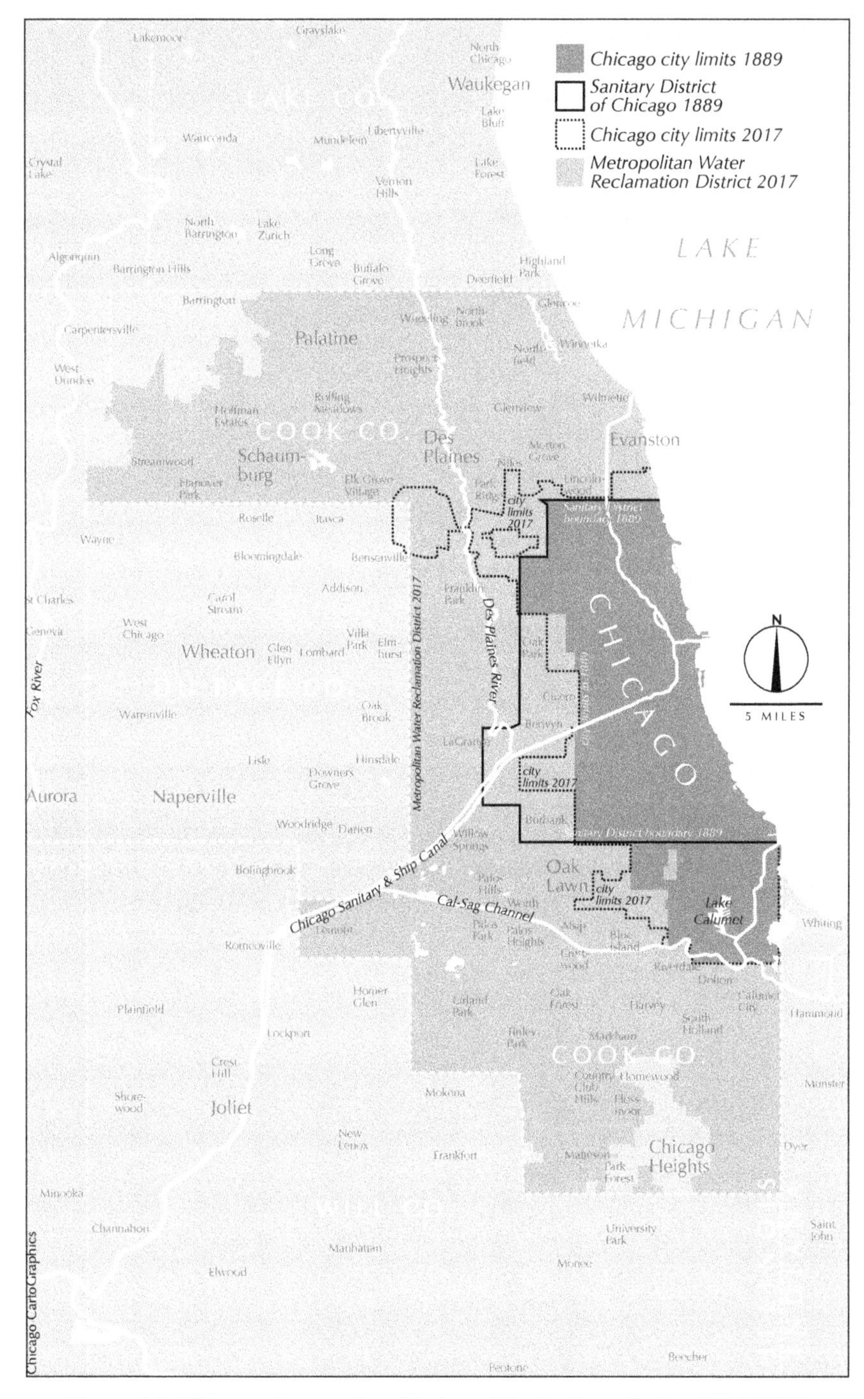

Figure 1.1. Chicago Annexations/Sanitary District Boundaries, 1889–2017 (Chicago CartoGraphics)

to the Republicans. Although the geographical boundaries of the city and the SDC were completely separate in law, the overlap of political power and taxing authority was equally undeniable. The politicians in the state capital made the new agency hold at-large elections for seats on its nine-member board of trustees, which were intended to increase the influence of the party bosses over the selection of candidates.[21]

Although reformers called for more scientific scholarship and less political partisanship, the hydrology of Chicago's prairie wetland played only a minor role in the debate over the outer boundaries of the sanitary district. Instead, the decision to include or exclude an area hinged on whether the primary purpose of the public-works project was a drainage channel or a supersize ship canal. The two uses conflicted, because draining the maximum amount of water into the Chicago River for the dilution of sewage would make its currents too fast for navigation. To reassure tugboat and barge operators, the lawmakers had set a limit of 3 miles/5 km per hour as the river's maximum rate of flow.

Engineers proposed two designs that could divert the maximum amount of water into the canal, which would keep the biggest vessels afloat at least to the junction of the Illinois and Mississippi Rivers. Their preferred plan was to dig more feeder canals to reduce the rate of flow sucked into the Main Branch, where the ships needed calm, harbor waters. The most ambitious plan called for a "Big Ditch" on the North Shore from the lake to the North Branch, another feeder channel on the Northwest Side to divert the upper Des Plaines River to it, and a third along either 39th or 87th Streets to the South Branch. A second more-expensive and more-disruptive option was to make the Main and South Branches much wider and deeper to slow the currents. This project would require replacing all the bridges along the route to allow for the passage of ocean liners moving upstream and downstream.

On the one hand, the ambitious promoters of a ship canal called for the inclusion of Evanston, the suburbs north to the Cook County line, the northwestern watershed of the Des Plaines River, and the city down to 87th Street, the southern border of Lake Township. On the other hand, the more-modest advocates of a drainage channel argued for restricting the district to the city's new borders, except for the exclusion of the low-lying watershed of the Calumet District below 87th Street.

The broad consensus in favor of drawing the southern border of the drainage district at 87th Street exposes the lack of importance both sides gave to the hydrologic geography of the metropolitan region. Despite some uncertainty caused by the extreme flatness of the topography, surveyors determined that the division between the watersheds of the Chicago and the Calumet Rivers was between 59th and 65th Streets. The "antis" pointed out that feeder canals and sewer systems in any area south of this hydraulic boundary line into the Chicago River would require pumping stations at great expense.

Only the former city engineer and now mayor Cregier, on the contrary, called for including the entire Calumet District. He reasoned on democratic grounds that its residents had a right to share in the benefits of the public-works project as citizens of the new, bigger city.

Furthermore, Mayor Cregier posited that improving the Calumet River and its harbor facilities would produce major advances for the entire metropolitan economy. Reversing the river's flow and turning it into Chicago's primary ship canal to the Illinois River was far more cost-effective than retrofitting the Chicago River. It was also the most politically promising way to gain federal funding for a navigable waterway because of the strong support for this plan by the U.S. Army Corps of Engineers. Almost as an afterthought, the mayor buttressed his political and economic positions by speculating that keeping the human and the chemical wastes of the industrial district out of the lake could benefit the city's public health. He remained a lone, isolated voice in the policy debate.[22]

The arguments in favor of including and excluding Evanston in the drainage district illuminate the absence of environmental values in the decision-making process. Like Chicago, the suburb used Lake Michigan as the sink for its liquid waste at no cost, except to the environment and the public health. Its network of sewer pipes serviced the main area of settlement, which occupied a narrow lip of land east of the hydraulic ridge (Ridge Avenue–Green Bay Road). The "antis" pointed out that the benefits of the so-called Big Ditch (later the North Shore Channel) would be limited to the sparsely populated, working-class sections of the community on the other side that drained into the North Branch. One of the leading spokesmen for a ship canal–size district, civil engineer Harvey B. Hurd, counterproposed that the sanitary authority could build an interceptor sewer to divert the suburb's raw sewage into this proposed flushing canal for the Chicago River.

He did not disagree, however, with the claim of Evanston "antis" that the sewage of its residents posed no threat to Chicago's drinking supplies. Village Board President H. C. Miller stated proudly that Evanston "had a pretty thorough system of sewerage."[23] The judges agreed, leaving the suburb out of the SDC. They had not been convinced that its inclusion on the grounds of either public health or environmental benefits outweighed the economic burden on its taxpayers. On the contrary, they ruled that adding its territory "is not seriously argued, or if it is, we do not find that the Chicago water supply is menaced or is in danger by reason of Evanston sewage. . . . If the Evanston sewage is not sufficient to pollute the Evanston water supply . . . how can it be regarded as endangering the Chicago water supply ten miles [16 km] away?"[24]

Judge Prendergast's panel used the same taxpayers' cost-benefit analysis for the far western, northern, and northwestern suburbs of Cook County, including the upper Des Plaines River. The jurists decided not to include them, because "they are farmlands and seem destined for such purposes for a long time." The scientific evidence had failed to persuade them that the drainage and storm-surge runoff of these upstream areas formed an integral part of the hydrology of Chicago's metropolitan region. Like Evanston, Riverside was omitted, because it had its own, tax-supported system of water management. The property owners in nearby suburbs along the Des Plaines River, the judges reasoned, should likewise be excused to give them equal treatment.[25]

The judges' map reflected their understanding of the political lay of the land, not its hydrologic geography. The three-member panel acknowledged that the organized opposition of the "antis" in Evanston, Riverside, and the northwestern suburbs had influenced their ruling. And they drew the southern line at 87th Street, because they knew that the residents living in the low-lying land of Lake Township supported the measure. Excluding the Calumet District also limited policy options to the Chicago River. The last step in the creation of the SDC was a special election in December, when the voters chose the first board of trustees. After the party bosses selected candidates unacceptable to civic leaders, they mounted a successful, independent campaign to elect a nonpartisan slate of best men. Judge Prendergast led the reform ticket, becoming the first president of the board. The *Tribune* declared a "glorious victory" for the reformers' modest $15-million drainage channel plan. "Honest men will rule," the newspaper forecast, ensuring that "more millions were saved to the people than any local election ever held in this state."[26]

Reforming a Crisis-Prone Society, 1885–1893

Chicago's expanded political and sanitary boundaries were manifestations of a community responding to the explosion of its population and size to metropolitan proportions. Following the flood of 1885, its civic leaders could claim credit for having solved the crisis of the physical environment—the drainage problem. As they also began to formulate plans for the celebration of the four-hundredth anniversary of Christopher Columbus's voyage, however, real and perceived crises of the social environment engendered a "new radicalism" among urban reformers. In their "search for order," this post–Civil War generation of college-educated experts and intellectuals helped guide American society toward a more-leisure-oriented culture of mass consumption, popular amusements, and outdoor recreation.[27]

As with their dread of an epidemic after the historic deluge, middle-class Chicagoans were held in the grip of fear of the social environment they had helped build. In a classic account of the "Age of Reform," Richard Hofstadter argues that this group of mostly white Anglo-Saxon Protestants suffered a loss of status in an increasingly multicultural society.[28] Their world seemed turned upside down by labor rising over capital, immigrants over native born, women over men, and blacks over whites. For them, the city had become a dangerous place in a perpetual state of crisis from outbursts of domestic terror and labor unrest; erosions of religious faith and practice; shifts in masculine identities, women's roles, and family relationships; failures to supervise youth in the streets; and loss of open space for leisure activities and nature preservation. Other historians have shown that the 15 to 20 percent of the city's population in the affluent classes were not alone in organizing reform movements. On the contrary, virtually every social and neighborhood group responded to the urban crisis of the 1890s by engaging in efforts to make Chicago a better, more-livable place.[29]

The World's Fair of 1893 was historic, in part, because it took place during a fleeting moment of time, when the past and the future overlapped. "By that time," John Higham observes, "a profound spiritual reaction was developing. It took many forms, but it was everywhere a hunger to break out of the frustrations, the routine, and the sheer dullness of a highly industrialized society. It was everywhere an urge to be young, masculine, and adventurous."[30] What appropriately became called the "bicycle craze" reflected a universal demand of people living within the new metropolitan-scale, industrial city to spend leisure time in nature.

The evolution of this machine from a dangerous "big wheeler" for male displays of strength and bravery into a "safety bike" that everyone wanted to ride represents a seminal case study in the social construction of technology. The simultaneous processes of sprawl and concentration left many Chicagoans with little open space within walking distance of their homes. Their desire to go mobile, to move freely around and beyond the city limits, led bicycle makers to design more-affordable machines with two equal-size, rubber-tubed tires; brakes; and a chain crank linked to the back wheel. As these models became available in the early 1890s, the number of middle-class buyers soared from a few thousand a year into the hundreds of thousands. Manufacturers developed the prototypes of modern assembly-line methods of mass production, while shopkeepers offered installment plans, trade-ins, used bikes for sale, and repair and rental services.[31]

Despite the polygenesis of distinct forms of reaction to each crisis, they converged on a single therapeutic remedy: get back to nature. And in sharp contrast to the Victorian, agrarian myth of picturesque, pastoral landscapes, moreover, the Progressives idealized a different imaginary of nature. "Simply put," Peter Schmitt states, "this urban response valued nature's spiritual impact above its economic importance; it might better be called 'Arcadian.'"[32] In other words, it was not a back-to-the-land movement. As spending time outdoors became commodified as a leisure activity, the benefits of saving the natural environment gained ground against the costs of its maximum economic value. Revival preachers, women's emancipators, medical experts, and a host of others advocated getting outside and moving about as the single-best way to heal the mind, body, and soul from the stresses of daily life in the city. In fact, Chicagoans needed no extra encouragement; they were bursting with the desire to get outside.[33]

On May 4, 1886, Chicago's working class gave violent expression to its pent-up demand for more leisure time. The Haymarket bomb, the first such terrorist act in the United States, caused a civic trauma greater than the natural disaster of the great flood less than a year earlier. Many employers imposed ten- to twelve-hour shifts, five-and-a-half days a week, leaving workers little free time. Sabbatarian prohibitions further restricted their options for outdoor recreation on their only day off. Sunday blue laws closed most easily accessible popular amusements. The workers' exhausted bodies and overstressed lives formed the backstory to their rallying cry: "eight hours for what we will."[34] A week of labor strikes for a shorter workday led to the fateful meeting on the western fringe of the CBD. The dynamite explosion among a charging phalanx of police and their retaliatory gunfire eventually left seven officers dead in addition to four fatalities and scores of wounded among the spectators.[35]

The Haymarket bomb was a watershed event that changed the course of Chicago's society and culture. "In the popular imagination," according to cultural historian Carl Smith, it represented "social and political protest, class warfare, and cataclysmic violence, all set against the backdrop of the industrial neighborhoods of American cities, as a single phenomenon. This nightmare, after all, had been inscribed in the public mind for some time."[36] The shock of the terrorist act forged a consensus among civic leaders. They not only cracked down on political radicalism but also upgraded local organizations devoted to municipal improvement, social welfare, moral order, and cultural philanthropy. The transformations of the YMCA and the Woman's Christian Temperance Union (WCTU) illustrate the first responses of established institutions of reform.

The now-obvious failure of past efforts also opened opportunities for novel approaches to bridge the gap of inequality between the rich and the poor. An outstanding example of the new radicalism was the campaign of Jane Addams, Graham Taylor, and Mary McDowell for public playgrounds and parks with sports facilities in every city neighborhood. These social settlement house workers took a hands-on approach, getting directly involved in improving the condition of the working class. Park proponents, Bachin underscores, "equated natural space with public space and sought ways to infuse nature into the city."[37]

The contrast between traditional and novel forms of urban reform has been characterized as "moral environmentalism" and "social environmentalism," respectively.[38] The Victorians, on the one hand, believed that poverty, crime, and mental distress were caused by failures of the individual.

They were moral determinists. In Chicago, an exemplar of this approach was the YMCA's guiding spirit, Dwight L. Moody, who worked tirelessly to convert the fallen one by one. From the start of the Civil War to the end of the World's Fair, the successful businessman turned charismatic preacher would lead a virtually endless evangelical revival meeting in an effort to create the sanctified community.[39]

The Progressives, on the other hand, believed that unhealthy slums, industrial pollution, and hazardous workplaces were the root causes of urban disorder. They were environmental determinists. Improving the physical conditions of the city, Addams thought, would bring about commensurate gains in the social welfare. Decent housing, clean streets, and neighborhood parks would lead the behavior of the masses toward higher, middle-class standards of morality, citizenship, and patriotism. Three years after the Haymarket bomb, she and her friend Ellen Gates Starr moved into a near West Side river ward. Close to the jobs in the CBD, it was the home for many European immigrants. At Hull-House, they began their postgraduate educations by applying an approach typical of the new radicalism: a scientific, door-to-door survey of the neighborhood. Addams and her allies also worked alongside traditional civic and charity groups as well as with labor unions, which were trying to close the class divide.[40]

The difficult transition of Chicago's YMCA from Moody's born-again vision of individual redemption to the more-corporate, social gospel of its "Metropolitan Plan" illuminates the period's "crisis of masculinity." Coming from New England to what he called a "wicked city" in 1856, the nineteen-year-old Moody, a salesman of boots and shoes, got involved in a second attempt to establish a YMCA a year later and immediately became its inspirational leader.[41] "[His] evangelical spirit," biographer Justin Pettegrew highlights, "with its emphasis on aggressive outreach and proselytizing, provided the distinguishing character that dominated the early Chicago YMCA from its inception and into the 1880s."[42] Although Moody's influence made the YMCA's reading room more important than its gymnasium, compared to those of other big cities, he embraced contemporary ideals of "muscular Christianity" in the basic sense of bodily health and manliness.[43]

The Haymarket bomb of 1886 made a profound impression on Moody and his longtime financial backers, including several of the city's richest men. He warned them, "Either these people are to be evangelized, or the leaven of communism and infidelity will assume such enormous proportions that it will break out in a reign of terror such as this country has never known."[44] His previously unheeded appeals for funds to establish a training school for urban missionaries—"gapmen"—to reach out from the churches to the streets were answered the following year. The graduates of the Chicago Evangelical Society (later the Moody Bible Institute) would play an active role in Moody's last great revival during the World's Fair.[45]

The terrorist act also marked a watershed event for the YMCA, ushering in a second generation of leadership by Chicago's wealthy families. In 1888, the sons of Cyrus H. McCormick, John V. Farwell, and other elite businessmen pushed out Moody's loyalists and installed a new chief executive from New York City. His "Metropolitan Plan" considered the new scale of Chicago by facilitating the formation and governance of YMCA "branches" throughout the region. More importantly, he responded to the crisis of masculinity by excluding women from the organization and instituting programs that stressed character building through team sports and athletic contests. In 1894, the YMCA opened a new building, which provided more floor space for playing games, swimming, and participating in other leisure activities than for prayer sessions in the auditorium.[46]

The YMCA responded to the gender crisis by creating a homosocial environment that was intended to foster what its most famous champion, Theodore Roosevelt, called "the strenuous life."[47] In the Gilded Age, the feminization of American urban culture, especially within the Protestant Church, added to perceptions of declining social status in these men. "The ideals of the strenuous life," according to historian Clifford Putney, "[were] physical hardness, vigorous action, and the rejection of various genteel constraints."[48]

The new generation of YMCA leaders also reacted in the public sphere by reasserting the power of men over women. "If the immigrant, the city, and rural depopulation constituted the first three horsemen of the nativist apocalypse," Putney confirms, "many considered the 'modern woman' to be the fourth."[49] Besides expelling them, the YMCA's emphasis on structured team sports—baseball, football, and basketball—was designed as an exercise in character building. Still appealing mostly to young office workers and salesmen, these reformers hoped to inoculate them with the right stuff to become community leaders. They were on a mission not only to restrict women from the public sphere but also to restore a proper Christian order of moral control over urban society and American culture.[50]

Chicago's women had their own crisis of gender relationships to deal with at the time of the World's Fair. The controversy over their use of bicycles helps expose the struggle of women for emancipation and equality. Evanston's Frances E. Willard, the national president of the WCTU, was outspoken not only as a prohibitionist but also as an advocate of the gospel of health. Linking the two causes, she claimed that bicycle riding was "the greatest agent of temperance reform." In her 1895 book, *A Wheel within a Wheel*, Willard provides a fascinating account of a member of the only generation who learned how to ride bicycles as adults. "I found a whole philosophy of life," she writes, "in the wooing and winning of my bicycle." It was not easy, but eventual mastery gave meaning to her two-wheeler named "Gladys" as a freedom machine. Composing from between a quarter and a third of riders, women were an important group of customers in a highly competitive marketplace.[51]

Nonetheless, women had to peddle uphill to overcome male resistance to their desire to enjoy the freedom of auto-mobility outside in public view. Ellen Gruber Garvey shows that "both defense and attack took medicalized form: anti-bicyclers claimed that riding would ruin women's sexual health by promoting masturbation and would compromise gender definitions as well, while pro-bicyclers asserted that bicycling would strengthen women's bodies and thereby make them more fit for motherhood."[52] In addition, Willard argued that it was a unique form of recreation, because it took men as well as women back to nature as an alternative to sitting inside, drinking alcohol in a saloon or nightclub. In a compromise typical of Progressive reform, women kept riding but on segregated terms of inequality. Despite adding to its price and weight, for example, manufacturers had to redesign the diamond-shaped frame of the safety bike to facilitate the continued wearing of long dresses. Some feminists wore bloomers in protest, but rules of proper decorum were prescribed in minute detail in mass-circulation magazines and romance novels.[53]

As a part of the back-to-nature movement, the bicycle craze became swept up in another crisis of modern urban life: the desecration of the Sabbath. The men had already deserted the church, and now the women were also spending their Sundays enjoying leisure activities. By 1893, whether the bicycle was God's or Satan's invention became absorbed in a major controversy over whether to close the World's Fair on the Sabbath. Like Willard, many ministers came to the defense of getting back to nature. Consider a sermon titled "The Church and the Wheel," delivered by Frank Lloyd Wright's uncle, the Reverend Jenkin Lloyd Jones. The Unitarian clergyman declared that bikes were machines from God's factory that built good character, restored physical health, and "made optimists out of gloomy dyspeptics."[54] And, like the "Metropolitan Plan" of the YMCA and other forms of the new radicalism, Jones's sermon sought to reform sports, not repress them. In a similar spirit of compromise, the gates of the fairgrounds remained open on Sunday, but many buildings were closed for the day.

Taking a parallel, albeit separate, line of attack on the urban crisis of the 1890s were Addams and her female supporters in a campaign to provide supervised playgrounds in every working-class neighborhood. "Organized play," sports historian Dominick Cavallo underscores, "was a means of shaping mental and moral faculties: it had little in common with the escapism of either sport or casual recreation."[55] Drawing on Darwin's biology, social scientists posited evolutionary theories of child development. To turn little savages into American citizens required experts to design the proper physical environment and social controls of early childhood. In short, child's play was too important to be left to children.[56]

In the year of the fair, Addams persuaded a philanthropist to donate a city lot near Hull-House for a playground. Taylor's settlement house on the North

Side and McDowell's on the South Side would soon add similar facilities for organized play. In 1897, the campaign would begin to pick up momentum when the first school playground was built with private donations, followed a year later with public funding by the school board for more open land and professional supervisors. The next year, New York City's child crusader Jacob Riis gave a lecture at Hull-House on the importance of organized play, which, in turn, inspired the local Municipal Science Club to lobby city hall to create a Special Parks Commission (SPC). As Addams had done to create maps of the neighborhood, the club conducted a scientific survey of public green space in Chicago. What they found was that additions of land to the system of parks and boulevards created just after the Civil War had fallen far behind increases of population to the city census.[57]

More alarming, the investigators discovered that the economic crisis of class conflict was compounded by a geographic crisis of environmental inequality. In the new quantitative jargon of social science, the SPC reported that affluent residents living in the eleven lakefront wards were provided with 1,814 acres/734 hectares (ha) of park space, or 234 residents per acre/578 residents per hectare, in contrast to 228 acres/92 ha of park space, or 4,720 residents per acre/11,697 residents per hectare, in the other, working-class wards. Women formed the backbone of the movement to reform Chicago's society and culture by enlarging the power of government. "Women," Bachin points out, "played a central role in linking recreation with municipal control."[58]

In response, all three of Chicago's park authorities on the South, West, and North Sides, respectively, would sponsor successful referendums to fund vast expansions of their recreational space and sports facilities.[59] What the SPC failed to report on was the crisis of ethnic and racial conflict over access to the limited public green spaces and semipublic amusements available to working-class residents within reasonable distance from their homes. Only in 1922, after a citywide race riot had erupted three years earlier from conflict over the public beaches, would another special commission admit that the most densely packed ghetto of the city, Bronzeville, was the least served by its park system.[60]

Shocked into action by the great flood and the Haymarket bomb, Chicago began to undertake a reorientation of its culture and society several years before the World's Fair of 1893. These two traumatic events not only exposed multiple layers of physical and social crises from the failure of past reform efforts but also opened the way for a new radicalism of direct action and environmental determinism to gain support from the civic, business, and political establishments. The double blast of their shockwaves set into motion the YMCA's promotion of muscular Christianity, the WCTU's crusade to get everyone back to nature on a bicycle, Hull-House's campaign for children's playgrounds, and many other new-style reform movements. For visitors to the World's Fair, the layout of the grounds and the buildings filled with stuff seemed comfortably familiar, because they were all meant to showcase the culmination of progress during the Age of Industry.[61]

At the same time, Chicago's modern downtown and its great fair displayed the promise of a dawning Age of Reform. "From its opening day in May 1893," historian William Wilson asserts, "the World's Columbian Exposition has exerted a major cultural influence."[62] The beauty and cleanliness of its grounds; the high technology of its electrical displays; and the efficiency, order, and safety of its day-to-day operations created a utopian image, a "Dream City" of possibilities.[63] Given Chicago's perpetual state of crisis, it is no surprise that the exposition's organizers and managers applied its lessons in formulating their plans of urban reform. For instance, the upgrading of the entire 686-acre/278 ha site of the fair, Jackson Park, taught the lesson that beauty pays. Creating an Arcadian landscape of lagoons and lakes, its planners set off a housing boom along its borders. Real-estate developers became major supporters of public parks, because they increased residential land values while attracting tourists and business to the city. Chicago's architects and landscape gardeners took away two additional insights from building the fair: the first was the utility of comprehensive planning on a metropolitan scale, and the second was the need to include environmental values in any future analysis of the costs and benefits of city building.[64]

Building the American Dream, 1893–1900s

Except for the outbreak of epidemics, the housing crisis was the greatest challenge facing the advocates of the new radicalism. As environmental determinists, they believed that overcrowding in the slums was the seedbed of crime, immorality, and family dysfunction. During the 1890s, the Chicago metropolitan area of Cook County gained 650,000 more people, rising to a population of 1,840,000 by the end of the decade (see Figure 1.2). For the first time, the census recorded a faster rate of growth in the suburbs of 65 percent, compared to a 54 percent rate in the city. Nevertheless, 92 percent of the increase took place within its expanded borders.

Keeping up with this demographic explosion posed a Herculean task in city building that chronically lagged behind needs for affordable housing and essential infrastructure. The success of the annexation movement was itself an accurate indicator of pent-up demand for all the modern conveniences of urban life. While social-settlement workers improved existing conditions in the inner city, other reformers sought to move its residents to healthy homes in garden suburbs.[65]

To a large extent, the movement for better housing in Chicago during the Progressive Era was driven by a series of public health crises. As in the past, the fear of imminent suffering and death from a contagious disease united its citizens and policy makers like no other cause of reform. The worst epi-

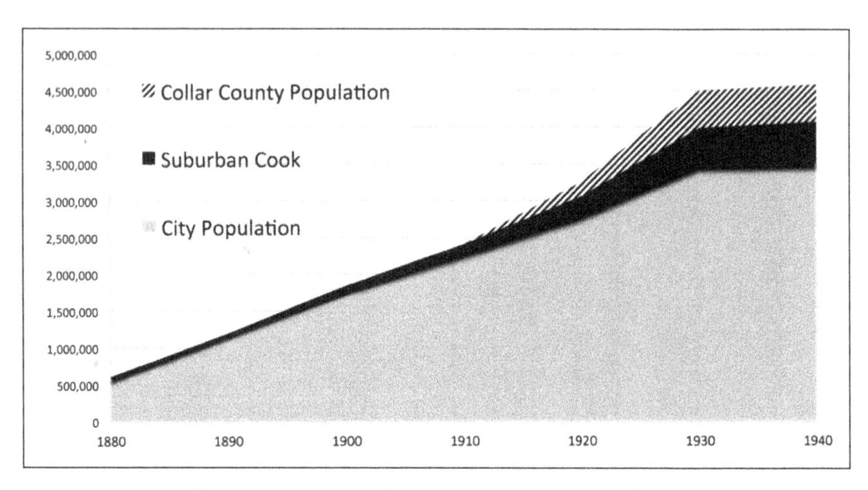

Figure 1.2. Chicago Area Population, 1880–1940 (Data from U.S. Census)

demic outbreak since the 1850s began in 1890 and lasted for three years. Causing 4,494 deaths, this plague of typhoid fever was followed during the World's Fair of 1893 by the spread of smallpox and the resulting vaccination of a half million school children and their parents. Medical experts led the campaign for the modernization of the Health Department and the appointment of the Pure Water Commission. They were also active supporters of Addams's efforts to clean the neighborhoods and provide them with playgrounds.[66]

The Progressives also believed in the promise of technology to solve the housing crisis. In postfire Chicago, advances in public transit from horse railways to cable and electric traction gave rise to "streetcar suburbs."[67] Like the railroad suburb of Evanston, Harvey was planned to create not only an ideal physical environment but also a sanctified village of temperance and moral order. And like the industrial suburb of South Chicago, this hybrid project in community building was designed to bring manufacturers and their blue-collar workers together. In more-upscale suburbs, such as Riverside and neighboring Oak Park, Frank Lloyd Wright and his followers were designing high-technology dwellings. Their prairie-style "organic architecture" was planned to reconnect the members of the household with nature and each other in a child-centered family.[68] Other anonymous homebuilders, inventors, and artisans collectively contributed to a masterpiece of domestic technology and vernacular architecture: the bungalow. Setting off a sustained, thirty-year housing boom, this model of modernity would come to represent the American dream for almost a hundred thousand families.[69]

One of Moody's early supporters, Turlington W. Harvey, advanced a plan to solve the housing crisis by building a suburban oasis for blue-collar workers. Making a fortune supplying lumber after the fire of 1871, the businessman also continued to be a leader in the evangelical work of saving souls. To create a sanctified community of beauty, health, and order, he formed the Harvey Land Association in 1891 and purchased 700 acres/283 ha on the far South Side, where two railroad lines crossed paths. Harvey, according to historian James Gilbert, "offered rural or small-town [white] Americans a safe urban haven and manufacturers a sober and orderly working force."[70] To be sure, the lumberman was not the first Chicagoan to build a model industrial suburb. Eight years earlier, the railroad carmaker George Pullman had constructed an ideal physical environment for his workers, but they had to live under his oppressive regime of puritanical controls. In comparison, Harvey offered more-liberal land policies and greater freedom to create autonomous community groups. The widely advertised venture became an instant triumph, a "Magic City" with five thousand inhabitants during the year of the fair.[71]

But, like that of Riverside twenty years earlier, the Harvey land company fell into bankruptcy in 1893, when the nation plunged into another depression. Pullman suffered a worse fate from a revolt of his workers in May 1894 that led to his company's occupation by the U.S. Army. Just after the Columbian Exposition closed, it suffered the worst fate of these utopian responses to the crisis: on July 5, 1894, arsonists burned its remains to the ground. With military troops guarding against class warfare breaking out in the CBD, and medicine men fighting against epidemics spreading in the neighborhoods, Chicago stayed gripped in a psychological state of emergency.[72]

If the Pullman strike revived ghosts of the Haymarket bomb, the high anxiety related to drinking a glass of water from the tap never went away after the deluge of 1885. Despite the victory of reformers in the battle to create a metropolitan-scale sanitary authority in 1890, the ongoing crisis of public health from contagious outbreaks of water-borne diseases got steadily worse. While engineers were constructing the Sanitary and Ship Canal, scientists were building the case in the public arena that Lake Michigan's water was unsafe to drink. Rather than a fountain inexhaustible of everlasting purity, as the engineers proclaimed, the lake was permanently contaminated, according to public health experts. Even after the Chicago River was reversed, they pointed out, the sewer outlets of 250,000 people living along the lakefront would still be emptying into the lake. Their new biological metrics convinced them that dangerous microbes in this liquid waste were not only reaching the waterworks' intake cribs but also pouring out of everyone's household faucets. In the face of raging epidemics, the Health Department's advice to boil the water was less than reassuring to a public living in fear for their lives.[73]

For those who could afford to escape from the city, the suburbs offered safe havens with all the amenities of modern life. What the middle class wanted, according to Wright, was a "city man's country home on the prairie."[74] In

1889, the aspiring, twenty-two-year-old architect got married and moved to Oak Park, which, like Evanston, rejected annexation. Over the next decade, he found a way to express this twofold desire to get back to nature and to revive domestic life in an original form of three-dimensional design. "The prairie house," architectural expert Robert C. Twombly points out, "appealed to an apprehensive upper middle class by emphasizing in literal and symbolic ways the security, shelter, privacy, family mutuality and other values it found increasingly important in a period of urban dislocation and conflict." Wright's "golden years" of success spanned from 1901 until his escape to Europe with a client's wife eight years later.[75]

Set within a controlled and beautified suburban nature, Wright's "organic architecture" and reconfigured interior spaces appealed especially to those seeking a resolution to the crisis of gender relationships that differed from muscular Christianity. Reflecting the flip side of the strenuous life, or what historian Margaret Marsh calls "masculine domesticity," the houses' open family areas fulfilled the desires of some men and women for more-intense, equalitarian marriages. "The new domestic ideal, centered firmly in the suburbs," she states, "represented family pride, family identity, and togetherness in the face of an urban society that promised individual achievement, anonymity, and excitement."[76]

Tellingly, many of Wright's clients came from the new middle class of corporate technocrats and managers. Enjoying unprecedented job security, they turned muscular Christianity into "male domestic responsibility" to teach their sons to become manly. These fathers did not mind that the hunting clubs of their father's generation had morphed into country clubs.[77] Besides the house façade's stylish blend of arts-and-crafts and prairie-landscape aesthetics, Wright's use of a new generation of building materials met women's demands for healthy interior spaces that were easy to keep clean. Although his suburban designs often included quarters for domestics, they pointed to a future of the servantless home. During this period of steady demand for new housing, access to such railroad suburbs as Oak Park and Evanston opened as electric street railways and paved roads linked them more conveniently to downtown.[78]

In many respects, the bungalow emerged as Chicago's most numerous and enduring type of single-family dwelling, because it represented a scaled-down prairie house. It was diminished in size and cost, but not in design or comfort. And it served in similar ways to reduce the tensions of the environmental and social crises of the 1890s. Conjuring images of California's land of sunshine, "the idea that the bungalow was especially suitable for outdoor living appealed to a generation of Americans who were rediscovering the joys of outdoor life," according to Clifford Edward Clark.[79] What emerged during the 1890s from the workingman's cottage was an arts and crafts–style, one-and-a-half-story brick bungalow. It was the combined product of new mass-production techniques and older traditions of individual craftsmanship and invention. The plumbing fixtures, electrical systems, and central heating units, for example, became increasingly standardized and manufactured on an assembly line. At the same time, the brickwork, limestone trim, art glass, and especially the interior woodwork remained in the hands of skilled workers, who continued to display the artisan skills of their forefathers.[80]

One important difference between Wright's upscale market homes and the mass-production bungalow (and "two-flat" apartment building) was the latter's need for a basement for the central-heating furnace and coal-storage bin. Wright had grown up on the prairie in southwestern Wisconsin; he solved the problem of a flood-prone environment by building on a slab foundation and raising the living spaces above ground level. He had the luxury of larger construction budgets and property lots to stretch out his land-hugging designs to bring the inside closer to the outside. At the end of his golden years, he would reach the ultimate expression of this integration of interior and exterior space, the built and the natural environments. In Riverside, he designed a house that was broken apart into modular units set within a landscaped garden.[81]

In contrast, the bungalow was set on long, narrow city lots. It needed a full basement to support the brick construction and to take advantage of up-to-date hot-water and steam-radiator heating systems. Moreover, storing the coal, tending the furnace, and getting rid of the ashes were dirty business, best kept away from the living quarters, which sat a half story above ground level. Homebuilders also put the laundry sinks downstairs from new-style kitchens designed for the modern housewife and the scientific management of her family's diet of increasingly mass-produced, processed foods.[82]

The spreading appeal of living in a suburban nature can be put in a context of the rise of metropolitan-scale industrial cities, but the "Bungalow Craze" of homeownership among the middle class requires a more-complicated explanation.[83] When Wright set up his own practice in the year of the World's Fair, about three out of four of these households lived in rental units with long-term leases. The uncertainties of the economy's booms and busts, the lack of long-term mortgages, and a host of other shortcomings of the housing market had kept people from buying property. They also remained skeptical about making an investment in real estate, especially highly speculative land on the crabgrass frontiers of a half-submerged marshland. Although counterintuitive, it was the immigrant worker who valued property rights in homeownership, because they established a sense of American identity, social respectability, and economic independence. In postfire Chicago, the response to this demand had been wood construction, balloon-frame cottages in fringe areas, and the rise of large-scale homebuilders. Samuel Eberly Gross became the single-largest developer by integrating the main steps of purchasing and subdividing the land, erecting the homes, and financing their purchase. The real secret of his success, however, was a genius of marketing his products by playing on the fears and dreams of his customers.[84]

In the 1890s, Gross made a transition from promoting low-cost cottages to more-expensive bungalows. In her brilliant history of homeownership in Chicago, *City of American Dreams*, Margaret Garb exposes how the mass-circulation advertisements of this super salesman tapped into the decade's sense of apocalyptic doom and its idealistic movements of reform. Building on the workingman's message, "the emerging middle class attraction to property rights in housing," she reports, "drew new attention to home ownership as an investment that would enhance status, secure gender order, safeguard family health, and prove, as builders asserted, a 'useful investment' for aspiring and more affluent families." From 1880 to 1900, Gross sold forty thousand lots and built ten thousand homes. Amassing a personal fortune worth $5 million, he was one of the masterminds of the reorientation of Chicago's culture toward consumption and its social geography toward suburbanization. "By 1910," social historian Clark confirms, "the ideal bungalow had emerged as the all-American family house."[85]

In the 1900s, Gross and his fellow homebuilders expected to enjoy rising profits from the seemingly endless flow of people moving to the again-booming metropolis of the Midwest. The city absorbed an additional 486,000 newcomers, a 29 percent gain in a total population of 2,185,000 inhabitants. Suburban Cook County grew at a rate twice as fast, increasing by 80,000 new residents to a total of 220,000 inhabitants. Together, the metropolitan region would hold more than 2.5 million people by the 1914 outbreak of war in Europe.[86]

While the CBD resumed building skyscrapers, the ring of land extending 3 miles/5 km around it experienced a more-checkered development. The victory of the downtown real-estate and manufacturing interests in routing the Ship and Sanitary Canal through the Chicago River meant that it remained the industrial heart of the region. And the river wards remained the most densely packed, with recent immigrants and others at the bottom of the economic ladder. In a 1-square-mile/2.6 square km area hugging the river on the near South Side, land-use economist Homer Hoyt exclaims, "were packed 73,400 people, or twice as many as lived in 88 square miles [228 square km] inside the outer edges of the city limit."[87]

This housing expert also marks this period from 1893 to 1914 as the time when spatial segregation by ethnicity/religion and race joined class divisions in sorting out the region's social geography. He identifies the emergence of African American, Chinese, and Eastern European Jewish ghettos within the 3-mile/5 km inner ring of postfire settlement. At the same time, older residents made an exodus to the crabgrass frontier of the subdivisions and the community builders. Most of the new construction of single-family houses and two-family apartments took place in streetcar suburbs along the routes of the expanding rapid-transit lines. Beyond the city limits, established suburbs, such as South Chicago, Riverside, and Evanston, sprawled outward toward their boundaries, while many new places cropped up all around them. Like the city's diversity of neighborhoods, Hoyt concludes, "these outlying settlements . . . ran the entire gamut of the social scale from the squalid quarters in South Chicago . . . to the spacious estates of the millionaires in Lake Forest."[88]

By making homeownership in the suburbs the American dream for the middle class, urban culture and nature itself became commodified. The price and location of your dwelling came to represent society's primary measure of family life, health, status, child welfare, personal autonomy, and, perhaps, happiness itself. The children of Chicago's first generation of big businessmen set the fashion, moving to the outskirts of town. They duplicated the exclusivity of their parents' residential districts within the city by decamping to a string of suburban enclaves beyond Evanston along the North Shore all the way up to the twin pinnacles of success: Lake Forest and Lake Bluff. For the growing mass of middle-class consumers, a bungalow in a suburban subdivision provided a route of escape from the real and imagined perils of living in the inner-city ring of postfire settlement. Mass-production techniques of residential construction brought the price of ownership within their reach. But including all the conveniences of modern technology increasingly excluded the working class from being able to afford to own property. The social construction of the American dream in Chicago, then, went hand in hand with the physical creation of a spatial order segregated not only by class but also by gender, race, and ethnicity/religion on a metropolitan scale.[89]

Preserving a Prairie Wetland, 1899–1909

The same crises and responses fueling suburban flight to save the family also gave birth to a reform movement to save nature from the city. As we have seen, one reaction to the crisis of open space was the 1899 formation of the SPC, which became the starting point of comprehensive city planning in Chicago. Five years later, its work would lead to a report calling for a "Metropolitan Park System" of 35,000 acres/14,164 ha to safeguard the city's remaining forests along the six river watersheds from the coming crisis of urban sprawl. In 1909, Burnham and Bennett would incorporate these ideas about conservation into their city of the future. Planning on a grand scale, their "outer belt" of preserves was twice as large.[90]

Yet the swelling ranks of the back-to-nature movement did not lead to a political consensus behind plans to expand public parks and outdoor recreation centers. On the contrary, the near-universal demand for more open space turned into a struggle among a wide range of place-based groups, economic interests, and political parties. After 1893, questions of site selection, funding, administration, and use became chronic sources of partisan contestation and social conflict.[91] The decade-long battle to establish a Forest Preserve District of Cook County illustrates the growing influence of a metropolitan point of view in the public sphere. To save the river watersheds in the suburban periphery, supporters had to overcome constitutional hurdles and organized opposition from civic elites and labor unions. The reformers' success in rallying a majority to keep voting in favor of setting aside a nature reserve for future generations marked the ascendency of environmental values in the formation of the city's building plans and policies.

The multiple crises of public-health emergencies from contagious diseases that literally drove people out of the inner city also engendered the first injection of environmental values in policy debates over the future of the city. In 1896, the Pure Water Commission had reiterated the accumulating scientific evidence that disease-causing microbes and toxic chemicals were present in the water supply most of the time. Echoing Chesbrough's and Hering's recommendations, these experts called for interceptor sewers along the lakefront to divert the contents into the soon-to-be reversed Chicago River. But in sharp contrast to the engineer-led report from the Drainage and Water Supply Commission of 1887, this panel's report expressed the views of the practitioners of the new public health. Going beyond previous planners, they argued vigorously that the social benefits of the conservation of nature outweighed the economic cost of saving it. To solve the crisis of deadly epidemics, they insisted that modern technologies of water purification and sewage treatment must be installed to solve the larger crisis of a polluted urban environment.[92]

The importance of their report cannot be overstressed, because it represents the first time environmental values were given consideration in official policy calculations of the costs and benefits of city building. The insurgency of the conservation movement, however, met stiff resistance from the professional politicians in control of the sanitary authorities. Conceding to the irrefutable evidence of contamination of the water supply by the lakeshore sewer outlets, the city council approved funding to begin the construction of interceptor sewers. It could well afford them for two reasons directly related to its management of Chicago's water. In the annexation of 1889, on the one hand, the ward bosses had shifted the costs of extending the sewer system from the municipal budget to a method of taxation that levied special assessments on the affected property owners. On the other hand, the council ordered the Water Department to increase the number of hookups as fast as possible to engorge its already obscene "surplus fund." In 1893, for example, the ratio of the department's costs compared to the receipts that poured into this slush fund was 400 percent. Free to spend this money any way they wanted, the power holders in charge of the formation of public policy had no intention of earmarking any of it for the conservation of the city's aquatic environments. On the contrary, the city councilmen continued to insist on exploiting them as free and unlimited economic resources. A cynical policy of wasting water fed the growth of not only their "surplus fund" but also their patronage system of jobs and contracts.[93]

For this reason alone, the opportunities in 1887 and in 1896 to combine the city's drinking supply, sewerage, and drainage into one, more-efficient agency of water management were lost. Moreover, the ward bosses had taken fewer than two years to wrestle control of the SDC from Judge Prendergast and civic reformers. Its top engineer, Lyman E. Cooley, had secretly made a deal with the Democratic leadership to fulfill his and other boosters' dream of surpassing New York to become the nation's largest port. In exchange for helping him take control of the SDC's governing board in 1891, he had pushed through a plan that called for hiring a maximum number of workers and employing a minimal number of modern technologies. In next nine years of constructing the 28-mile/45 km channel, Cooley's patronage army of thousands dug out more tons of ground than would soon be removed in building the Panama Canal.[94]

Although the advocates of a conservation ethic of environmental values had reached the limits of reform of Chicago's waters, they were able to regain the initiative by mobilizing public support behind the back-to-nature movement. After 1893, the logic of setting aside little patches of open space for children to play snowballed into an avalanche of expert and layperson advocacy behind proposals to provide similar opportunities for teenagers, adults, and whole families. While children's playgrounds required less than 1 acre/0.4 ha, according to SPC member Graham Taylor, "recreational parks" needed from 2 to 60 acres/0.8 to 24 ha each. Besides outdoor playing fields and swimming pools, these larger green spaces were designed with year-round community centers, which acted like settlement houses. A Chicago invention, the public-park "field house" contained gender-segregated gyms and showers and gender-integrated meeting, lunch, and reading rooms in addition to "well-educated social director[s]."[95]

Prairie-style architect Dwight Heald Perkins, landscape gardener Jens Jensen, and public official Henry G. Foreman envisioned saving an even larger system of forest preserves and prairie wetlands on a metropolitan scale. In the battle to save this "outer belt" of open space, each of these spearheads of reform came to represent a basic reason to give currency to environmental values of nature and human conservation in the calculus of the costs and benefits of urban growth. Born during the Civil War period, they reached the peaks of their careers at the turn of the century. Perkins personified the visionary plans of the future metropolis; Jensen, the spiritual values of the natural environment; and Foreman, the civic ideals of business leadership. The three men were a part of an interlocking network of Progressives associated with Addams and Hull-House. Short biographies shed light on how this inner circle of new radicals built and sustained the momentum of their cause for more than a decade, turning it into an enduring legacy of the Age of Reform.

The story begins with Perkins, because he was one of those rare creative artists who were also first-rate organizers and managers. Born in 1867, he was raised in Chicago by a single parent, Marion Heald. A member of one of its pioneer families, the widow and her son were taken under the wing, so to speak, of her childhood friend, Charles Hitchcock. One of the city's most successful lawyers, Hitchcock and his wife funded Perkins's education at the Massachusetts Institute of Technology in Boston, where he became an architect and met his future wife. When Perkins returned to Chicago in 1888, his family connections got him a job in the fast-rising firm of Burnham and Root. For the next five years, he served as the supervisor of several skyscraper projects while helping plan the World's Fair. With Burnham's support, Perkins established his own firm, becoming a leader in the prairie-style school of architecture.[96]

His metamorphosis from designer of tower blocks to community centers and public schools took place at Hull-House. His mother had become one of its social workers, linking Perkins to Addams, the arts-and-crafts aesthetic, the children's playground movement, and the plight of the poor. An inexorable organizer, Perkins brought together a group of young, new radicals, including Wright, by providing them studio space in the attic of one of his first commissions, Steinway Hall. Perkins and his inner circle of the Chicago Architecture Club collectively worked out an "organic architecture" of the Midwest. At Hull-House, he also met Jensen. They collaborated professionally and as activists in the Municipal Science Club's initiative to establish a system of public parks on a metropolitan scale. After serving as the aptly named "compiler" of the landmark report of 1904, Perkins would be appointed the architect of the public schools, designing forty buildings over the next five years. Considered a "masterpiece" of the prairie style, his Carl Schurz High School presents "a brilliant exhibition of virtuosity that marks the high point of non-commercial architecture in the Chicago tradition," historian Carl Condit observes.[97]

Perkins's contributions to the plan for a "Metropolitan Park System" went far beyond an editorial role. In fact, he was the main author of what in retrospect was a crucial stepping-stone or precursor to another masterpiece, the *Plan of Chicago*. Like others involved in building the World's Fair, Perkins had learned the lesson of comprehensive planning. In effect, the park plan represented a topographical layout of Chicago fifty years in the future, minus the buildings (see Figure 1.3). Based on scientific calculations of population growth, its projected ten million people would cover the entire region. The city's present shortfall of parkland acreage per person, Perkins reasoned, underscored the need to act now to save nature from the city before it was too late.

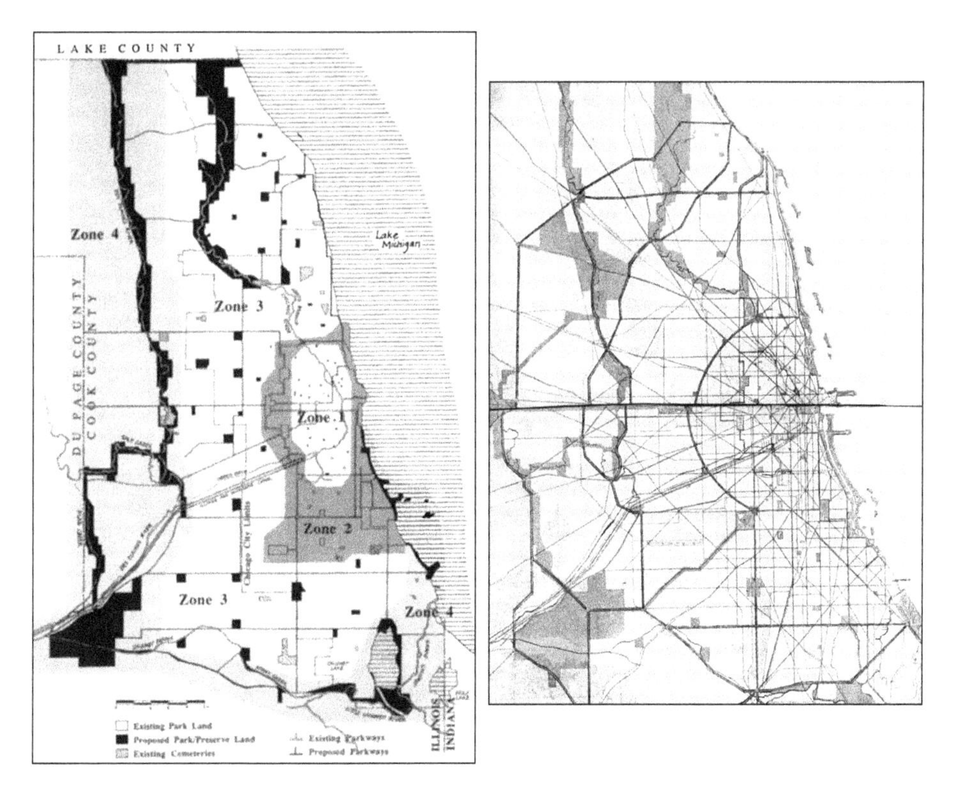

Figure 1.3. J. Jensen Outer Belt Park Plan, 1902, and *Plan of Chicago* Preserves Plan, 1909 (Dwight Heald Perkins, compiler, *Report of the Special Park Commission to the City Council of Chicago on the Subject of a Metropolitan Park System* [Chicago: City of Chicago, 1904], following p. 44; Daniel H. Burnham and Edward H. Bennett, *Plan of Chicago* [Chicago: Commercial Club, 1909], fig. 64, available at http://www.archive.org/details/planofchicago00burnuoft)

The report also presents an outline for the beautification of the downtown area. "What that district needs," it declares, "is 'open space.'" Giving credit to his fellow reformers, Perkins summarizes current ideas for a remake of the lakefront and the Chicago River. Its reclamation called for replacing warehouses and rail yards with boulevards (the future Wacker Drive) and river walks.[98] In addition to setting aside recreational space for future generations, the contribution of the commission's landscape architect, Jensen, provides two other reasons to save an outer ring of river watersheds and prairie wetlands from too much urban growth: some of these still-unspoiled areas contained historic sites from the pioneer days, while others were virtual biology laboratories for scientists. Jensen served as the city's guide to its wild places, where he pointed to examples of human-caused destruction and the restoration of nature. Like his kindred spirit, Henry David Thoreau, Jensen became an expert naturalist through careful and systematic observation.[99]

Jensen, a twenty-four-year-old native of Denmark, had moved to the United States with his bride in 1884. Like Perkins and Foreman, he came from a privileged background, but unlike these native sons, the immigrant started at the bottom rungs of the economic ladder as a street sweeper when he arrived in Chicago two years later. Getting a job with the West Park District, Jensen took only two more years to work his way up to superintendent of Union Park, where he had planted an indigenous "American Garden." He continued to move up the ranks of the park authority until he was fired in 1900 for refusing to cooperate with a corrupt administration. He formed a private company, attracting some of the city's richest families, who were moving to suburban estates. In 1905, his influential friends would have him reinstated at the West Park District, where he would hold the position of consulting landscape architect for the next fifteen years.[100]

In effect, Jensen was Chicago's counterpart apostle of the prairie wetlands to Wright, the leading exponent of an "organic architecture" of the built environment. Although they worked together on several residential projects, the outspoken iconoclasts clashed. In contrast to Wright, Jensen was embraced by the city's elite families, becoming a member of their most prestigious clubs, a sought-after guest speaker at their dinner parties, a well-paid retainer of their patronage, and a chaperone of enlightenment in their own backyards and surrounding countryside. A charismatic figure, Jensen blended the contemporary philosophies of Gifford Pinchot's utilitarian conservationism and John Muir's transcendental preservationism into a functionalist approach to everyday life. "Jensen's primary tenet," according to scholars William H. Tishler and Erik M. Ghenoiu, "was always that environment engenders human character, that the places where we live help determine who we are."[101]

A weekend bike ride or stroll on the beach, Jensen believed, was not enough. City dwellers needed constant contact with nature to restore their minds, bodies, and spirits. Starting from the bottom in the immigrant neighborhoods of the West Side, he was especially sensitive to the environmental conditions of the working class living amid intense concentrations of commerce and industry. His report describes a stunted urban nature, overwhelmed by too much steel and bricks, bad air and water, and overcrowding and disorder. To counteract the debilitating effects of an overbuilt environment, he helped the SPC visualize a hierarchy of public open space on a metropolitan scale.[102]

Nonetheless, under the influence of Addams and her playground movement, Jensen kept the neighborhood park at the center of the system. He realized that few members of the working class could afford to get out of the inner city to the proposed forest preserves. In his report on the region's natural history, Jensen envisions what planners would call the "neighborhood unit" of community development a quarter century later. Not surprisingly, this template of the modern suburban subdivision would be the brainchild of one of those "well-educated social directors," Clarence Perry. The outdoor recreation expert would base community size on the number of children needed to fill an elementary school, which would be located at the heart of a central green space. In contrast, the landscape gardener called for a maximum 1- to 2-mile/1.6 to 3.2 km walk for anyone to reach a recreational park that would act as an incubator of health, democracy, and Americanization.[103]

In addition to foreseeing the metropolitan scale of the 1909 plan, the 1904 report was groundbreaking in its effective use of photography. In the 1890s, technology turned methods of graphic reproduction into new forms of mass communication. Taking advantage of this popular novelty, such reformers as Jacob Riis had shocked the middle class by exposing "how the other half live[d]" in deplorable slum conditions.[104] About one half of the 120-page SPC report includes black-and-white photographs Jensen had taken in the proposed reserves. Of course, Burnham and Bennett would take this new medium of mass persuasion one step further by reproducing in stunning colors detailed science-fiction images of what Chicago could look like a half century in the future. Jensen's photographs support the report's finding that detrimental influences were already putting these peripheral areas in imminent danger of irreparable, ecological harm. Consider his titles: "Dead Walnut Trees—the Effects of Sewers," "River[banks] Destroyed along Desplaines River," and "Destruction of Woodlands Now Going On," which shows stacks of timber amid a landscape stripped bare.[105]

At the same time, Jensen strove to demonstrate the promise of science in solving the crisis of an overbuilt environment. Under the stewardship of such experts as Jensen, nature could be not only nursed back to health but also lifted to a more-elevated state of beauty. For example, he lists several schemes to remake the vast marshland surrounding Lake Calumet, because he considered it visually boring. For the most part, however, this group of illustrations shows riverbanks and forest groves in the process of recovery, when protected from further exploitation. In addition, Jensen includes a gallery of images of some of his favorite old trees, river scenes, and views of prairie landscapes. Concluding the report with an upbeat message of the benefits of nature, the reformers anticipated a rough road ahead in persuading policy makers and voters to approve their proposal for a metropolitan-scale system of public open space.[106]

The heavy lifting needed to overcome the opposition to the creation of a park authority embracing a territory larger than the sanitary district fell on Foreman's shoulders. The oldest of the three reform leaders, he was the son of a German Forty-Eighter, who had left the homeland in 1848 with significant resources. The Jewish banker set up shop and, in 1857, became the father of a newborn son. Henry Foreman dutifully went into the family business but in the mid-1880s broke away from his father, Gerhard, to become a land speculator and stock trader. Another social magnet, he helped organize the Chicago Real Estate Board and the local stock exchange. Despite exclusion from some social clubs because of his religion, he was accepted by the business community as one of its most prominent spokesmen. In 1902, the Republican Party rewarded the civic activist with an appointment to the South Park District, where he served as its president for the next ten years. In addition, he was slated by the party and elected by the voters as the president of the Cook County Board of Commissioners, another job he would hold for a decade.[107]

Of the three park districts, Foreman put the South Side in the vanguard of the conservation movement. True, this district had always been the best funded and had benefited from playing a major role in the World's Fair. Yet Foreman took the lead in translating Addams's appeal to save the children into a network of ten recreational parks. Three years later, on May 13, 1905, he attended the opening ceremonies of one of these green spaces, Davis Square. The 8.5-acre/3.4 ha site was equipped with a field house designed by Edward Bennett. Located in the stockyards district, or "Packingtown," a park for its immigrant workers was the inspiration of Mary McDowell, the University of Chicago's settlement house director. She mobilized four thousand local residents "dressed in holiday attire" to attend the event, where Foreman officially dedicated the park.[108]

He told them that it would serve their community as a springboard to achieving the American dream. An environmental determinist, the urban reformer promised that the park would become a seedbed of the nation's values of individual self-help and local self-government. "The south park commissioners," he declared, "believe they can help you to help yourselves, and have built you this neighborhood center to give you the means to do so." A long list of petitions from other neighborhood associations for playgrounds and recreational facilities supplies a good indication that Foreman had no problem convincing his audience that Davis Square would improve their lives.[109]

Over the coming months leading to the November elections, the businessman Progressive faced a much harder sell in persuading a majority of voters to approve the formation of a new countywide agency to save nature from the city. Following up on the 1904 report, Foreman had lobbied a bill through the state legislature to let the people decide whether to implement its plan to create an outer belt of open space. Although everyone agreed that a reserve of forest and prairie for future generations was a good idea in the abstract, his proposal and character came under attack from every side of the political arena. On the conservative side, several businessmen's organizations, such as the Civic Federation and the Citizens' Association, undercut his support by pointing out the lack of checks and balances to safeguard democracy from corruption by the party bosses. In the middle, the school board mounted a vote-no campaign, because the new agency could potentially sap funds for children's education in the city to pay for pleasure grounds in the suburbs. And on the liberal side, the labor unions lambasted his proposal as class legislation, because blue-collar families could not afford the expense of a trip to a remote location. Moreover, his political enemies got personal, accusing him of corrupt, insider deals that allowed his silent partners to buy most of the land around the borders of a big new park before its location was made public. A tireless campaigner, Foreman beat back his opponents and rode to victory on election day.[110]

Despite Foreman's victory at the polls, political contestation over a metropolitan park system continued for ten more years before winning constitutional approval. The first court challenge convinced the state high bench that the enabling act required not only a majority of those voting on the proposal but also a majority of all those voting on election day. After a second successful referendum campaign in 1910, opponents again found a procedural flaw in the legislation. A third vote in November 1914 passed with 235,000 votes in favor of and 131,000 votes against the creation of a forest preserve under the authority of the county government. After waiting a year for a third ruling by the state supreme court, Foreman, Jensen, and Perkins were finally able to see their visionary plan to save a prairie wetland come to fruition.[111]

Conclusions: The Metropolitan Legacy of Progressive Reform

From the Great Chicago Fire of 1871 to the great flood of 1885, Chicago became the fastest growing city in the Atlantic world. Its center and periphery became more interdependent and tightly integrated into a single metropolitan region. This transformation changed not only physical landscapes but also social relationships and cultural values. In the act of city building, Chicagoans underwent a reorientation in their evaluation of the costs and benefits of urban growth. In 1889, the creation of the SDC and the success of the annexation movement marked the conceptual turning point in their perceptions of this transition of place from city to metropolis. Over the next two decades, the explosion of new forms of private suburban natures and public open spaces would represent the ascendency of environmental values in the popular imagination.

A new generation of reformers was able to mobilize this near-universal demand to get back to nature into a political groundswell of support for visionary plans for building the metropolis of the future. In the act of turning Jackson Park into the fairgrounds of the Columbian Exposition, Chicago's architects and landscape gardeners became planners with increasingly comprehensive perspectives on the urban environment. Like other new radicals, they were driven by fears of impending urban doom as well as hopes of establishing a sanctified community. They believed that making patches of nature accessible to every city dweller was critical to restoring minds and bodies in addition to uplifting the immigrant masses to American standards of moral order and democratic patriotism. In 1896, the Pure Water Commission outlined a complete plan to save Chicago's aquatic resources from pollution, but the politicians in charge of city hall defeated it. Turning from public health to public parks, such men as Perkins, Jensen, and Foreman joined forces with such women as Perkins, Addams, and McDowell to mount a more-successful campaign to create a system of recreational centers on a metropolitan scale for future generations. A straightforward progression can be seen from Jensen's hand-drawn map of an outer belt of nature preserves in 1902 to its adoption in the 1904 report and then in the *Plan of Chicago*'s brilliant images of parks, boulevards, and an "emerald necklace" of prairie wetland framing the city center.

The injection of environmental values in the formation of public policy became another source of social and political conflict. Public open space became another limited resource in a long list of infrastructure improvements that chronically lagged behind the development of the built environment. The ensuing struggle among communities to get access to more parks joined the competition over the provision of sewers and drains, paved roads and sidewalks, and all the other conveniences of modern life. Moreover, the establishment of new agencies of environmental management created additional sources of contestation among the politicians and their parties for control of their funds and administration. After 1905, the staying power of both sides in the battle to establish a forest preserve for a decade reflected the enduring importance Chicagoans now gave to the cause of saving nature from the city.

2

The Defeat of Conservationism, 1910–1920

The Year of Decision: 1910

For a city built on water, 1910 became a year of decision. With the Commercial Club's presentation of the *Plan of Chicago* to the city, it reached a milestone as well as a crossroads in the remaking of its flood-prone environment. Daniel Burnham and Edward Bennett's masterpiece represented the culmination of a generation of urban reform. While portraying an inspiring vision of the metropolis of the 1960s, it also represented the distillation of the businessman's point of view on the best way to solve the crisis of the 1890s.[1] Over the course of the next two years, critics ranging from leading figures, such as Jane Addams and Jens Jensen, to grassroots movements hotly contested its proposals. The urban economy as an *entrepôt* and manufacturing center seemed to depend on the modernization of its nautical infrastructure to metropolitan proportions. At stake were pivotal decisions on water management that would determine the fate of the city's lakefront and the suburbs' river watersheds.

The need to set policy directions in city building was driven not only by the business community's grand plan but also by unresolved local, national, and even international crises. Despite opening to great fanfare in January 1900, the Sanitary and Ship Canal failed to prevent either damaging floods or deadly outbreaks of water-borne epidemics. Moreover, speeding up the currents in the Chicago River had made navigation by ships and barges unsafe. The federal government immediately had to order a reduction of more than 50 percent in the diversion of water from Lake Michigan into the Main Branch, restricting it to 2.7 billion gallons per day (bgd)/10.2 billion liters per day (bLd).

Nine years later, an association of the states bordering Lake Michigan concluded that the human and industrial wastes being generated by the Calumet District posed a clear and present danger to the public health. The continual contamination of the lake was compounded by a growing crisis of sewage from Chicago polluting the inland waterways of the region and downstate. In 1910, moreover, the Canadian government protested under a new, year-old treaty with the United States against any diversion of the Great Lakes' waters by the Sanitary District of Chicago (SDC). The city's hydraulic and sanitation engineers agreed that instead of depending on dilution methods of sewage disposal, they needed to start erecting a network of modern treatment plants to keep up with projected increases in population.[2]

As the comprehensive plan of 1909 underscored, defining new sanitary strategies was inextricably tied to making decisions regarding how to resolve Chicago's long-coming economic crisis as an *entrepôt*, where ships and railroads transferred goods and people. In 1906, the Port of Chicago reached a turning point when the tonnage of shipping entering the Calumet River harbor exceeded the amount moving on the Chicago River. A technological revolution from wooden sailboats to iron and steel steamships had turned the city center's entire riverine infrastructure of bridges, tunnels, and wharfs into an obstacle course for these ever-larger vessels. Adding to their operating costs from lost time were fees for tugboats to pull these ships through a river bottom clogged with raw sewage and factory "trade wastes." After a year's study, in 1909, a special harbor commission recommended that the central business district (CBD) should handle only ships and trains carrying passengers and packages, while most manufacturing and bulk cargo facilities for grain and coal should be shifted to the Calumet industrial district. In the downtown area, the exponential growth in street congestion from automobiles, streetcars, teamsters, cyclists, and pedestrians gridlocked by river traffic reinforced the plan's and the report's call for the area's redevelopment into an attractive, lakefront hub of commerce and leisure.[3]

Conspicuously absent from the businessmen's plan, but not from other reform agendas, was any mention of the crisis of decent housing for the mass of city dwellers. It was purposely excluded, because the members of the Commercial Club believed this crucial issue of urban life should remain in the "invisible hand" of the marketplace. For the middle class, the real-estate industry's production of suburban subdivisions and model homes seemed to offer a solution to its crisis of gender and family relationships. Yet a self-proclaimed comprehensive plan without reference to the management of residential sprawl and land use was a fatal flaw, which threw Burnham and his sponsors on the defensive.[4]

Consider, for example, the opposition of Chicago's women-led reform groups. They advocated for more-proactive regulation of the land, sponsoring the enactment of stronger municipal health and building codes. Furthermore, they protested against top-down planning by patriarchal elites, as opposed to democratic decision making by communities engaged in local self-government. In addition to linking epidemic outbreaks to slum conditions, they continued to organize support behind the open-space and recreation movements, promoting conservation as the best public remedy to the social injustice of the environmental conditions of the working class. In 1909 and 1910, they mounted a second campaign to pass an enabling act for a referendum on the creation of a forest-preserve district.[5]

In sum, long-standing and immediate crises accumulated to a point of intense political pressure on public officials to make decisions that set in motion the next major steps in the rebuilding of Chicago on a metropolitan scale. Besides the universal constituency of its inhabitants in favor of conserving the purity of Lake Michigan's drinking supplies, the large number of conflicting stakeholders invested in the fate of its riverine environment further intensified the contestation over the formation of public policy. The commercial interests fought against the industrialists; the Chicago River property owners against the Calumet River harbor owners; the businessmen's civic groups against the women's reform clubs; and the City Profitable against the City Humane.

During this psychological state of political warfare, a series of four damaging floods in a year's time added urgency to the universal demand to resolve the crisis of water management. The first storm, on August 12, 1908, was the worst since 1885, dumping 4.33 inches (in.)/11 centimeters (cm) of rain in a twenty-four-hour period. Sewer backflows ruined five thousand basements, lightning bolts set several buildings on fire, and high winds killed thousands of birds. The following spring, on April 19, the surge of surface runoff from a deluge again overwhelmed the sewers in the downtown area. Only sixteen days later, tragedy struck the city when a tornado touched down on the far South Side, killing five people, seriously injuring another thirty-nine, and causing at least $2 million in damages.

Desperate efforts by the sanitary authorities to keep the Chicago River from flowing into the lake failed, sending millions of gallons of untreated sewage toward the water-supply intake cribs. The Des Plaines River overflowed, sinking Riverside and other suburbs on the western periphery. Just a year after the first downpour, on August 15, 1909, another historic rainfall of 3.5 in./8.9 cm led to the drowning of a baby left unattended in a submerged basement and a little girl found underwater in her front yard. Thousands of homes and businesses throughout the metropolitan area suffered property damage. Storm-runoff surges and public-health threats supplied reformers with an irrefutable argument that neither the old combined sewer system nor the new sanitary canal was adequate to the metropolitan scale of the city.[6]

The formation of water-management policies in 1910 set the course of city building until the onset of the Great Depression. This process of planning took place within a political framework of tension between the rivalries of intraparty factionalism and the pressures of popular demands for bipartisan unity. In that year of decision, policy makers had to grapple with the modernization of the city's hydraulic infrastructure. The Chicago River was too small to handle the ever-bigger ships, which influenced the choice of whether the lakefront would be developed into recreational space or a commercial harbor. Equally important, policy makers had to cope with the Calumet River's contamination of Lake Michigan and the SDC's pollution of inland waterways.

In contrast, land management remained in the private sector for the most part. During the 1910s, the opposition of the city's manufacturers to a zoning ordinance sponsored by the Chicago Real Estate Board (CREB) limited public-sector intervention. The proposal to divide the city into residential, commercial, and industrial areas sought to reinforce the board's long-standing policy of inserting restrictive covenants in property deeds to segregate the neighborhoods by race and ethnicity/religion.[7] Perhaps the most important role of the public sector in the steady growth of the metropolitan region was its multi-million-dollar incremental program of infrastructure improvements in newly built-up areas in the city and the suburbs. The other significant intervention in regulating the land was the result of the reform movement for open space, which proved to be an unstoppable political force. Opening in 1918, the forest-preserve district made steady acquisitions of property in pursuit of Jensen's original vision of saving Chicago's river watersheds. This relinking of water and land-use management proved to be the exception.[8]

From the 1890s to 1920, a political culture of government fragmentation and intraparty factionalism stymied the resolution of the environmental crises facing policy makers. Yet they enjoyed a holiday from coping with a flood-prone environment, because relatively few damaging storm surges occurred during this extended period of drought conditions. But an equally long period of metropolitan growth made them especially vulnerable to corruption by the party bosses, because their large budgets for construction contracts and unskilled workers were the lifeblood of machine politics. One sewer contractor, Patrick A. Nash, would build a personal fortune and Chicago's first citywide machine, while an engineer in the sanitary district, Edward J. Kelly, would work his way up the ranks to become the mayor. The corruption of the sanitary authorities sabotaged effective planning to deal with the problem of increasing amounts of human and industrial liquid wastes polluting the lake and the rivers. Resisting reform and demanding the diversion of more water from Lake Michigan into the ship canal, Chicago fought a rearguard action against the Great Lakes states, Canada, and the federal government until 1922, when the U.S. Supreme Court intervened to force it to adopt a sanitary strategy of nature conservation.[9]

The Politics of Urban Growth

As we have seen, the Progressives' "search for order" engendered a vast expansion in the scale and scope of the public sector over the physical environment: annexation, sanitary district, public-health authority, and parks and playground facilities. The expansion of Chicago to metropolitan proportions, moreover, required a commensurate coordination among overlapping government agencies that only the two major political parties could provide. With tax revenues lagging behind urban growth, the rapid suburbanization of middle-class homeowners and manufacturing firms kept outpacing public-works programs to meet people's demands for all the modern amenities. Choosing which outlying residential neighborhoods, shopping districts, and industrial zones got their improvement projects listed in the annual budget and which contractors were awarded the winning bids gave the politicians tremendous discretionary power as well as tempting opportunities for the bribery of votes and contributions to their campaign coffers.

An extreme fragmentation of government authority and party factionalism forestalled a full-fledged regime of one-party boss rule until after World War II. Chicago has always had a strong council and weak mayor system of municipal government. In a city of privatism, where "property ruled," constitutional and structural limitations kept the public agencies in charge of building the urban environment—the streets; Water and Sewer Departments; the SDC; park commissions; suburban municipalities; and counties—chopped up and semi-independent of one another.[10] Political power devolved to the ward level, where party leaders ruled like benevolent kings or petty tyrants over their fiefdoms. A city of deep class, ethnic/religious, racial, and gender divides also fostered a political culture of seething, intraparty rivalries among its contending groups. Traditions of elite civic reformers battling corrupt ward bosses further fractionalized party cohesion and public administration.[11]

Only after the turn of the century, Robin Einhorn notes, did "Chicago began to exhibit some of the characteristics of 'machine politics' at the ward level, though not those of a citywide 'machine.'"[12] Historian Roger Biles explains that they were "built largely on the votes of diverse immigrant populations, dispensed [patronage] jobs and assorted welfare benefits while offering avenues of social mobility at a time when local governments provided a paucity of such [social] services."[13] Over the course of the decade, political bosses contributed to the defeat at the polls of charter-reform and municipal-ownership movements. In addition to reaching the limits of structural reform to strengthen the public sector, the Progressives' moral crusades for prohibition and Americanization backfired among voters with working-class and ethnic/immigrant identities. Only near-universal support in favor of more open space in the city and forest preserves in the suburbs mustered the majorities needed to maintain the momentum of reform.[14]

The roots of machine politics in Chicago can be traced to the corruption of sanitary authorities. In 1910, the SDC board of trustees would confront the same basic decision that it had faced at its origins twenty years earlier: was the primary mission of the agency sanitation and public health or shipping and commerce? If the former, then the goal of the agency was to optimize the conservation of nature. The trustees needed to devote its resources to stopping wastewater from polluting Lake Michigan and safely disposing of it in treatment plants before it entered the inland rivers. If the latter, then the goal of the agency was to take full advantage of nature as a freely expendable economic resource. The agency's policies needed to focus on diverting the maximum 6.5 bgd/24.6 bLd of water from the lake without disrupting navigation on the Chicago River. The trustees faced options that included digging additional feeder canals to the main channel and enlarging the Chicago River by making it wider and deeper. Of course, they could choose to divide the budget between the two goals, but, given fiscal limitations relative to metropolitan growth, they were under tremendous political pressure to set a priority of one over the other.

On the sanitation side, the formation of policies to end the endangerment of the public health from water-borne diseases remained at the top of the reform agenda. Fewer than three years after the completion of the 28-mile/45-kilometer (km) canal, the flare-up of a typhoid-fever epidemic in Addams's neighborhood on the near West Side smashed hopes that the big-technology project had ended Chicago's state of emergency. In contrast to the citywide spread of the outbreak in the early 1890s, this and subsequent epidemics tended to cluster in more-localized areas. Reflecting the regional scale of metropolitan Chicago, lakefront sewer outlets and water-supply intake cribs stretched down from Milwaukee, Wisconsin, to the North Shore suburbs, the city, the Calumet District, Indiana's industrial cities at the bottom of the lake, and around the other side to the resort communities in Michigan as far up as Benton Harbor. With each new panic over the water supply, political demands mounted on the sanitary authorities to take whatever steps were necessary to end the contamination of Lake Michigan and save this precious natural resource.[15]

In favor of giving shipping priority, the commercial stakeholders in the CBD lobbied with relentless determination to overcome the crisis of technological obsolescence. Since Lyman Cooley had taken charge of the SDC board of trustees in 1891 for the Democrats, it had followed his single-minded plan to make the Chicago River's harbor the nation's largest seaport. To achieve this goal, the agency's engineers needed to divert a maximum amount of water from the lake into the ship canal to keep deeper-draft vessels afloat all the way to the Gulf of Mexico. In one expert's estimation, this amount of water was equivalent to the flow of the Mississippi River at the twin cities of Minneapolis and St. Paul. As soon as the shipping lane opened, however, the federal government thwarted Cooley's plan, because the river was too narrow and shallow for safe navigation at an accelerated rate of flow.[16]

In the 1890s, the downtown lobby also had to overcome the resistance of the municipal government and the U.S. Army Corps of Engineers to dredging the Chicago River and lowering the tunnels that went under it. City hall did not want to bear the cost of replacing the tunnels. Moreover, the local branch of the federal government in charge of navigable waterways protested paying to remove sedimentation from the city's untreated sewage that was clogging the bottom of the river. Overriding the experts, Chicago's congressional delegation had secured funding for deepening it. The lawmakers had also ordered city hall to remove the tunnels, because they were obstructions to navigation. But the boats kept getting bigger, making a resolution to the crisis of moving thousands of vessels each year in, out, and through the Chicago River more and more urgent. The Commercial Club and other downtown businessmen's groups put all of their considerable political clout to bear on the SDC trustees to upgrade the waterway to accommodate ocean liners.[17]

In the middle of these contending forces was a political faction that can be called the Irish Quartet: Roger Sullivan was the power broker; Thomas A. Smyth, the public servant; Kelly, the expert engineer; and Nash, the connected businessman. Their story puts a human face on the "machine." Their personal histories show the importance of ethnic/religious identities and class solidarity in making the party organizations into vehicles of upward social mobility. These were ambitious men on the make; they used the business of government as a springboard to the top rungs of respectable society. At the same time, however, their story reveals that the self-serving goals of the Irish Quartet undercut the ability of government to fulfill the public sector's role in planning and building the urban environment.[18]

The Democrats had followed Cooley's ruse to gain control of the SDC board of nine trustees in 1891 with a full slate of candidates four years later. Under the election law, an accumulative five-vote rule virtually guaranteed victory for four of the at-large candidates from each party. The party that won a majority of the total votes meant victory for its fifth man, giving the Democrats or the Republicans control of the board. In 1895, Sullivan sponsored the wannabe public servant, Smyth, for a seat on the board. The oldest of the group, at forty-seven years old, he and his brother, John A., were successful merchants and furniture salesmen. Victorious, the civic activist continued to rise in society and the SDC, eventually becoming its president. During that year, the hyperactive Sullivan became the "gray wolves'" leader of the pack. With the help of his sponsor, Mayor John Hopkins, he secured a public-utility franchise from the city council worth $8 million to $10 million for the Ogden Gas Company despite the fact that it existed only on paper at the time.[19]

The members of the Irish Quartet were all sons of immigrants, although their births spanned a generation. They were all born in Chicago, with the exception of Sullivan. In 1879, the eighteen-year-old had moved to the city from a farm about 75 miles/121 km to the west. Like many rural youth, he had found work taking care of the horses of a street railway company on the West Side. He also had joined Hopkins's Irish faction within the Democratic Party, landing a lucrative job in 1890 as a clerk in the probate court. After cashing in on the Ogden Gas deal, the instant millionaire no longer needed to work a steady job. In addition to becoming a party boss with statewide influence, he invested in other ventures doing business with the government, such as the Great Lakes Dock and Dredge Company. Created in 1891 in response to the technological revolution in lake vessels, it sought contracts with the city, the SDC, and the U.S. Army Corps of Engineers for projects to improve navigation on the Chicago and Calumet Rivers.[20]

Another loyal member of the Hopkins/Sullivan West Side organization was the immigrant Thomas Nash, who was a sewer builder. Born in 1863, Patrick and his brother Richard took over the family business as the Nash Brothers. "P. A.," as he was called, had learned from his father to get along with the Republicans as a private businessman but to join the Democrats as a public advocate for Irish/Catholic pride and progress. Moreover, he had continued the age-old practice of crony capitalism that exchanged government contracts for political sponsorship of the workforce he hired for the job. In fact, the city's utility companies and building contractors supplied far more patronage jobs than the government's public-works departments. For Nash, this kind of sweetheart deal proved to be a win-win proposition that generated extra profits and political power. In 1892, Sullivan rewarded the up and coming young businessman with an official position as a ward delegate on the party's slate-making committee. The Nash brothers also began winning bids for minor contracts from city hall.[21]

Activism within the Irish/Catholic faction of the Democratic Party also brought Nash together with Kelly, an engineer working for the sanitary district. The youngest member of the quartet, he was born in 1876 in the Irish/Catholic neighborhood of Bridgeport on the South Side. Like Horatio Alger's *Ragged Dick*, his rise in society was built on hard work and lucky breaks.[22] He began at the bottom as a newsboy before getting a job with the new agency as an axman when he was eighteen years old. Kelly quickly climbed the ranks through on-the-job training to achieve the position of supervising engineer while Cooley's army of patronage workers dug the canal. He thus became a manager of workingmen in the field rather than a designer of technological systems in the office. Kelly was also put in charge of inspecting and signing off on the work of private contractors before it was literally buried underground.[23]

Following the money, Nash had sought to become a builder of the new generation of giant interceptor sewers, which were turning the Sewer Department into a big business. In August 1899, however, his membership in the Hopkins/Sullivan faction became a liability when it clashed with the more-reform-oriented Carter Harrison II wing of the party. Two years earlier, it had taken control of the Democrats, putting the son of a mayor into his father's office. Now Harrison rejected the low bids of the Nash brothers for two contracts worth almost a million dollars and demanded that the work be completed by the city.[24]

The "gray wolves" had corrupted the very first of these big-technology projects to protect the public health. According to Harrison's allies on the city council, the scheme involved lowball bids to get contracts for the 39th Street interceptor sewer, followed by claims for extra work that ended up costing far more than higher, albeit more-realistic, estimates of the costs of the project by other bidders. "I thought," exasperated Commissioner of Public Works Lawrence E. McGann testified, "that with the surety company's bond and other safeguards the city was protected, but now the contractors are in with claims for extras, which . . . somehow or other [the] courts always do allow."[25] But the mayor's proposal to create hundreds of "day labor" jobs to finish the work sounded to the Republican minority on the council like a call to create a patronage army for the 1900 elections. Their spokesman railed against a repeat of the "payroll-stuffing and other scandals [that] have been brought to light" in the Water Department. Together with the Hopkins/Sullivan faction, they blocked the mayor's plan, creating a stalemate in the completion of this important project to protect the water supply on the South Side.[26]

During this political standoff, the labor unions exercised their power by going on strike against the contractors and the city. In November, more than two hundred men stopped work on the 39th Street project. Their representatives demanded a pay raise and a closed shop. They argued that digging these deep, underground tunnels was dangerous work that required trained miners and skilled masons. The building contractors' council retaliated with its own ultimatum that demanded complete control over the work rules as well as the payroll. After a month off the job, the workers ended the strike without a raise. Although their unions lost this round of the battle between labor and capital, they continued to fight on the economic and the political fronts for a better life.[27]

Their success in pressuring the faction-ridden Democrats to unite behind a plan to help them achieve the American dream sheds light on the reasons why working-class voters became loyal members of political machines. Rather than having no patronage jobs to hand out, the Hopkins/Sullivan faction made a deal with Mayor Harrison that led to a straight party vote in January 1900 in favor of his "day labor" plan. Three weeks later, the majority party passed another ordinance granting these workers a 33 percent pay raise, from \$1.50 to \$2.00 a day. These jobs provided a foothold on the bottom rung of the ladder of success for many newcomers to the city. They also supplied a form of welfare or charity for what the new breed of efficiency experts described as workers who were unfit, incompetent, and simply too old for this type of hard physical labor. Forming the base of political machines, patronage workers believed that their personal and family gains from allegiance to the ward bosses outweighed the cost of corruption in terms of diminishing the city's capacities to promote the general welfare. In the case of the 39th Street sewer, the experts put the price tag at $52,000 in extra costs for labor compared to the private sector's pay scales.[28]

While men on the bottom exchanged votes for jobs, men on the make, such as Smyth, paid the ward bosses to get their names on the party tickets, including as candidates for the judiciary. In 1900, for instance, the slate-making convention of the Democrats renominated the two-term trustee of the SDC for the upcoming November elections. Party bosses also levied an "assessment" of a year's salary of $3,000 on the candidates for this office. Two other incumbents joined Smyth in a revolt, leading to a compromise with Sullivan and the other power brokers on a $2,000 levy. In part, these office holders prevailed because they had added at least five hundred extra jobs to a workforce of fifteen hundred men. As the chair of the engineering committee, Smyth had rejected Chief Engineer Isham Randolph's claim that these patronage workers added $31,000 to labor costs. He listed other wasteful spending practices on contracts with overbuilt design specifications. Under Kelly's supervision, Randolph also charged, no more than a quarter of the workforce was ever on the job at the same time. He also called out Smyth, because he was "a candidate for re-nomination by the Democratic Party and took care of his friends." And they took care of him, ensuring his victory on election day.[29]

The Politics of Water Management

Under the incoming administration of SDC President Smyth, corruption penetrated all the administrative departments, which employed at least 150 workers. Frank Wenter, an original trustee and former president, had hired his son not only as an absentee clerk but also at a higher rate of pay than his coworkers. Smyth would give virtually all the sanitary district's insurance business to his son. The trustees also recouped their party "assessments" by paying back their friends with contracts for legal, printing, security, and consulting services. Besides nepotism and crony capitalism, building contractors were expected to pay the trustees so-called commissions, or kickback bribes, if they were awarded winning bids. These businessmen, in turn, formed secret cartels to ensure that their lowball figures were high enough to make exorbitant profits, even before charging for "extras." During Smyth's first five-year term, illegal conspiracies were exposed among the dredge contractors and the sewer builders, including the Nash brothers.[30]

Under his leadership, the SDC continued to give priority to shipping over sanitation. In 1903, Smyth published an article in the *Chicago Daily Tribune* titled "The Present and Future of the Drainage Canal." He started with a promise that "we are passing through the boiled water period in local history." In his view, city hall's anticipated completion of the interceptor sewers by the end of the year would fulfill the agency's original mission to protect the public health. His vision for the future turned the drainage canal into a "part of a great national waterway, which will connect the [G]reat [L]akes with the [G]ulf of Mexico." For a half century, he noted, this has been what "the industrial interests have been clamoring for."[31]

To achieve this goal, Smyth outlined a two-part plan to divert the maximum amount of lake water into the ship channel. Claiming to save the lake, he announced the annexation of the Calumet District below 87th Street and a public-works project to dig a navigational-size canal to the main channel along the Sag Slough. Reversing the Calumet River, the proposed "Cal-Sag Canal" would become a second major drainpipe to tap Lake Michigan. The second part of the plan called for reengineering the Chicago River to handle bigger ships and more water. Smyth took credit for dredging parts of it from 17 to 20 feet (ft.)/5 to 6 meters (m) deep and widening sections to 200 ft./61 m across from Van Buren Street south to the canal. He also committed the SDC to replacing all the center-pier, swing bridges with bascule-style models. Other ship-canal visionaries, including Burnham and Bennett, added grand plans to fill the riverbed along its meandering path through the rail yards on the southwestern side of the CBD and to dig a new route straight through them.[32]

Missing from Smyth's report of the drainage district's future was a third plan to annex the North Shore to build the "Big Ditch." The 8-mile/13 km project turned a clay bed into a riverbed, adding 650 million gallons (gal.) per day/2.54 bLd of the lake's water to the ship canal. Like the Cal-Sag project, the North Shore Channel was designed as a workaround of the federal government's restrictions on the amount of lake water the sanitary authorities could drain into the Main Branch of the Chicago River. Along with the Calumet District, the SDC expanded its boundaries to the edges of Cook County, increasing in size from 185 to 358 square miles/479 to 927 square km and serving a population of more than two million inhabitants.[33]

The public did not need Smyth's boosterish article to convince them that draining more fresh water into the Chicago River would help improve the urban condition and restore the health of nature. On the North Side, the city's diversion of the raw sewage from the lakefront areas into interceptor sewers made much worse the pollution of the North Branch. This sluggish waterway had already been degraded by the organic wastes of humans and horses, breweries and attached feedlots, and tanneries and leather-goods factories. In addition, the utility companies used coal to manufacture gas that generated carcinogenic by-products. They polluted the river with chemicals and the air with black smoke, which cast a pall over the neighborhood: it was known as "Little Hell."[34] The policies of the sanitary authorities made those forced by poverty to live in the riverfront wards suffer the social injustice of environmental inequality in their relationship to those living nearby in the more-affluent lakefront wards.

In November 1905, the voters in the expanded territory of SDC were given a clear choice of priorities between deep-water navigation and safe tap water. Wenter's supporters within the Democratic Party prevailed over Sullivan's faction, strong-arming Smyth to the sidelines of the campaign. Running for president of the board, Wenter put the improvement of shipping on the South Branch at the top of his platform, followed by digging the North Shore Channel and building bridges. Running against him was Robert R. McCormick, a twenty-four-year-old graduate of Yale University's Law School and an heir of the International Harvester Company and the *Chicago Daily Tribune*. A protégé of boss Fred Busse, he had served less than a year as an alderman before being slated for the presidency of the trustees. A self-proclaimed civic reformer, McCormick called for "aggressive honesty" to root out corruption, install efficiency, and protect the water supply. His platform pledged not only to complete the two canal projects but also to connect them to intercepting sewers along the North Shore and the Calumet District. Uniting the factions of the Republican Party behind him, he saw his victory as a stepping-stone to becoming "one of the foremost public men of the [M]iddle [W]est."[35]

But he did not stop the political machine—it ran over him. Declaring that appointing honest department heads was the critical first step of reform, the ambitious, albeit inexperienced, public official took a stand against the other four Republican trustees. They expected him to conduct a "clean sweep" of the district's administrative jobs, and they threatened to sabotage his entire agenda unless he agreed to their demands for the spoils of victory. In January 1906, McCormick caved on the test vote to appoint the Busse-sponsored candidate for the chief clerk's position.[36] As president of the board, McCormick also found himself swept up in a bipartisan booster campaign at the local, state, and national levels to create a deep-water canal from Chicago to New Orleans. In 1908, he helped win statewide support for a $20-million public-works project to turn the Illinois River into an extension of the ship canal.[37] Meeting stiff resistance in Washington, however, the agency's top spokesman had little time to follow through on his platform pledges.

During McCormick's stormy term in office leading to the crossroads of 1910's year of decision, the two feeder canal projects advanced neither the goal of navigation nor of public health. The contracts for the $3-million North Shore Channel had gone to Sullivan's Great Lakes Dock and Dredge Company. In November 1910, the SDC opened the channel on the eve of the election, promising that its flow of lake water was "sufficient to dilute all the sewage that may be turned into the canal." But a newspaper reporter observed, "[The] Evanston ditch fails to purify." At best, it was "making the water a little clearer in the [N]orth [B]ranch of the Chicago [R]iver." At worse, taxpayers had wasted their money on a project that had done nothing to reduce the amount of sewage being dumped into the lake, contaminating their water supply.[38]

In the Calumet District, the SDC immediately ran into a roadblock set up by the federal government to stop it from exceeding the limit set in its 1901 permit, which allowed Chicago to take 2.7 bgd/10.2 bLd from Lake Michigan. After rejecting its application to drain an additional 2.6 bgd/9.8 bLd through the proposed Cal-Sag Canal, the nation's top regulator of navigation on inland waterways, Secretary of War William Howard Taft, announced his intention to sue the SDC for violating its permit. In 1907, he sought an injunction in the federal district court to halt the construction of the project, because it would take additional water from the lake to reverse the Calumet River and keep the canal filled. The issue would remain moot for the next fifteen years. Without the permit, the SDC would drag out the pace of construction of the 16.2-mile/26 km channel, and a recalcitrant, local judge would refuse to issue the inevitable ruling upholding the national authority. The U.S. Army Corps of Engineers also turned a blind eye while the sanitary authorities broke the law, gradually diverting twice as much water into the Main Branch as the permit allowed in a vain attempt to dilute all the district's organic waste to safe levels.[39]

In 1910, the conservation of Lake Michigan continued to dominate civic debate in McCormick's reelection campaign, because typhoid fever remained at epidemic levels. In the ten years since opening the drainage canal, the SDC had done little to end "the boiled water period in local history." The territory south of Hyde Park served by the 68th Street water-supply intake crib was repeatedly stricken with outbreaks of the terrifying disease. South Chicago's forty thousand working-class residents were suffering typhoid fever rates 18 to 126 times greater than those of people living near the city center.[40]

Public-health officials, civil engineers, and civic reformers led the revolt against the board of trustees' priority on shipping over sanitation. In April, for example, the elite City Club sponsored a forum to rally support for building the Cal-Sag Canal. As we have seen, Chicago's business community was well aware of the growing importance of the Calumet District to the urban economy. In 1910, the combined Port of Chicago was the third largest in the nation, handling more than thirteen thousand vessels carrying 15 million tons of freight.[41]

The city's commissioner of health, Dr. W. A. Evans, stated that there was no longer any doubt about the direct linkage between the quality of the water supply and the rate of typhoid fever. "The district south of Seventy-[F]irst Street," the public-health crusader remarked, "over to a point about 10 miles [16 km] east of Indiana Harbor is a district in which all the shore water is polluted and will be unfit for city use until there is a comprehensive effort for preventing it in this district."[42] Following up in May, newspaper editorials and businessmen's tours of the Calumet River kept the public's attention focused on a problem that put between two hundred thousand and five hundred thousand people at risk of death every time they drank a glass of tap water.[43]

Chicago's three top engineers also spoke at these events to protest the lack of progress toward adding modern, biotechnologies of water purification and sewage treatment at each end of the pipes. Of course, they had to pay lip service to the businessmen's and politicians' mantra of 6.5 bgd/24.6 bLd as a necessity for the public health. But leading the call for alternatives to dilution methods was George M. Wisner, Randolph's handpicked replacement as chief engineer of the SDC. Joining it in 1891, the college-trained former assistant chief drew an imaginary line of the metropolitan region's population growth that crossed the limits of dilution from a maximum diversion in 1922. Wisner and the other experts from the city and the U.S. Army Corps of Engineers argued that any "comprehensive effort" must immediately start with the building of a modern infrastructure of urban conservation.[44]

Although unappreciated in the upcoming elections, McCormick belatedly sponsored two of the chief engineer's proposals that set important precedents for the future. A year earlier, the board president had persuaded his fellow Republican trustees to approve funding to set up an experimental testing station at the end of the 39th Street interceptor sewer. The preliminary report of the engineers and the scientists in charge of the project stated that septic tanks were removing the solid waste from 200,000 gal. per day/757,000 L per day, or 0.5 percent of the city's sewage. As the November 1910 elections approached, moreover, McCormick got the board to establish a blue-ribbon panel of experts who were charged with drawing a long-range plan of water management for a metropolitan region with a population of three million inhabitants.[45]

The Politics of Public Administration

During this civic debate about engineering the urban environment, an explosion of corruption scandals derailed reform politics, dragging Chicago's era of Progressivism to a standstill. Coming under scrutiny were the city's Water and Sewer Departments and the SDC. Leading this crusade against the predatory "gray wolves" on the city council was the alderman from the University of Chicago's Hyde Park neighborhood, Charles E. Merriam.[46] In August 1909, the founding member of the school's political science department initiated a special committee to root out waste and inefficiency. He knew he would have to work undercover if he hoped to expose the perversion of public administration by the professional politicians in charge of managing the city's water resources. Rather than hire private detectives, his "municipal sleuths" were trained accountants and engineers armed with measuring rods, boring tools, and testing laboratories.[47]

In December, Merriam created a diversion by publicly disclosing one criminal scheme while his agents were literally uncovering another nefarious plot to defraud the city. The bookkeepers, he announced, discovered that the Sewer Department had approved a $46,000 claim by a contractor for removing nonexistent "shale rock" from an interceptor-sewer project on the North Side. At the same time, his technical experts were secretly inspecting another one of these recently built big sewers on the South Side to determine whether it met the city's contract specifications. Although an inside diameter of 18 ft./5.5 m and walls five bricks thick were required, they found only a 15 ft./4.6 m bore and two bricks held together with substandard mortar. In both cases, the alderman lectured, such an organized conspiracy among the building contractors, department inspectors, and trade unions could take place only with the complicity of the political bosses.[48]

On January 31, 1910, Merriam drew public attention to this latest shocking revelation to provide another smokescreen for a covert reconnaissance of the work routines of the Sewer Department. His investigators trailed the teams of men assigned to clean the catch basins and measured their activities with mathematical precision. In a council meeting, he charged that the superintendent of the sewers, William E. Quinn, was "unsuited for his job," because he was responsible for enrolling the department's army of unqualified patronage workers. They "wasted time" at an average rate of 46 to 50 percent of their shifts, according to Merriam's sleuths. Making matters worse, many of the workers were "poor and decrepit, rendering poor service[,] . . . and another [was] ready for the boneyard." The bottom line, the experts calculated, was a departmental efficiency score of 30 percent, or a cost of $230,000 a year to maintain the patronage system of the political parties.[49] In effect, this was wasted time and money that were dearly needed to keep the city's inadequate flood-control system at maximum capacity. As if to punctuate this point, 0.5 in./1.3 cm of rainfall two weeks earlier had submerged the streets on the western side of the CBD "from six inches to a foot [15 to 30 cm] deep[,] and pedestrians had to walk blocks in some places to reach a crossing, or wade." The sewers backed up because they were clogged with the buildup of decades of greasy sedimentation.[50]

The outcomes of the shale rock and the sewer loafers scandals reveal the far-reaching extent to which the professional politicians had centralized control over all the branches of local government. The assistant chief engineer in charge of keeping on-site records of the excavation in a field book,

Ralph A. Bonnell, came under withering interrogation by the Merriam committee's legal counsel, Walter L. Fisher. The fast-rising future U.S. secretary of the interior had just completed writing a special appendix to the *Plan of Chicago*, a constitutional treatise on the power of the states to create forest-preserve districts and other metropolitan-scale public-works projects.[51]

In a courtlike setting, Fisher got the supervising engineer to admit that he had broken the rules. He took the field book home after work rather than return it each day to the department. Bonnell then claimed that burglars had broken into his house and stolen the book. Fisher sprang his trap, catching him in what could now be interpreted only as a complete fabrication. The attorney produced the book in question, which had been recovered after investigators raided the department's offices. Causing a sensation, Fisher dramatically held it up, showing that three pages had been ripped out. The lawyer shut the trap, declaring that these were the very pages that had documented Bonnell's approval of the contractor's vouchers for removing the extra, unanticipated bedrock.[52]

Based on these revelations, the administrative and judicial agencies responded swiftly with investigations of their own. The civil service board suspended the supervising engineer for misconduct, and a grand jury indicted him as a co-conspirator in a criminal scheme. In a similar way, outrage over the sewer slackers led Commissioner of Public Works McGann to fire Superintendent Quinn for dereliction of duty. Other dismissals and forced resignations followed. Merriam's good-government campaign seemed to be making headway in cleaning out the crooks from city hall.[53]

But the unraveling of his two leading cases of corruption and the restoration of the status quo in the Sewer Department left no doubt that the Progressives' movement for municipal reform had been stopped dead in its tracks by the corrupt aldermen. At first, the prosecutor in the shale-rock conspiracy trial seemed to prove beyond a reasonable doubt that Bonnell and the contractor had attempted to defraud the city. A parade of expert witnesses testified that Chicago's subsoil contained dense, hardpan layers of clay, but no shale rock. The prosecutor also called to the stand a private detective who had posed as the owner of a construction company seeking business in Chicago. The undercover agent had followed Bonnell to his vacation home in Florida, where he had sought the suspended engineer's advice on getting a sewer contract with the city. Letting his guard down, Bonnell confessed that the field book's pages had to go missing: they actually recorded another $20,000 worth of approved vouchers that the contractor prudently had not submitted for payment after the announcement of Merriam's investigation.[54]

In May 1910, however, the judge tore the case apart and ordered a directed verdict of not guilty for all the defendants. According to this elected officer of the court, the prosecutor had failed to connect the dots of a conspiracy among them, regardless of any individual wrongdoing. On the basis of the court's ruling, the members of the mayor's appointed civil service board dutifully reinstated Bonnell and all the others who wanted their jobs back, including Quinn. The patronage regime of the Sewer Department snapped back into place, while Merriam was defeated a year later when he ran as the Republican candidate for mayor.[55]

The unraveling of Merriam's anticorruption campaign suggests that machine politics on a citywide scale was beginning to emerge. In the coming years of the twentieth century, this story's three-act trajectory of corruption exposed, investigated, and rolled back would be repeated with tiresome regularity. A narrative motif of civic tragedy like this one plays out only when a political party gains control of all the agencies of local government, including its judicial and administrative watchdogs. Then its loyalists can act with impunity for their own self-serving interests at the public's expense.

Another indicator of the growing power of machine politics was the defeat of McCormick by his Democratic opponent, Smyth. His party united behind Sullivan's faction, while the Republicans were distracted by an intraparty struggle between the Taft and the Roosevelt factions. For McCormick, defeat meant the loss of his stepping-stone to celebrity as a leading public figure. Turning to the private sector, he set up the Illinois Conservation Association. Protesting machine politics, the humiliated civic reformer organized a lobby dedicated to the preservation of nature and included Wisner as one of its directors. "The policies for which we stand," McCormick lamented, "constantly have been defeated in the interests of seekers after special privilege."[56]

Wielding his veto power, President Smyth was equally demonstrative in signaling the rejection of a conservation ethic in the formation of public policy. At issue was the demand of the North Shore for intercepting sewers to protect its lakefront water supplies. A committee representing Evanston and Wilmette complained that these communities had already paid $423,000 in taxes but had received no benefit from the "Big Ditch." The committee pointed to precedents, moreover, that the district had helped pay for the 39th Street sewer and was currently operating its pumping station.[57]

But in April 1911, Smyth announced the intentions of the board to backtrack from conserving natural resources in favor of returning to the path of exploiting them to the fullest possible extent. To stop the planning commission in its tracks, he withheld the pay of Wisner and a second district expert serving on the panel until they discontinued their participation. "This board," he declared, "is not in the business of building sewers and is not going into that business." The McCormick-run *Tribune* attempted to shame the civic leader into reconsidering his position against the experts' call for a comprehensive plan of water management. "It is to be hoped," an editorial chided, "that President Smythe [*sic*] will not destroy his standing in the community by further obstinate opposition to the public good." The newspaper blast had no noticeable effects. The district confirmed his policy directives in September, when it sent official letters to the North Shore's communities, advising them to build these costly public works at their own expense.[58]

As in the past, the one and only force powerful enough to change the sanitary strategies of Chicago's policy makers was a natural catastrophe in the form of either a great flood or an epidemic disease. After recording twenty-six cases of typhoid fever in Evanston in two months, the local branch of the Chicago Medical Society warned residents in November that they were drinking "raw sewage." Water quality along the lakefront had become so offensive that swimmers were being forced to retreat to the beach. Unexpectedly, the board of trustees was thrown into the harsh glare of public scrutiny, facing blame for not preventing this frightening plague from attacking some of the wealthiest and most powerful residents of the metropolis. In this unique case, the board reversed course, holding out an offer of help. To save face, it contended that building the sewers presented a special case, because these suburbs were contaminating the city's water supply on the far North Side. Casting blame on others, the trustees held the federal government responsible for preventing them from doing anything similar to safeguard the public health of the Calumet District.[59]

In February 1912, Smyth reaffirmed the board's decision to prioritize the interests of the political machine over the public good. He disbanded the blue-ribbon panel and buried its comprehensive, thirty-year plan of urban conservation. To protect the drinking supply, the experts recommended that the sanitary authorities should shift from the dilution of wastewater to biotechnological systems of purification and treatment. To prevent damaging floods, the report proposed that they should construct a completely new, separate system of storm drains and sewers, starting with the replacement of the overtaxed network of postfire pipes in the CBD.[60]

Instead, Smyth turned the sanitary district into an engine of upward mobility for the Irish Quartet. For the remainder of the decade, each member of this self-reinforcing mutual-aid society prospered. The most obvious example of machine politics was the nepotism enjoyed by Smyth's twenty-six-year-old son, whose flourishing insurance business was located in his former office. The most notorious case of corruption involved Sullivan's Great Lakes Dock and Dredge Company. It had secured a $400,000 contract to remove 1.5 million cubic yards/1.15 million cubic meters of sedimentation from the Chicago River and deposit it in the lake more than 8 miles/13 km out from shore or in the landfill project to expand Lincoln Park on the North Side. Instead, a whistle-blower informed Chief Engineer Wisner that the company was filling the barges in the daytime and then dumping them in the same spot at night. Because every company tugboat had a sanitary district inspector aboard, only a conspiracy among the work crews, businessmen, and public officials could account for what Commissioner of Public Works McGann called "nothing new." Yet he offered only a lame excuse for not stopping this illegal practice in the past: he had simply failed "to get any evidence," possibly because the company was being tipped off regarding his detectives' surveillance plans.[61]

Until the opening of the Cal-Sag Canal ten years later, Smyth's "obstinate opposition" to the federal government left the drinking water of hundreds of thousands in the metropolitan region polluted with deadly germs and toxic chemicals. In desperation, Chicago's sanitation experts became among the first in the nation to recommend the chlorination of the water supply. Wisner's test station at 39th Street had been using bleach as a disinfectant for some of the stockyard's discharges into the city's infamous open sewer, "Bubbly Creek." This and other so-called trade wastes, representing the equivalent of a million people's raw sewage, were already overloading the dilution method beyond its (illegal) maximum limits to render harmless the water in the Chicago River. The experiments led scientists to the counterintuitive discovery that adding minute amounts of this poison to the water supply made it safe to drink. The chlorine killed the dangerous bacteria, but not the ultimate consumer. After a hard sell to the city council in 1912, the Water Department began installing chlorination machines, beginning at Hyde Park's 68th Street intake crib. The number of cases of typhoid fever and other water-borne diseases dropped dramatically, finally ending this crisis of public health.[62]

Moving through the crossroads of decision making on water management, the sanitary authorities discarded comprehensive planning and adopted a policy of incrementalism. This approach gave the board of trustees a maximum amount of power over the queue line of improvement projects and the winning bids for building contracts. Evanston and the North Shore got interceptor sewers; South Chicago and the Calumet District got nothing. The South Branch got more new bridges; the North Branch got more raw sewage. The investors in commerce and industry got priority over the stakeholders in sanitation and conservation. At least until the end of World War I, public-works projects in Chicago supplied the political machines with the fuel of patronage and contracts to continue their consolidation of citywide control of the two major parties.

The Politics of Land Management

In contrast to the management of its water resources, the job of building Chicago from the ground up was left largely to the private sphere. An open prairie and a profusion of speculators, such as Samuel Eberly Gross, gave some truth to Adam Smith's "invisible hand" of the marketplace. But the sale and rental of property was not blind to race, ethnicity/religion, gender, class, and immigrant/citizenship status. In particular, the CREB exerted a powerful influence over the metropolitan area's social geography. In the 1890s, it began sponsoring the universal adoption of restrictive covenants, which not only made discrimination legal but made segregation mandatory as well. The appearance of the "Black Belt," an overpacked ghetto of African Americans, made visible the human agency that was indeed playing a formative hand in drawing the map of Chicago's settlement patterns.[63]

As with the contestation over the city's waters, the stakeholders in the value and the use of land included the full range of society, from elite businessmen's groups at the top to neighborhood/parish organizations at the grassroots. And, as with the politics of water management, the public sphere was turned into a battleground of competing interests. From 1912 to 1922, for example, the industrialists stymied the CREB's grand plan to zone every parcel of property within the city limits. The housing reformers continued to confront the slumlords by keeping pressure on city hall to enforce the health regulations and building codes. At the same time, direct confrontations over urban space were constantly taking place in public spaces, such as parks and beaches, and semipublic places of amusement, such as dance halls and movie theaters. Victims of environmental injustice, blacks and Latinos fought back, refusing to be expelled from the city to the periphery. These ongoing skirmishes would culminate in the city's worst race war in 1919, when blacks held their ground against the assaults of white mobs.[64]

In part, Chicago's uninterrupted attraction to foreign immigrants and rural migrants alike accounted for conflicts over urban space at the street level. The new arrivals helped add another half million people to the city during the 1910s, representing a 24 percent increase in population to 2.7 million. Suburban Cook County increased at a faster rate of 60 percent, or 131,000 new inhabitants, rising to a total of 351,000. The momentum of residential construction also increased to unprecedented levels, gaining thrust from the "Bungalow Craze" (see Figure 2.1). Subdividers responded by mapping out more than 110,000 lots inside the county and almost 90,000 more in the surrounding metropolitan region. Housing demand kept real-estate developers busy with new construction, which also created work for all the providers of utility services and public infrastructure. The Sewer Department, for example, was extending its system of pipelines about 50 miles/80 km each year.[65]

Human agency shaped sprawl's dual process of deconcentration and concentration to produce new patterns of settlement. On the one hand, homebuilders took advantage of rapid advances in automobiles, public transit, telephones, and other modern technologies to turn farms into neighborhoods. Ironically, the clay excavated from the North Shore Channel was turned into the bricks that were paving over the land on both sides of the city's borderline. Despite a halt in construction for two years following the American entry into World War I in 1917, they erected 30,000 houses and four times as many apartment units to produce an impressive 155,000 new places to live during the decade. Most of them were concentrated in a crescent-shaped arc located

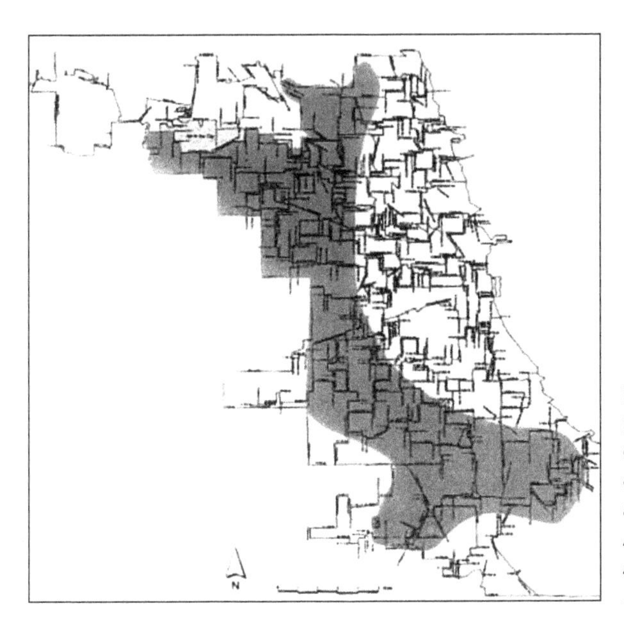

Figure 2.1. The Bungalow Belt, circa 1930 (City of Chicago, Department of Housing, available at http://tigger.uic.edu/depts/ahaa/imagebase/buildings/Files/Bungalow.html)

4 to 7 miles/6 to 11 km from the city center. It ballooned threefold to 1.1 million residents. The next arc, from 7 to 10 miles/11 to 16 km out, also enjoyed the steady growth of networked, residential suburbs and manufacturing districts, doubling in number to 332,000 inhabitants.[66]

On the other hand, the one million people living in the central zone were squeezed more tightly together in deteriorating conditions, especially in the Black Belt. The expansion of the CBD replaced housing with offices and stores. Sociologists from the University of Chicago described this process as "piling up . . . an increase in Negro population without a corresponding increase in living space. It is reflected in the heightening of . . . residential density, the creation of additional dwelling units by one or another kind of conversion, and an increase in room crowding in dwelling units."[67]

In 1910, reformers mounted a campaign to ameliorate slum conditions in the emerging black ghetto, South Chicago's "Bush," and other neighborhoods overcrowded with recent arrivals. Led by social-settlement workers, this movement resulted in what turned out to be a hollow victory: a revised building code for virtually all multifamily dwellings. Like its predecessor, the landmark "New Tenement" ordinance of 1902, according to Thomas Lee Philpott, "the biggest boodlers [corrupt officials] in the City Council backed it." Full of grandfather clauses and other loopholes, the hijacked legislation reinforced mutual bonds of support between the political machines and the real-estate industry. By the outbreak of World War I in 1914, the author of *The Slum and the Ghetto* concludes, "the housing movement in Chicago was in tatters."[68]

While living in the center became claustrophobic, the opening of the periphery offered a wide variety of opportunities to move into a contemporary home set in a suburban landscape. The rapid buildup of Rogers Park/West Ridge illustrates the transformation of semirural villages into networked suburbs after being consolidated within the city borders. The residents of outlying areas, such as this one on the far North Side, bordering Evanston, voted for annexation because it promised political empowerment and technological modernization. Rogers Park/West Ridge's thirty-five hundred residents got telephone wires in 1889, gas service in 1893, electricity in 1895, trolleys in 1906, and street improvements under the special assessment system. The area's rapid growth would influence the sanitary authorities to make it the first to have its sewage diverted from the lake to the North Shore Channel.[69]

By 1910, Rogers Park/West Ridge was well equipped with the physical infrastructure and social institutions to facilitate a metamorphous into a thriving, suburban community. Already attracted here were the Jesuits, who bought land on the lakefront for a new college, and the brickmakers, who set up their factories along the banks of the sanitation channel. In between was a new country club with an eighteen-hole golf course. A year later, a newspaper article titled "Suburban Gains Market Wonder" extolled Roger Park's "remarkable development and reasonable prices." Appealing to the American dream, the reporter observed, "'own a piece of land' has become the slogan of the clerk as well as the employer."[70]

This triumph of suburbanization, however, meant displacement for the small-scale farmers living in an ethnic community long known as Germania. They lived west of the appropriately named Ridge Avenue. The sand dune–like hydraulic divide between the Great Lakes and the Mississippi River basins paralleled the shoreline about 1 mile/1.6 km inland. In the late 1890s, real-estate speculators exploited reforms that allowed for the creation of special-purpose tax districts to build parks and boulevards, which they used to increase the value of their developments at the public's expense. They bought large parcels of land on both sides of Ridge Avenue. Setting off the "Cabbage Head War," the real-estate men maligned the farmers and defeated them in the enabling referendum. With the paving of this road that connected the city center to the North Shore, their fields were planted with new crops: rows of single-family bungalows and two- and three-flat apartment buildings.[71]

The founders of the Argo manufacturing district avoided conflict by selecting a green space in the village of Summit, about 14 miles/23 km southwest of downtown. Its name came from the fact that, at 31 ft./9.4 m above the water level of the lake, it was the highest point in Cook County. Reducing the risk of a damaging flood, the investors in the development, the owners of the Corn Products Company, chose this location for three additional reasons. First, they built a $15-million complex designed to employ up to three thousand workers there because it was close to the major switching yard of several railroads and the ship canal in addition to being a safe distance from the gridlock in the Chicago River harbor. Second, the company exploited the sanitary district's policy of free disposal of industrial pollution into the rivers and the ship canal. Like a malt-liquor brewery, the cornstarch and food-additives factory consumed massive quantities of clean water and produced equal amounts of liquid waste. This single, albeit giant-scale, plant became second only to the entire stockyards district in polluting the inland waterways with industrial organic-waste by-products.[72]

The other reason for choosing the Argo District was its advantageous location for the pursuit of the American dream of upward mobility through homeownership. Accessible to and from the city center by several streetcar lines, the developers of this "Factory Belt" modeled their project on South Chicago and the instant, steel-mill city of Gary, Indiana, at the bottom of the lake. In 1912, a promotional piece published by the *Chicago Tribune* ran under the banner "It Pays to Live Where Factories Never Bank Their Fires." Aimed at luring working-class families to the suburbs, it promised to save their children from the saloons and sin of the inner-city slums. The Argo District developers also pleaded for wannabe homeowners and investors to start building two-flats because of the acute housing shortage for its factory workers. Yet the advertisement-like article was conspicuously silent in boasting about the modern amenities of a networked suburb, with the single exception of electric service. Instead, it hinted that the availability of six thousand homesites on the open prairie was keeping prices down. One measure of the success of this appeal was Summit's mushroomlike growth from 950 inhabitants in 1910 to more than 4,000 a decade later.[73]

Nonetheless, class conflict moved to this suburb with the factories and the workers. A strike less than a year earlier in August 1911, of course, did not fit into the newspaper's narrative trope of progress and prosperity. The two-page illustrated article did not report the story of the walkout of the building tradesmen in the Argo District over wages and conditions. When the factory workers threatened to join the strike, the companies gave in to the unions' demands. Attracting recent arrivals from Eastern Europe, Summit became a place where residents had forged bonds of solidarity as workers and neighbors by the time they staged a walkout five years later. The company hired three hundred guards "armed with rifles, shotguns, and axhandles" to attack the picket line of five hundred strikers, about one-half of the workforce.[74] While remaining peaceful, the workers defied the pleas of the Polish parish priest of St. Joseph's Catholic Church to return to their jobs. Only after the company offered a 13 percent pay raise did they go back to the factory a week later.[75]

On July 10, 1919, the struggle between labor and capital in the Argo District turned violent, when three strikers were killed and seventeen were wounded. But bonds of community among workers living in Summit strengthened their resolve to resist the intimidation tactics of the company. In addition to engaging literally in class warfare, the company threatened to fire all of its foreign-language-speaking immigrant workers. Although arbitrators from the national office of the American Federation of Labor were the official strike negotiators, local organizers were in charge at the street level. They targeted company foreman and strikebreakers living in Summit for revenge. A strike leader from the neighborhood, Frank Gogwell, plotted to dynamite one of their houses, but he and his companion were ambushed in a gunfight with five sheriff's deputies. The eventual settlement of the strike left unsettled the post-war future of the American dream: the promise of upward social mobility.[76]

For African Americans and other easily identifiable racial/ethnic/religious groups, the road to personal success and social status through homeownership narrowed during the 1910s. Ghetto formation in Chicago is a well-told story by generations of artists, writers, and scholars. The slums of its Black Belt (and Packingtown) probably have been the most intensely investigated urban spaces in history. For our purposes, what should be emphasized is how the institutionalization of racism led African Americans to engage in machine politics. Vice lords and ministers of the lord organized a swelling base of voters into the ranks of the Republican Party in return for patronage and protection from police crackdowns. In *Freedom's Ballot*, Margaret Garb shows how the informal economy of corruption and crime lifted the fortunes of a few and helped everyone else by bringing money and resources from outside the ghetto into the community. But "the tragic paradox," she underscores, "was that in claiming power within the machine, activists in effect legitimized the machine brand of politics and, ultimately, helped construct the gilded fortress that prevented more radical demands for social change and civil rights from gaining a foothold in city government."[77]

In 1915, the Republicans claimed victory in not only electing the mayor, William Hale "Big Bill" Thompson, but also the first African American alderman, Oscar De Priest. In the midst of the Great Migration, the symbolic importance of making legitimate a voice of the race in the formation of public policy cannot be overstated. In addition, the new mayor paid back the black community for giving him a margin of victory in the primary and in the general elections. He gave African Americans a much larger share of the spoils of victory than his predecessors, integrating several city departments. Making a clean sweep of Democratic placeholders, Thompson hired ninety-two hundred "temporary civil service appointments" to replace them in his first five months in office.[78]

He became the first political "boss" of Chicago by leveraging control over patronage and contracts to hold together a winning coalition of intraparty factions. He became the first mayor to gain the moniker "the Builder" by redressing the Burnham Plan as his own designs for lakefront beautification and riverfront modernization. During the war years, moreover, he exploited tensions among the city's European ethnic and religious communities to broaden his base of support while further disrupting the already fragmented Democrats. Shifting his appeal from the middle to the working class, the patrician turned politician would sweep past his primary opponent, Merriam, as well as Sullivan's handpicked candidate in the next general election four years later.[79]

By 1919, however, the "piling up" of African Americans within a ghetto had reached the bursting point. As private living space grew smaller and more depressing, demands for access to public space became greater and more insistent. On the street, residents took pride in patronizing black-owned stores, restaurants, and nightspots along an expanding strip of State Street known as "the Strand." But as the race-riot commission reported belatedly three years later, this most densely packed area on the South Side was the least served by public parks, beaches, and playgrounds. What has also gone unacknowledged for too long is the full extent of working-class attraction to these outdoor spaces of urban nature for recreation and leisure activities.[80]

Recent research reveals that foreign immigrants and rural migrants brought with them rich and diverse cultural traditions of relating to nature from their homelands. New scholarship also records the contestation of African Americans and other nonwhite groups for use of the parks and beaches on their own terms. Locked out of the "gilded fortress" of decision making regarding land management, they took their struggle for civil rights directly to the public and the semipublic spaces of the city in confrontations with whites attempting to enforce the color line. Encouraged by Mayor Thompson's demagoguery, tensions also mounted between white groups during wartime demonstrations by Irish nationalists, Polish protesters marching against Germany, Zionist activists rallying Jews to stop anti-Semitism, and so on. In 1919, this virtual state of emergency became a powder keg of civil disorder as the war economy's abrupt end set off a nationwide crescendo of strikes like the pitched battle at Argo. Set within this new, broader context of seething political, social, and economic rivalries, the intensity of the violence that white mobs unleashed on the American scapegoat—the black man—becomes more understandable, if no less tragic. A fuller account of the formative hand of African Americans in building a city within the city during the 1910s also helps explain why they fought back, ending any further frontal assaults in the future.[81]

Conclusions: The Politics of a Flood-Prone Environment

In 1910, public-policy decisions regarding the future of Chicago had dire consequences for the remaking of its flood-prone environment. This year became a crossroads in its history of water management, because the publication of the *Plan of Chicago* created an opportunity to reform sanitary strategies within new frameworks of metropolitan scale, scientific knowledge, and conservation ideas. Ten years after the opening of the Sanitary and Ship Canal, the city's engineers, doctors, and biologists reached a consensus that Chicago had to change course and take immediate steps to save the lake and the river watersheds as valuable natural resources. Under the influence of machine politics, however, the sanitary authorities continued on their well-trodden path of exploiting water as an unlimited, free economic resource for commerce and industry.

The SDC gave first priority to building the ship canal, leaving little money for its other primary mission of public health and flood control. Adopting an incremental approach rather than a comprehensive plan, it left the western suburbs along the Des Plaines River unprotected against severe rainstorms. Rejecting the McCormick commission report, moreover, resulted in a lost decade in erecting the treatment plants that inevitably would be needed to supplement dilution methods in such a fast-growing metropolitan region. By the end of World War I, only the complete installation of chlorination machinery saved the city from its drinking supplies.[82]

The defeat of conservationism also meant time was wasted in reversing the degradation of Chicago's waters from untreated sewage and toxic chemicals. In fact, the sanitary authorities became one of the biggest polluters of the lake. A price paid for machine politics was corrupt practices, including conspiracies that allowed contractors dredging the river bottoms to break the law against dumping their poisonous loads within 8 miles/12.9 km of shore. And as the experts predicted in the McCormick report, the region's inland waterways became dangerously overloaded with organic waste. Another sweetheart deal with the stockyard companies indefinitely put on hold an agreement for them to help pay for the installation of treatment equipment. Instead, the SDC broke the law, taking twice as much water from the lake as its federal permit allowed to cope with increasing loads of pollution pouring into the Illinois River. In return, the federal government and the communities downstream of the ship canal were suing Chicago by the time Thompson moved into the mayor's office.[83]

Three years later, in 1918, the SDC made another appeal to Congress to authorize it to take even more water from Lake Michigan, or 7.8 bgd/29.5 bLd. But after more than a decade of rejecting recommendations to get a start on reducing the city's pollution problems, Chicago's delegation stood alone in the House of Representatives. All members from the Great Lakes states now joined the executive branch and the Canadian government in opposition to the city's barefaced claims on behalf of public health. They rejected its demands for the use of this valuable economic and natural resource at everyone else's expense.[84]

In environmental terms, the costs and benefits of machine politics in giving priority to shipping over sanitation remained marginal during this decade of constant intraparty factionalism. The fact that the sanitary authorities did not deviate from the long-standing policy of exploiting water as an economic resource provides strong, if ironic, evidence of the ultimate dependence of the party bosses on the business community. Their inability to consolidate a citywide machine limited their freedom of action against entrenched vested interests, which were profiting from advantageous locations on the Chicago River. Even as Mayor Thompson became "Big Bill, the Builder," he adhered closely to the recommendations of the downtown-oriented plan commission for civic improvements, such as the Michigan Avenue Bridge and the landfill projects to expand lakefront parks and beaches.[85]

Another irony of machine politics was the stream of beneficial scientific and technological innovations coming out of Chief Engineer Wisner's experimental station. Because the trustees had no intention of implementing his recommendations for conservation, they gave him the freedom to build a first-rate team of experts to conduct tests on the latest methods of sewage treatment and water purification. The single-most-important contribution to sanitary science and public health undoubtedly was the chlorination of the drinking supply. With the help of University of Chicago bacteriologist Edwin Oakes Jordan, Wisner's team also conducted research to find the best alternative to dilution methods of sewage treatment. In 1916, the sanitary district trustees officially adopted its recommendation for the activated-sludge method, yet it turned out to be a meaningless victory for the environment—they took no further steps to actually build such a facility.[86]

The defeat of a conservation ethos was also the result of the failure of reformers to attract the support of working-class voters. Certainly, they had outstanding candidates from the Harrison faction of the Democrats and the Merriam/McCormick wing of the Republicans. The roots of this political failure lay in the inability of Chicago's civic elite to bridge gender, racial, and class divides. On the contrary, the men's groups did their best to discredit the engagement of women's groups in civic affairs. In addition, some settlement workers had little faith in Eastern European immigrants and southern migrants to achieve the American dream on their own. The elite's patronizing and condescending attitudes toward those forced by poverty and prejudice to live in slum conditions alienated the very constituencies they were trying to assist in helping themselves.

Furthermore, the poor paid most of the costs of the tenement-reform acts, which led to higher rents and new opportunities for corruption. Published in 1910, Addams's *Twenty Years at Hull-House* contains her admission that "the mere consistent enforcement of existing laws for their advance often placed Hull-House into strained relations with its neighbors."[87] Philpott confirms that "the cost of graft, when the laws were not enforced, came out of the pockets of the poor as well." These kinds of unforeseen consequences of reform reinforced sexism, racism, and classism that the party bosses exploited to their advantage. After incoming Mayor Thompson stuffed the Building and Health Departments with patronage workers, enforcement came to an end, especially in the Black Belt.[88]

In 1915, voter approval of the forest-preserves district was the one major class-bridging effort that represented a victory for the conservation of Chicago's natural and human resources. In this case, support for the outdoor recreation movement came as much from grassroots insurgency as from top-down reform. In the ten years since the first referendum, access to the periphery of the metropolitan area improved significantly for the masses of people who could not afford cars. Expanding steam and electric rail lines in addition to building better roads for special express buses helped open fringe areas, such as Roger Park/West Ridge and Argo. At the same time, the interrelated construction of the bungalow belt brought a continuous flow of upwardly mobile, blue- and white-collar workers and their families into more-natural, albeit networked, suburban environments. When the first forest preserve opened three years later, the county's elected officials made sure to inform the public about its multiple, public-transportation links to the city center. The ward bosses also took credit for the steady expansion of the park system in the neighborhoods, with the exception of the Black Belt.[89]

Perhaps the greatest expression of appreciation of Chicagoans for the benefits of machine politics during this period was the turnout for Roger Sullivan's funeral. On April 17, 1920, seven thousand to eight thousand mourners marched in the procession from his home on the West Side to Holy Name Cathedral, where another fifteen thousand people gathered. All government offices and courthouses were closed; everyone from the state governor to the street sweepers was there to pay their respects. President Woodrow Wilson sent flowers. After services that recounted Sullivan's fulfillment of the American dream and extolled his devotion to the church, an honor guard of five hundred policemen and five hundred firemen flanked his cortege on its way to Mount Carmel Cemetery, passing through "crowds, [which] were lined twenty deep in the streets, heads uncovered."[90]

A similar story of success for each of the surviving members of Sullivan's Irish Quartet reinforces the conclusion that the working class believed that the benefits of the political machine outweighed the costs. Counted among Sullivan's best friends, Nash was one of the pallbearers. His sewer-construction business continued to grow with the expansion of the metropolitan area, boosting his annual income into the ranks of the top-ten moneymakers during a coming decade of prosperity. Inheriting Sullivan's role as a powerbroker, Nash continued to promote Kelly as the public face of the Democratic Party. In 1920, he reached the top job in the sanitary district and "the pinnacle of his profession," according to his biographer.[91] By then, he no longer needed the support of Smyth, who had retired four years earlier as the president of the SDC's board of trustees. The sixty-eight-year-old Smyth could take pride in his record of public service. Reduced to a duet, Nash and Kelly would continue to work tirelessly to bring their party's rival ethnic/religious factions together in an era of Republican rule. They faced their most formidable opponent yet, Thompson, who as mayor was busy consolidating power as a citywide boss by becoming "Big Bill, the Builder" of public-works projects on an unprecedented scale of corruption.

3

The Rise and Fall of the American Dream, 1920–1945

The Dry Years of Depleted Waters

In 1925, a landmark decision by the U.S. Supreme Court hinged on climate change. Since Chicago had started diverting water from the Great Lakes into its Sanitary and Ship Canal in 1900, their surface levels had fallen 21 inches (in.)/53 centimeters (cm). For the shipping industry, each decline of 1 in./2.4 cm represented an annual loss of $5 million in revenue. Shallow water meant that its vessels had to carry less weight to avoid running aground. Its customers ultimately paid in higher freight bills. With the exception of Indiana, the other five states bordering the Great Lakes also blamed Chicago's river-in-reverse for damaging their economies and shorelines. In *Sanitary District of Chicago v. U.S.*, they asked the court to revoke Chicago's permit and to force it to return all the water it removed from the lakes in the future in an equal state of purity. Only then, their lawyers contended, would the Great Lakes begin to rise to more-normal, historic levels.[1]

In this case, the decision depended as much on the court's understanding of science and technology as the law. Nevertheless, the expert witnesses—mostly engineers—did not engage in a discourse on long-term climate change. Instead, they provided a consensus of testimony on how the rain and the temperature affected lake levels on a seasonal basis. Although sixty years of records were available, neither side in the controversy thought about analyzing the interrelationship between weather and hydrologic cycles. Since the planning of the Sanitary and Ship Canal in the late 1880s, the experts had predicted that taking 6.5 billion gallons per day (bgd)/24.6 billion liters per day (bLd) would cause a 6 in./15 cm drop in all the Great Lakes, except Lake Superior, which was higher than the others. By the 1920s, they concluded that their calculations had badly underestimated the impacts of the divergence.[2]

The city's defense rested on its right to safeguard the public health from typhoid-fever epidemics, which it claimed had been virtually eliminated by reversing the river. Chicago's advocates were willing to accept responsibility for the damages caused by a 6 in./15 cm drop in lake levels, but they cast blame on everyone else, including the Canadians, for taking the rest of the lakes' depleted waters for their own selfish benefit, especially the hydroelectric power at Niagara Falls. Fighting back, Chicago argued that the lawsuit was the last resort of a conspiracy of its urban rivals to stop it from becoming the nation's leading seaport.[3]

In making its decision, the justices of the high court also had to take into account the alternatives to Chicago's sanitary strategy of diluting its wastewater and sending it into the Mississippi River basin. In 1906, they had ruled in the city's favor against downstream communities that asked for an injunction to stop it from polluting their drinking supplies. In *Missouri v. Illinois*, Justice Oliver Wendell Holmes Jr. reasoned that because St. Louis and other riverine places were equally guilty of dumping untreated sewage into the waterways, each had to take its own steps to protect the public health.[4] Since then, however, biologists had been making rapid advances in the study of fresh-water lakes and streams that led to a scientific revolution: ecology.

Considered the founder of the new paradigm of nature in the United States, Stephen A. Forbes was also a pioneer in establishing a laboratory in the field on the Illinois River. Leading the state's natural history survey, his team was finding more and more stretches of dead zones. The human and industrial effluents from Chicago and cities downstream were the cause of these empty voids of aquatic life. Even more permanently disruptive of nature's balance was an ongoing, massive transformation of the river's floodplain into agricultural land. Their research, albeit preliminary, left little doubt that dilution methods were not sufficient to deal with a problem that could only get worse in the future. In coming to a decision, the justices had to consider Chicago's sanitary strategy in terms of not only the Great Lakes but also the rivers of the Mississippi Valley.[5]

In retrospect, they were left to their deliberations without a brief on a second crucial aspect of the case: Chicago's flood-prone environment. In part, the dry years of lower-than-average rainfall and fewer damaging events since 1885 account for the failure of both sides in the controversy to draw attention to the inseparable connection between water quality and flood control in a combined sewer system. More important, this missing link can be attributed to the city's policy makers, who always insisted that reengineering a separate system of storm drains and sewer pipes was simply "impracticable." In the same way, they dismissed several storm surges that sent millions of gallons of polluted water in the Chicago River into the lake as inescapable acts of God.[6]

Although uncontested by the city's out-of-state antagonists, local reformers had been calling for this alternative infrastructure of wastewater management for at least a decade.[7] While admitting that retrofitting the built-up areas of the city center would be expensive, their technical experts argued that these costs would be offset by savings in flood damages and sewage treatment within the Sanitary District of Chicago's (SDC's) expanding territory of undeveloped land at the periphery. Furthermore, the city's conservationists drew connections from the massive volume of lake water needed to dilute the city's sewage due to the profligate waste from a water-supply system without meters. They estimated that 75 percent of the 1.1 bgd/4.2 bLd pumped by the Water Department every day leaked underground, exacerbating the combined system's dual problem of flood damages and sewage disposal. This reform alone—universal metering—would go a long way toward reducing the total amount of lake water diverted from the Great Lakes, according to their calculations.[8]

This chapter examines the remaking of Chicago's flood-prone environment from the point of view of the engineers, doctors, and scientists who worked to solve its problems of water management. During the interwar years, the life sciences advanced rapidly, which in turn led to impressive gains in public-health applications, such as epidemiology and biotechnology. Researchers made parallel progress in limnology, the science of the ecosystems of rivers and lakes. Their accumulating knowledge, of course, needs to be placed within larger frameworks of an evolving urban society and culture. The responses of the grassroots, politicians, and policy makers to the authority of experts and their recommendations for reform ultimately determined the outcome of the Supreme Court case on the fate of the Great Lakes and the Mississippi River basin.

The Machine Age of Outdoor Recreation

The place to begin constructing this framework is six years earlier in 1919, when virtually every Chicagoan had a plan for the anticipated end of the conflict in Europe. Fueled by the wartime boom, personal and civic expectations of progress were great. Better housing and consumer goods were high on the wish list, along with infrastructure improvements, such as better roads, suburban electrification, and more public space for outdoor recreation. In part, the ethnic/racial and class tensions leading to the great race riot was the result of hopes dashed during a period of economic chaos. The national government had failed to prepare and carry out a transition to peacetime. Nonetheless, corporate profits and individual savings amassed during the war soon restored the economy, which then expanded to unprecedented levels for the next ten years. Although large groups, including farmers, women,

African Americans, and Hispanics, were left out of this dawning Machine Age, the United States became an urban nation of mass consumption and leisure.[9]

The primary engine driving Chicago's prosperity during the 1920s was real estate. More than the healthy revival of either manufacturing or trade, according to economist Homer Hoyt, the buying and selling of residential property set into motion a self-perpetuating climb in land values. The war had turned a shortage of modern housing into a crisis, encouraging landlords to raise rents far beyond their expenses: between 1919 and 1924, they doubled. "With rapidly advancing rents, practically no vacancies, and almost stationary operating cost," he finds, "the owner of an apartment building in Chicago had a bonanza." These fat profits, in turn, attracted not only easy credit from the banks but also large pools of investment capital seeking productive use. Building bungalows also proved lucrative, because average construction costs of $5,000 were about one-third less than the dwellings' selling price of $7,500. Hoyt estimated that all this new construction was equivalent to 730 miles/1,175 kilometers (km) of street frontage, or 25 square miles/64.8 square km of paved-over prairie wetland. The seemingly insatiable demand for better housing came from a combination of upwardly mobile residents and newcomers.[10]

As the metropolis of the Midwest, Chicago continued to act as a powerful magnet for people in search of a better life. In sheer numbers alone, the forces of concentration continued to outweigh the forces of deconcentration. The national census year of 1920 found a majority of people living in cities. For our purposes, it also marks the point in time when suburban sprawl beyond Cook County became significant in the five surrounding collar counties: Lake, McHenry, DuPage, Kane, and Will. Chicago's population rose by 674,000 during the decade, reaching a total of 3,375,000 city dwellers; its periphery gained 383,000 people, boosting their number to 818,000 suburbanites. While McHenry and Will Counties retained a majority of farmers, DuPage County was overrun by fifty thousand new settlers, a 120-percent increase that halved the proportion of its rural population to only 20 percent. Established middle-class communities, such as Riverside and Evanston, captured most of the suburban housing boom.[11]

Becoming a stampede, the land rush of homebuilders, subdividers, and speculators turned the city's manufacturers from opponents to supporters of the Chicago Real Estate Board's plan of 1919 for comprehensive zoning. Their objections to being restricted to certain areas of the city became fears of being squeezed out of it entirely and into the suburbs by the skyrocketing price of real estate. By adding their support to the passage of a zoning ordinance four years later, business owners gained a voice in dividing the land into residential, commercial, and industrial uses. The resulting map shows that their primary goal was saving the Chicago River as a sacrifice zone for the free dumping of their liquid waste. The official designation of an industrial corridor also helped remove the legal roadblocks that had frustrated plans for straightening the South Branch for at least ten years. Despite the steady shift of shipping to the Calumet Harbor, city boosters refused to give up their dream of a deep-water channel from the Great Lakes to the Gulf of Mexico.[12]

In early 1919, the real-estate board also joined the Association of Commerce and other business and civic organizations in drawing laundry lists of public-works projects to make a smooth transition to a postwar economy. The power holders as well as the community organizers expressed their faith in Chicago's future by calling for reforms to raise more money to pay for them. In January, they added their support to city officials in approving $21 million in special assessment taxes for street and sewer improvements that would create hundreds of jobs for returning soldiers. In March, policy makers presented a more-ambitious $100-million plan for installing street lighting, building fire and police stations, straightening the river, expanding open space, and creating a new solid-waste disposal system. A month later, the property owners of South Chicago voted to tax themselves $1 million to build a collector sewer that would help drain an area of about 3 square miles/7.8 square km.[13]

The extent to which the "worse springtime flood conditions . . . than they have been for years," according to the *Chicago Daily Tribune*, influenced the referendum cannot be determined. However, the saturation of the soil by almost 4 in./10 cm of rain during the first two weeks of March, followed by a deluge of more than 2 in./5 cm, inundated the entire metropolitan area. In some far West Side neighborhoods, the water was 4 to 5 feet (ft.)/1.2 to 1.5 meters (m) deep. Overtaxed sewers from Evanston to Lake Forest on the North Shore backflowed into homeowners' basements. And despite the disclaimers of the SDC's spokesman, engineer Horace P. Ramey, the Chicago River poured its polluted contents into the lake. "If any river water did get into the lake," he equivocated, "it did not get to the [drinking supply intake] cribs."[14]

Despite nature's reminders that flood control was still on the agenda of unfulfilled infrastructure projects, the record of civic discussion is silent on this topic. The metropolitan-wide flooding of March could not be written off as an exception, because two more damaging storms occurred by the end of the year. In April, a cloudburst on the North Side forced the streetcars to keep their doors closed and basement-apartment dwellers to evacuate their homes. In October, another severe downpour of 1.5 in./3.8 cm in an hour overwhelmed the sewers in the city center. The fire department received hundreds of calls to rescue the old and infirm from drowning in their underground homes and to pump the polluted water out of them. Most of the sewers in the business district dated to the pre-skyscraper era, when 1 to 2 ft./0.3 to 0.6 m diameter pipes were laid with inadequate grade to keep them from clogging up with sedimentation. And as Charles Merriam's investigators had revealed, the corruption of the Sewer Department left the job of cleaning them undone. Experts put the price of building a new, separate system of sewage and drainage for the downtown area at $5 million to $6 million. Their pleas again fell on deaf ears.[15]

Instead, the press and the public at large were mesmerized by the announcement of the biggest plan ever, a $141-million package of improvements to complete Daniel Burnham and Edward Bennett's vision of the lakefront from the central business district to the Calumet Harbor. In July, the city won a three-year court battle against the Illinois Central Railroad Company, which confirmed the city's authority to regulate its operations. Forcing the railroad to convert its lines on the South Side from coal-burning to electric engines was a prerequisite to upgrading the lakefront for parks and recreation. Spearheaded appropriately enough by the plan commission's Charles Wacker, a bond issue to start this massive project sailed through the city council on a vote of 62–2. "When these improvements are completed," Chicago's most prominent civic leader boasted, "this city will have passed from the provincial town class to a real metropolitan city." Among the highlights of the plan were erecting a sports stadium seating 75,000 people, reclaiming 1,500 acres/607 hectares (ha) of lakefront for parks, adding four swimming beaches to accommodate 120,000 bathers, and building a new generation of harbor facilities for freight and passengers.[16]

The near-unanimous vote of the policy makers was emblematic of an equally universal support of Chicago's grassroots for more open space and recreational facilities. But this consensus quickly resorted back to contestation over the use of the parks and beaches among the various racial/ethnic and socioeconomic groups living within the city. The race riot of August 1919 began as a fight between whites and blacks over access to the lakefront. Violence on this scale was not repeated during the interwar years, yet an endless guerrilla war for control of these public spaces was fought on a daily basis throughout the period. On the one hand, park supervisors, settlement workers, and schoolteachers were able to channel some of this aggression into competitive (male) team sports, especially baseball, football, and basketball. On the other hand, the almost exclusively white police force added to the racialization and segregation of urban space by enforcing the color line against African Americans as well as a new group of immigrants: Mexicans.[17] A series of thumbnail sketches stretching down the South Side from the city center to the Calumet District illuminate the frameworks of social and cultural conflict that fought over the meaning of urban nature.

The Grant Park (renamed Soldier Field) Stadium was the signature project of the city's establishment, its business, civic, and political leadership. It was built to outshine all of its rivals in other cities. Officially opening on October 9, 1924, this great achievement was credited to the president of the South Park District, Edward J. Kelly. Two years earlier, the Democrats slated him for its board of trustees as an expert in public works. Keeping his position as the chief engineer of the SDC, Kelly became a household name as he cut ribbons to open sections of Lake Shore Drive, parks, beaches, and other recreational facilities. Behind the spectacle presented to the sixty thousand people at the new sports arena, Kelly and his sponsors, Patrick A. Nash and George Brennan, had built a new kind of political machine. It combined ethnic-based bosses from both parties to defeat their respective reform factions. In this case, Kelly cut a deal with the Charles Deneen wing of the Republican Party on an evenly divided board to be elected president. Each of these coalitions was based on secret agreements to split the patronage of jobs and contracts. Under Kelly's leadership, the proportion of politically appointed jobs jumped from less than 5 percent to more than 25 percent. The public paid the price; the new stadium cost more than $8 million, which was about four times as much as comparable facilities built during the same period.[18]

For the crowd assembled at the city's largest outdoor arena, the spectacle of the opening ceremonies cast aside any questions about the corruption of the South Park District. They cheered with delight as Robert McCormick's polo team competed against the one led by Frank Bering of the Sherman Hotel. And they applauded enthusiastically when Kelly recalled the city's rise like a phoenix from the ashes of the Great Chicago Fire of 1871. "There is no doubt in the world," he boasted, "that Chicago will some day not so long hence be the largest and leading city of the world." Two years later, the park district president would take center stage again before an even larger, standing-room-only crowd of a hundred thousand at the stadium's formal dedication. He gave them something worth coming to see: the national college football championship game between the Army and the Navy. As attendance continued to accumulate over the course of the decade at shows with a widening variety of appeals, Kelly's standing in the community as a successful and popular public servant rose in kind.[19]

For the same reasons, Mayor William Hale Thompson pledged in March 1919 that he would redouble his efforts to be "the Builder" during his second term in office. Although he would orchestrate the most corrupt administration yet in the city's history, his adherence to the *Plan of Chicago* sustained support from the commercial-civic elite, at least until the next election cycle. For the Main Branch of the Chicago River, Burnham and Bennett had envisioned a Parisian Beaux Arts façade that would combine beauty, efficiency, and order. A double-decker roadway would separate passenger and commercial vehicles, and a riverside promenade would complement the new Michigan Avenue Bridge and replace the chaotic and unsightly, albeit lively, fresh-food market on South Water Street. Financial and legal complications delayed the beginning of the area's demolition until October 1924, but Thompson deserves credit for what Wacker proclaimed was "'the most important event in Chicago history.'"[20]

The renamed Wacker Drive illustrates the political benefits of civic-improvement projects. "In as much as this plan was organizationally and aesthetically refreshing," historian Amy D. Finstein highlights, "it was also strategically shrewd. The projects recommended in the Plan of Chicago were so numerous and far reaching that they had to be tackled individually as the political and economic climates allowed. The completion of one project often anticipated the construction of another, effectively encouraging the continued implementation of the Plan. This was especially true of South Water Street."[21]

The widespread appeal of the city beautiful, the events at Soldier Field, and the expansion of urban open space were representative of the commodification of nature in two ways. First, they embodied the metamorphosis of the back-to-nature movement into a consumer product. Along with the rise of professional sports, Chicago was becoming a "society of the spectacle."[22] But even record-breaking occasions, such as the Army-Navy game, paled in comparison to the number of people directly involved in outdoor recreational and leisure activities. Consider that during the first year of Kelly's reign, 1922, the South Park District alone recorded 11.3 million participants (not different individuals) in supervised play, including more than 2.4 million in team sports, 1.5 million using the gymnasiums, and a million each at children's playgrounds and swimming pools. The number of park users engaged informally in pickup games and individual training went uncounted. In this second way, then, city dwellers turned urban nature into a product of consumption. Reflecting these trends was the creation in 1919 of the "recreational engineer," whose job was to package pieces of physical space into a predicable experience. For the most part, these new experts replaced the social reformers of the Progressive Era.[23]

Our next snapshot of the back-to-nature movement in the city can be called "the Battle of Washington Park." This case study of low-intensity warfare was a three-way contest among the working-class Irish/Catholic (and Polish/Catholics) living west of the 372-acre/150 ha park, the African American residents to the north of it, and the white middle class of Hyde Park/Woodlawn bordering its East and South Sides. Although the report of the post-riot commission highlighted the contrast between Bronzeville's deficiency of open space and high population densities, the South Park board completely ignored its recommendations. On the contrary, the expansion of its facilities everywhere but the black ghetto and its stretch of lakefront can only be seen as a conscious policy of discrimination and neglect.[24]

Reinforcing this view is the vivid portrait of racism revealed in native son James T. Farrell's autobiographical novel, *Studs Lonigan*. He felt dispossessed by blacks, Jews, and other ethnic/religious groups despite the coming of age of Irish/Catholics in society during the 1920s, such men as Kelly, Nash, and William E. Dever, who was elected mayor in 1923 as the clean-government, reform candidate. In the prewar period, the teenage Studs and his gang could claim personal triumphs in the turf war by beating up African Americans. But he was fighting a losing battle against what he perceived as an invading horde of barbarians. After 1919, a dissipated Studs got knocked down by Jewish opponents on the football field as well as in the boxing ring. Both his fictional and real family moved out of the parish, leaving behind the new church, which it had taken such pride in helping build.[25]

In contrast to the parish-centered Catholics living near Washington Park, more-affluent Anglo/Protestants were better able to insulate themselves from the neighborhood's demographic succession from white to black over the course of the decade. They did not have to resort to violence, because they could afford lawyers to enforce restrictive covenants in property deeds. They also had the financial resources and political clout to privatize the lakefront, creating the exclusive South Shore Country Club. They also took over Hamilton Park at 72nd Street and Normal Boulevard in the all-white Englewood neighborhood. Because the area's largely Catholic adult population rejected the social-reform mission of secular society, the Protestant well-to-do expropriated the 30-acre/12 ha site and field house for their own use. Called the "most ideal park in the system" by a University of Chicago sociologist in 1928, it was the home of the Hamilton Park Women's Club, which had six hundred paid members. They organized a symphony orchestra; a theater group; and literature, art, applied-education, and public-welfare departments. The men formed the South Side Male Chorus, sponsored Boy Scout and Campfire Girl troops, and fielded a championship tennis team. As a reflection of the experts' own ideals of urban nature, Hamilton Park was cast as the epitome of the "community service" model of recreation and leisure.[26]

For black Chicagoans, the institutionalization of racism in housing and government meant drawing on their own resources to get back to nature. The one major exception—the funding of a Young Men's Christian Association chapter by Jewish philanthropist Julius Rosenwald—gives even greater emphasis to the significance of their self-help accomplishments. Based for the most part in the church, the creation of sports teams, scout troops, day camps, and outdoor socials became an important part of forging a sense of identity as an urban community. Although small in number, elite African Americans also carved out a few suburban enclaves and summer retreats at the periphery of the metropolis. At the same time, the grassroots never surrendered their rights to the city in countless individual acts of defiance that crossed the color line in Washington Park and other public spaces and semipublic places of amusement. Farrell's sense of loss and his family's move to South Chicago became typical responses that eventually gave the black community a park of its own by the end of the decade.[27]

For white racists, such as Farrell, living in the mill-gate communities of the Calumet District brought them into contact with a group they believed were even more undesirable than African Americans: Mexicans. Like other immigrants before them, they were imported by industry as strikebreakers. Arriving in 1919 to take the jobs of workers in the steel mills, their population grew to between ten thousand and twenty thousand by the mid-1920s. Their numbers were relatively small, but they were highly concentrated in South Chicago and Packingtown. As Hoyt observes, land values went down, and housing rents went up. This extra tax on the immigrants forced many families to double up or take in borders and reinforced demeaning stereotypes by whites. The newcomers also suffered forms of racialization and exclusion unknown to them in their homeland. And, like Chicago's African Americans, they had to rely on their own collective resources and individual acts of defiance to gain access to outdoor recreational facilities. Taking pride in their sponsorship of baseball teams and other competitive sports encouraged them to keep fighting for the use of Bessemer Park, the Calumet District's only public green space.[28]

Different groups played in different ways, but they shared a desire to get outside to enjoy recreational and leisure activities. The few microhistories of public open space sketched here do not take into account the vast expansion of commercial and private facilities devoted to the consumption of nature. To take just one example of the growth of the back-to-nature movement, consider the fifty thousand golfers who participated in this elite and expensive sport on a single holiday, Memorial Day, 1928. In the process of the commodification of nature into a product of mass consumption, more and more Chicagoans began to weigh the costs and benefits of its exploitation as an economic resource. Undercurrents of conservationism during the Progressive Era became more-mainstream efforts to save the Great Lakes and the rivers of the Mississippi Valley.

The Turning Point of Water Conservation, 1925

If almost everyone in 1919 had a plan for the postwar era, the most glaring exceptions were the trustees of the sanitary district. They had become so entangled in in-fighting that their meetings turned into actual fistfights. The combat had begun three years earlier, when a reform Republican, Charles H. Sergel, had superseded the scandal-ridden administration of Thomas A. Smyth. In his inaugural address, the new president had become the first member of the board to break its solid front in defense of the dilution method of sewage treatment. On the contrary, he had stood with the agency's top engineer, George Wisner, and the city's other technical experts, who had recommended a sanitary strategy of building treatment plants. "Our great metropolis," Sergel had declared, "which should march in the van, has been lagging in the rear."[29] Unlike previous disputes over the division of patronage among political factions, this one was about policy. The question remained moot, however, while the nation was engaged in a real war.[30]

At their May 29, 1919, meeting, the trustees' long-running battle of verbal fisticuffs turned into what one member called a "minstrel show."[31] Willis O. Nance, Sergel's only ally on the nine-member board, repeatedly interrupted Patrick J. Carr, who punched him in the face. At issue was a resolution sponsored by the majority denouncing the president's inaugural address, because it had "proposed that the sanitary district abandon the basic principles upon which the sanitary district works were constructed."[32] Tensions among the trustees were high, because the U.S. Army Corps of Engineers had just informed them of its recommended plan for Chicago. Together with Wisner, and the city's board of supervising engineers, this plan disagreed with all three of the agency's major "basic principles." Stating the obvious, the experts pronounced that the Chicago River was obsolete as a harbor for modern, deep-draft ships. This made the SDC's policy of building expensive and disruptive swing bridges rather than fixed bridges equally outdated.[33]

Moreover, the engineers did the math to come to a bottom line in accord with Sergel's call for a fundamental shift in sanitary strategy from the exploitation to the conservation of nature. They used the original formula of the ratio between water and population to show that the dilution method was already unsustainable. The current load of human and organic industrial effluents required twice the amount of water being drained from Lake Michigan, or more than three times as much as the legal limit set by the federal government. The amount currently needed—9.75 bgd/36.9 bLd—exceeded the canal's design capacity by 50 percent. At the end of May, the U.S. Army Corps of Engineers submitted its official report, which concluded that the SDC needed to create and implement a twenty-year construction plan for sewage-treatment plants. But the weekly meetings of the trustees during June degenerated into slugfests of criminal recriminations, personal defamations, and three more near fistfights. While they paid lip service to the engineers' recommendations, these men had no intention of fulfilling their pledge of compliance. Instead, the only plan concocted by the seven-man majority was an ongoing plot to depose the president and take over the two key committees: engineering (contracts) and employment (jobs).[34]

The U.S. Army Corps' engineers were not deceived by the SDC's smokescreen, because they finally began taking steps over the next year to enforce its federal permit, limiting Chicago's divergence of water from the Great Lakes to 2.7 bgd/10.2 bLd. In September, the Corps served notice that it would force the sanitary district to treat the liquid waste of the stockyards, the equivalent of a million people's sewage. Two months later, the War Department issued a report that openly criticized the SDC for the first time and pressed the federal district court to issue an injunction against Chicago from taking "an excessive amount of water for sewage disposal."[35]

More damning, the engineers accused the board of trustees of engaging in a long-running conspiracy of deception. "Chicago is," the report charged, "therefore, debarred from any claim for indulgence as to work done and expenditures incurred in recent years. If, in defiance of the opposition of the Government, and in open disregard of the law, the officials of the Sanitary District of Chicago have continued to expend the money of their constituents in the prosecution of unwise and illegal plans, these officials and their constituency are to blame, and they should expect no great indulgence from the general public whose government they have ignored and whose interests they have disregarded." Under pressure from Washington, the court ordered the sanitary authority to stop violating its permit the following July, in 1920.[36]

Over the course of the five years while Chicago waited for the Supreme Court to hear its appeal, biologists joined the engineers in studying the effects of Chicago's sanitary strategy on a flood-prone environment. What historian of medicine Christopher Hamlin terms the "science of impurities" had undergone its own revolution, from chemical to bacteriological methods of analysis. Although older indicators of water quality, such as oxygen and chlorine content, were still useful, new tests were showing how microscopic organisms interacted with other microbes and chemicals in the water. During the Progressive Age, scientists had made two important breakthroughs that resulted in fundamental changes in the technology of the treatment of urban drinking supplies and wastewater: (1) they learned that the layer of bacteria that formed on the top of the sand beds of filtration systems was the actual agent of purification, and (2) they discovered that other microbes acted in similar ways to eat the organic compounds in sewage and render them harmless. Working together, the engineers and biologists had designed several different systems of treating human and industrial effluents. By lagging in adopting these systems, Chicago stood to benefit by installing the most up-to-date technologies, including the activated-sludge method.[37]

Armed with microscopes and petri dishes, scientists were conducting a parallel series of investigations of the ecology of lakes and rivers. The opening of Chicago's Sanitary and Ship Canal immediately drew the attention of the Illinois State Board of Health, because the American Medical Association "ha[d] about come to the conclusion that no river is long enough to purify itself," according to the state's Dr. James A. Egan.[38] However, the chemical indicators he and other doctors used initially verified their dilution formula. A second group of scientists working for the state government, including such biologists as Forbes, also began to study the impacts of the massive increase of water flowing into the Illinois River on its thriving fishing industry. At first, they observed that the fish were eating the organic compounds in the pollution-laden water, resulting in a spectacular increase in their populations. Doubling the catch between 1900 and 1908 to 24 million pounds (lb.)/11 million kilograms (kg), this unanticipated consequence was deemed an economic benefit. But the boom turned just as quickly to bust over the next six years, when the catch had fallen to only 8 million lb./3.6 million kg.[39]

Forbes and his team of scientists attributed the decline to causes other than Chicago's sewage. This failure is perfectly understandable, because working downstream, they were witness to an equally dramatic change of the river from a natural state to a human-made channel. From 1904 to 1914, levee and drainage district projects below Beardstown alone had reclaimed 172,000 acres/69,606 ha of bottomlands. With the help of local fisherman, Forbes learned that their little lakes and marshy areas provided safe havens of food and calm water in which the fish could breed and grow before reentering the river. Channelization, moreover, increased its rate of flow, which had a similar effect of degrading the aquatic ecology, including the fish. Although he joined the townspeople in their fight against the privatization of the bottomland commons by land developers and sportsmen's clubs, the state government supported their levee-building projects.[40]

In making this policy choice, a flood-prone environment proved a decisive influence. In 1904, the Illinois River Valley suffered its greatest inundation since 1844, when the first such disaster was recorded. While acknowledging that levees increased the speed and height of the river's currents, state officeholders and experts put their faith in science and technology to tame the forces of nature. "It is doubtful," an official report of 1915 opines, "if anyone will seriously consider the abandonment of the investments in the valley and the reverting to conditions of nature which would be likely to correct the present difficulties. For obvious reasons it is out of the question that we go back to the days of the buffalo and the Indian." The experts outlined a plan for the establishment of commercial fish hatcheries and artificial storage areas to compensate for the loss of the bottomlands.[41]

The flood of 1904 also played a role in giving greater emphasis to the study of weather patterns and climate change to gain control over nature. In the same year, a legal dispute over the fate of the Des Plaines River provided additional incentives to conduct research on how human settlement had caused changes in the land. In *Economy Light and Power Company v. U.S.*, the national government sought an injunction to stop the electric company from building a dam just below Joliet, because it was a "navigable waterway."[42] In the power company's defense, its lawyers argued that the river was not used for this purpose; it had either little or no current most of the year. And its scientific experts built a strong case based on daily measurements of the river's flow rates, which had been recorded at Riverside since 1886.[43]

Their analysis of the data showed that very little water flowed in the 90-mile-long/145 km river for almost 60 percent of the time. Only 30 to 40 percent of the rainfall on the 400,000-acre/161,875 ha watershed reached the river. On average, no water flowed past the rain gauge on 92 days a year, and fewer than 24.8 million gallons (gal.)/93.9 million liters (L) per day flowed past it on 117 days. At the turn of the century, 40 percent of this land was tilled by farmers, and suburbanites paved over some of it as well. A recent retrospective comparison of these river-flow figures with the 1943–1990 period illuminates how rural drainage projects and urban sprawl during the interwar years changed the riverine environments of the metropolitan region of Chicago. The median flow of the Des Plaines River was four hundred times as great, and the number of flood-stage events more than doubled, from an average of 1.5 to 4 times a year. A full explanation of the complex factors causing the increase in the river's waters lay beyond the needs of this study. In summary, however, the researchers posit that "preliminary modeling suggests the following results: reduced infiltration, reduced evaporation and evapotranspiration, greatly increased runoff and hydraulic volatility, and increased sediment yields and in-stream water quality problems caused by destabilization of stream-banks."[44] None of the data impressed the U.S. Supreme Court.

In 1921, the high court rendered its ruling against the power company. Justice Mahlon Pitney dismissed the scientific evidence and delivered a history lesson. Going back to "the days of the buffalo and the Indian," he recounted the story of Father Jacques Marquette's mapping of the Illinois, Des Plaines, and Chicago Rivers route that was used by fur trappers and settlers for more than a hundred years. His opinion also quoted the passage of the Northwest Ordinance of 1787, which declared, "'The navigable waters leading into the Mississippi and St. Lawrence, and the carrying places between the same, shall be common highways, and forever free.'" Since the 1820s, Pitney admitted, the Des Plaines River had not been used as a navigable waterway. Taking a long-term point of view, nonetheless, it might again be put to this purpose in the future. "It has been out of use for a hundred years," he lectured, "but a hundred years is a brief period in the life of a nation."[45]

If the trustees of the SDC entertained any hopes of winning their appeal to the Supreme Court, this decision should have convinced them that they were illusions. All inland waterways that in their "natural state" were navigable came under the authority of the national government, which was empowered to remove obstructions to the movement of commerce upon them.[46] On the heels of the ruling, the other Great Lakes states joined the government's suit against Chicago on the grounds that its river-in-reverse was impeding commerce by reducing the amount of cargo that ships could carry. Realizing that it was on the losing side of this argument about the lakes' depleted waters, the SDC sought to bolster its cause on behalf of safeguarding the public health. In desperation, perhaps, it exploited Chicago's flood-prone environment to make a counterintuitive rationale for draining more water from the Great Lakes.[47]

Despite denying that severe rainstorms in March and May 1919 had resulted in backflows of the Chicago River into the lake, the sanitary district reversed course five years later. It now claimed that the only way to prevent the contamination of the drinking supply from these kinds of torrential downpours was by maintaining the maximum amount of flow, 6.5 bgd/24.6 bLd, into the Sanitary and Ship Canal. Its experts made a special report on the storm of August 11, 1923, which dropped 3.68 in./9.35 cm of rain in a twenty-four-hour period and "caused the [single] most serious reversal of the Chicago River."[48] The engineers reported that approximately half of the Chicago River's watershed of 307 square miles/795 square km had sewers that disposed five times as much runoff as the less-built-up and paved-over half. With the sewer system at full capacity pouring 7 bgd/26.5 bLd into the Chicago River and an equal amount coming down the Des Plaines River, the dam at Lockport was opened to release as much water as possible without causing flooding downstream.

This emergency measure was not enough: 25 million gal./95 million L of sewage-laden water backflowed into Lake Michigan. The SDC's experts calculated that heavy rains could cause similar reversals seven or eight times a year. They warned that the potential for disastrous epidemic outbreaks could only get greater with the growth of the metropolitan region. They also made a dire prediction of catastrophic flood damages, because "as a rule, the street inlets are too small and in some large areas the sewers should be about double the present capacity." Therefore, they concluded, a maximum amount of diverted water was needed to keep the Chicago River flowing in the right direction. This line of argument came too late to be added to the SDC's appeal and did not influence the Supreme Court's decision. However, it would play a significant role in subsequent rulings that determined the plan to bring Chicago into compliance with the ruling in the case.[49]

Given the secretary of war's scathing report and the constitutional precedents, the judgment of the court was a foregone conclusion in *Wisconsin v. Illinois*. Speaking for a unanimous bench, Justice Holmes took the SDC to task for twenty years of defying the national government, and he ripped to pieces its defense of saving the public health as "too futile to need reply." On the contrary, the opinion reinforced the War Department's position that the national government had no authority to grant a diversion of water from the Great Lakes for sanitation, only navigation. "The investment of property in the canal and the accompanying works," Holmes concluded, "took the risk that Congress might render it valueless by the exercise of paramount powers. It took the risk without even taking the precaution of making it as sure as possible what Congress might do." He gave Chicago sixty days to come into compliance with its permit and the War Department authority to grant temporary relief to it while a plan to achieve this goal was implemented.[50]

Professing shock and disbelief, Chicago's policy makers and civic boosters continued to cast the narrative of the city's sanitary strategy in tropes of victimization by its rivals. In the week following the court's decision on January 5, 1925, they portrayed its consequences as an apocalypse of deadly plague and economic collapse. The *Chicago Daily Tribune*, for example, called for the city "to rise in revolt against a tyranny which now threatens its very existence." At the top of its enemy list was the U.S. Army Corps of Engineers, "because . . . [Chicago's] vital welfare is left in the hands of a military organization which for twenty years has acted toward it as if it were conquered territory." Demanding action by its congressional delegation to overturn the court's ruling, the editorial warned, "If they fail so to act, many thousands of us must die of preventable [typhoid] fever in the coming years."[51]

Led by Chief Engineer Kelly, the SDC organized an uprising of the grassroots against the city's opponents. On January 10, he addressed the largest town meeting ever assembled at city hall. He told the overflow crowd of two thousand citizens that they would soon have to strain their drinking water to remove bugs and worms. They would also watch the river burn from thick layers of scum that would cover its surface. The plan commission's Wacker introduced a resolution addressed to President Calvin Coolidge and Congress to authorize the city to continue taking at least 6.5 bgd/24.6 bLd. By the end of the month, the trustees of the SDC had sponsored another forty-six meetings in the neighborhoods as well as several in the suburbs. While they were pumping up the public into a state of hysteria, they also engaged the Corps' engineers in more-practical terms to begin designing a system of public works to reverse Chicago's sanitary strategy from the exploitation to the conservation of nature.[52]

By the beginning of March, they had hammered out a four-year plan to bring the city into compliance with the law, but without endangering the public health (see Figure 3.1). Assembling a team of a hundred engineers, the secretary of war presented the SDC with an ultimatum: in exchange for a temporary permit allowing a divergence of 5.5 bgd/20.8 bLd into the Main Branch of the Chicago River, it would have to build enough treatment plants to eliminate completely the equivalent of 1.2 million people's sewage. To prevent backflows of the river during severe rainstorms, it would also have to erect at the river's mouth a controlling works. And it would have to surrender the management of daily operations and flood-stage events to the U.S. Army Corps of Engineers. The decree demanded, moreover, that the city government adopt a universal meter system within the next six months. Chicago was given ten years to complete this project, which had been the practical experience in other major urban areas. Failure to meet any terms of the overall plan or to make significant progress each year in constructing treatment plants and installing meters would result in a revocation of the temporary permit.[53]

While this blueprint of a sanitary strategy reflected a conservation ethic, it also had three curious blank spots. It called for 100-percent treatment of

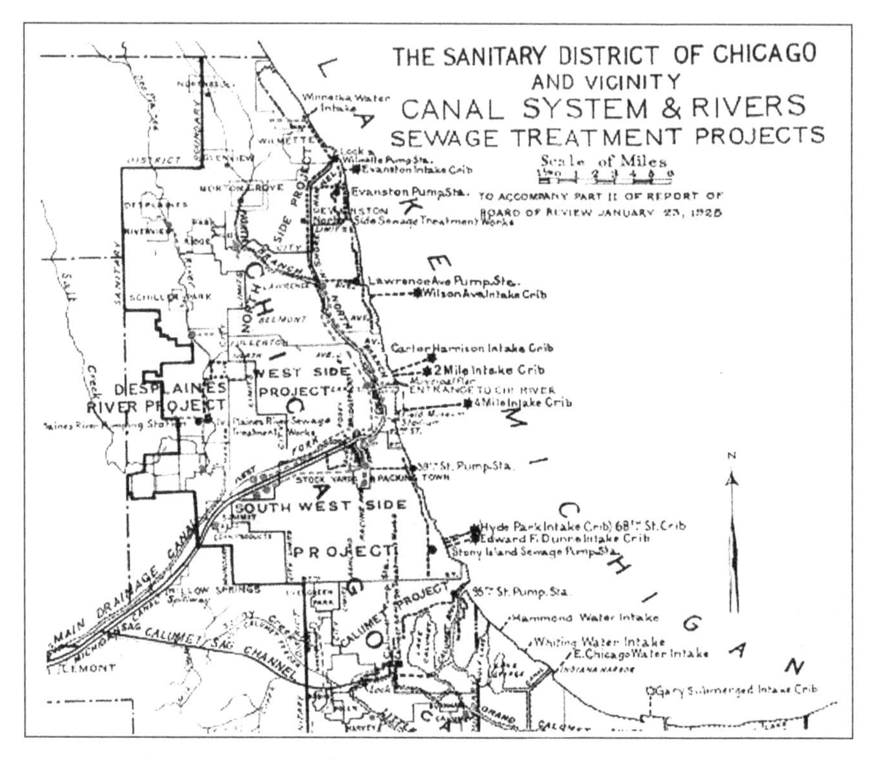

Figure 3.1. Sanitary District of Chicago Plan, 1925 (City of Chicago, Sanitary District of Chicago, *Report of the Engineering Board of Review* [Chicago: Sanitary District of Chicago, 1925])

one-third of the district's human population, but it dropped previous demands by the Army Corps' engineers for the SDC to treat the industrial effluents of the stockyards and the Argo corn-products factory, the equivalent of another 1.5 million people. Second, the amount of water that was permitted to be drained from the Great Lakes into the Main Branch was close to the amount currently being taken. However, the 650 million gal./24.6 million L per day flowing through the North Shore Channel and the 1.3 bgd/4.9 bLd diverted into the Calumet Canal since opening in 1922 were not included. And the decree also left out the Water Department's massive consumption of 825 million gal./3.1 billion L per day. Given that the bottom line of the plan allowed Chicago to take 8.3 bgd/31.4 bLd, or almost 2 billion gal./7.6 billion L per day *more* than it had asked for, the trustees had no problem coming to a unanimous vote of approval.[54]

The "Whoopee" Era of Machine Politics

In most respects, Chicago was well equipped to accomplish the goals of the federal government's plan for the conservation of the Great Lakes. The nation's second-largest city was in the midst of a real-estate boom of rising land values and new construction, which increased its tax base. The SDC not only covered a larger area but also kept increasing in size through a series of annexations. Since adding the North Shore and Calumet District in 1903, its expansion had brought most of the Des Plaines River watershed within its boundaries. In the six years following the end of the war, its assessed values had grown by $176 million, or 14 percent, to reach a record-breaking $2 billion. In addition, the state legislature responded to the secretary of war's decree by increasing the SDC's authority to issue bonds to cover the costs of building the sewage-treatment plants and a flood-control project at the mouth of the Chicago River. And in 1925, the end of the boom was nowhere in sight.[55]

The agency, moreover, was well prepared to begin this program of public works, although it had lost fifteen years since Wisner's call for a new sanitary strategy. Before resigning in June 1920, in the wake of Sergel's removal as president of the board of trustees, Wisner had set up three more temporary experimental stations to study the effluents of the tanneries, corn products, and pickle industries. His successor as chief engineer, Kelly, had been smart enough to choose an equally competent expert, Langdon Pearse, as his technical assistant. The Harvard University–trained and Massachusetts Institute of Technology–trained engineer, in turn, had supervised the construction of two full-scale treatment plants: one was located on the Des Plaines River, with a capacity of the equivalent of 58,000 people, and the other in the Calumet District processed the waste of the equivalent of 134,000 people. In a July 1926 special issue of the *Journal of the Western Society of Engineers*, Pearse outlined a $73-million plan to expand these facilities and build two much larger ones to handle the waste of the North and the West Sides by the 1929 deadline. In addition, he confirmed that the sanitary district was in the midst of designing the world's largest treatment plant on the Southwest Side. It would process the human and industrial wastes of the Chicago River, including those of the stockyards and the Argo corn-products company.[56]

Reinforcing Pearse's team of engineers was a small army of scientists, who kept constant surveillance over the water quality of Lake Michigan and the rivers of Illinois. The SDC's chief chemist, F. W. Mohlman, was not exaggerating when he stated, "Probably no stream in the United States has been sampled and analyzed as much as the Illinois River during the past thirty years."[57] By 1926, his colleagues had a much better understanding of the decline of the fish populations below the dam at Lockport. While some organic effluent enriched their food supply, too much depleted the dissolved oxygen in the water below a level that even microbes needed to live. Then, sedimentation coated the riverbeds, causing further ecological damage, because their plants could no longer generate oxygen from photosynthesis. Beginning at the city limits on the North Side, the entire length of the Chicago River and long stretches of the Illinois River had become virtual dead zones, where only "sludge worms" could survive.[58]

A final piece of preparation was widespread support from the public at large for the implementation of the federal government's sanitary strategy for Chicago. Everyone had a stake in preventing a "failure to act[, when] many thousands of us must die" from an epidemic of water-borne disease. And most Chicagoans shared an interest in saving the lake and the riverine environments for recreation and leisure. Each year, millions of people enjoyed the lakefront parks and beaches. Many others had a special interest in nature conservation, such as boaters, fishermen and fisherwomen, bird-watchers, hikers, and summer campers and picnickers making use of the forest preserves. Moreover, grassroots organizations formed to preserve patches of open space from destruction by polluted water. In 1913, for example, the residents living west of the North Shore created the Northwest Sanitary Drainage Association to protect the streams running through their fast-growing suburbs from sewage contamination. Along the Des Plaines River, too, community groups, local governments, and the Forest Preserve District of Cook County (FPDCC) lobbied the sanitary district to fulfill its often-repeated, albeit broken, promises to meet the goals of its public-works program.[59]

Yet the combination of all of these resources of money, expertise, and popular support was insufficient to overcome the resistance of machine politics. During the four years of its temporary federal decree, Chicago failed to meet every one of its terms. Even Mayor Dever's successful sponsorship of an ordinance in 1925 in favor of universal water meters was nullified two years later. Making a comeback, Thompson ran on a platform that called for its repeal, because "water should be as free as air."[60] Winning by a margin of eighty-five thousand votes, he immediately ordered a halt to the installation program.[61] Together with like-minded Democrats, he helped form coalitions that turned the SDC and the FPDCC into patronage armies that did little or no work. The money that should have been spent on a sanitary strategy of water conservation was wasted on bloated payrolls, sweetheart contracts, and frivolous projects, including the now-delusional dream of a deep-water shipping lane to the Gulf of Mexico.

After Sergel was deposed in April 1918, the sanitary district became the prototype of the evolution of machine politics from intraparty factionalism and interparty rivalry to citywide control. Compared to a city council composed of seventy prickly aldermen, the ward bosses had to get only five trustees to work together to gain control of an agency with a huge budget and little public oversight. They often slated themselves for positions on the board, including Republicans Lawrence F. King, Alex N. Todd, and Morris Eller and Democrats Patrick J. Carr, James M. Whalen, John J. Touhy, and Timothy J. Crowe. Under their leadership, the payroll more than doubled, from 619 employees in 1919 to 1,447 three years later. Although journalists and civil organizations, such as Citizens' Association, occasionally criticized the SDC for excesses, the city's commercial interests generally supported its sanitary strategy and the promotion of the Chicago Harbor over the Calumet District. Industrialists, too, approved of its policy of keeping the Chicago River a free dumping ground for their liquid waste.[62]

With the election of Crowe as president of the board in December 1926, the SDC entered its most notorious, "Whoopee" era of machine politics. This label comes from the extravagant parties, complete with illegal liquor and chauffeur-driven limousine service for prostitutes, that were paid for with public funds.[63] The fifty-one-year-old committeeman of the thirty-third ward had been a steelworker and a real-estate salesman before starting a new career in 1912 as a professional politician. Before being elected as a trustee in 1922, Crowe had held various appointed positions in the county and the state governments. In 1925, he had also become president of the Oak Park First National Bank. Together with two other Democrats linked to Brennan's faction, Crowe formed an alliance with the three Republicans from Thompson's wing of the party.[64]

After April 1927, the trustees acted with impunity. The War Department failed to follow through on its threat to revoke Chicago's temporary permit following Mayor Thompson's order to stop the meter-installation program. While he declared "victory," President Crowe interpreted the federal inaction as "vindication" of Chicago's right to divert as least 5.5 bgd/20.8 bLd to dilute its sewage.[65] He promised to fulfill all the other terms of the decree, but he did just the opposite. On the one hand, the SDC did not even bother to apply for the required permit from the War Department for the controlling works at the mouth of the Chicago River, let alone begin building it. Nor did the trustees take the first step of purchasing a piece of land for the treatment plant on the Southwest Side or provide the funds to complete the building of the one on the West Side. They did, however, violate the decree by increasing the diversion of water from the lake into the river to a peak of a little more than 6.5 bgd/24.6 bLd.[66]

On the other hand, they shifted $16 million from bond funds earmarked for infrastructure improvements to the corporate account for corrupt purposes. In the run-up to the November 1928 election, the payroll reached $9.3 million and four thousand employees. At least 1,147 of them were on the books as "water investigators," whose only official duty was to watch the lake. Nevertheless, they were unable to ensure victory for Crowe, who came in last. Even after the new board of trustees fired these patronage workers shortly after the election, Crowe's ward alone still employed at least 126 of the politically connected appointees.[67] The trustees had also spent millions of dollars on sweetheart contracts for projects that had nothing to do with either sewage disposal or flood control. In 1927, for example, they paid a company a million dollars for a cinder path along the entire length of the canal, and they awarded another contract of $198,000 to a wire-cable company to install a "life-line" on each side of it in case anyone slipped off the path and fell into the water.[68]

In March 1929, the party ended with the exposure of the trustees' scheme of raunchy merrymaking. After the fall of the Crowe regime, the state and the federal governments began investigating the SDC's corrupt practices. To a grand jury, the state's attorney presented evidence that the operation of its entertainment activities was run out of the Central Automobile Company, which had received $70,000 during the previous year. Some of the money had been laundered through Crowe's Oak Park bank. The company's bookkeeper perjured herself to protect her cousin, who was listed as the president. She had left town weeks earlier in anticipation of the exposure of this criminal conspiracy. Digging deeper, the prosecutor uncovered the names of seven other members of the family on the SDC's payroll. In May, the grand jury returned indictments on charges of graft against seven of the nine trustees and seven others involved in stealing $250,000.[69]

On the state level, a three-act play of corruption exposed, investigated, and rolled back was stretched out by the criminal court over a three-year period. Delaying the proceedings until two months after the March 1930 primary elections, the prosecutor obtained a second round of indictments against ten more officials of the SDC, including Kelly, for payroll padding and fraud. The court postponed the hearing for another year, when the charges against Kelly were dropped. In December 1931, a bench trial finally got underway, leading two months later to the acquittals of three trustees and guilty verdicts against four of them, including Crowe. However, only Frank J. Link went to prison, because two of the convicted trustees had died, and Crowe had also passed away while his lawyers were appealing his jail sentence. Link had made the mistake of cashing a forged check with his signature on it.[70]

In contrast, the federal authorities pursued the corruption of Chicago politics with zeal. Following the money, the U.S. attorney uncovered a pattern of crony capitalism between the SDC and its contractors. Like Al Capone, Kelly, Nash, and his nephews, the Dowdle brothers, were charged with tax evasion. Consider that Kelly's official pay during the postwar decade totaled $150,000, but his income amounted to $750,000. Nash underreported his take from a $2-million sewer contract, which was doubled for "extras," and then was awarded another $3 million "extra" for these "extras." The Dowdles also could not account for their share of $13 million in business with the SDC during Crowe's two years of leadership. Unlike Chicago's most notorious gangster, none of them was sent to jail. Instead, they were allowed to buy their way out by paying their back taxes.[71]

The Supreme Court also took a more-proactive role in forcing the SDC to comply with its decision in *Wisconsin v. Illinois*. In the face of the agency's flagrant defiance and shameful perversion, the justices issued increasingly stringent rulings between 1929 and 1933 in a series of three hearings. In the first case, the court agreed with the SDC's court-appointed Special Master,

Charles Evans Hughes, that the diversion of water for diluting the city's sewage was a temporary expedient that must end in the near future. Furthermore, Chief Justice William Howard Taft was equally dismissive of its arguments on behalf of safeguarding the public health. "The sanitary district authorities . . . ," he chided, "have much too long delayed the needed substitution of suitable sewage plants as a means of avoiding the diversion in the future. Therefore they cannot now complain if an immediately heavy [tax] burden is placed upon the district because of their attitude and course."[72]

A year later, the court set a five-year timetable for the reduction of the amount of water the city could divert from the Great Lakes to 3.2 bgd/12.1 bLd, and then three years later to 969,500 gal./3670,000 L per day, or just a little less than its original permit. Justice Holmes ridiculed Chicago's longstanding claim to having built a ship canal. Stripping away its defense, he protested, "There is no navigation in any practical sense coming into the Chicago River by way of the Illinois Waterway or the Illinois-Michigan Canal. Only a few canoes and small pleasure craft have passed through the little lock of the sanitary district." The court, however, agreed with the city that its water supply for domestic and industrial use should be excluded from the calculation of the amount of water released at the Lockport Dam, at least for now. At the same time, Holmes tightened it by including storm-water runoff from the Chicago and Calumet Rivers' watersheds. "It already has been decided," he concluded, "that the defendants are doing a wrong to the complainants and that they must stop it. They must find out a way at their peril. We have only to consider what is possible if the State of Illinois devotes all its powers to dealing with an exigency to the magnitude of which it seems not yet to have fully awakened. It can base no defenses upon difficulties that it has itself created."[73]

In 1933, Hughes, who was now the chief justice, took the SDC to task in harsh language of disapprobation. The delays in meeting the terms of the secretary of war's decree, he chastised, were caused by the SDC's "total and inexcusable failure" to act in good faith. In response to its new excuse of a lack of funds, an exacerbated court expanded its ruling to make the state government responsible for meeting the 1938 deadline to complete the building of all the planned sewage plants as well as the controlling works at the mouth of the Chicago River.[74]

Perhaps the only fully truthful disclosure made by the SDC to the Supreme Court was that its finances were in dire straits. On the eve of the exposure of the "Whoopee" era in February 1929, it could not meet its payroll despite firing thirty-five hundred of its forty-seven hundred employees, including Crowe's niece. In desperation, the trustees had to issue $12 million in "tax anticipation warrants," or IOUs, which were bought by a syndicate of local banks. In addition, the pilfering of its bond fund left the SDC without money to build infrastructure. Expecting a public blowback from the scandal, the trustees sought an exemption from the state legislature of the requirement of a referendum to issue more bonds. But with support from neither lawmakers nor taxpayers, the SDC's construction program came to a standstill. The people of Illinois had good reason to become distracted by pressing problems beyond its virtual bankruptcy: their personal lives had been turned upside down after October 1929, when the stock market crashed and the nation plunged into its greatest economic crisis, the Great Depression.[75]

In this case, the serendipity of history provided the way out for the SDC. The Supreme Court announced its third, conclusive ruling in May 1933, during the hyperactive, first hundred days of the New Deal. Congress had just passed a multi-billion-dollar bill for public-works projects. In charge of this massive spending program was Harold L. Ickes, one of Chicago's foremost reformers. Under his sympathetic leadership and because of President Franklin D. Roosevelt's growing friendship with Ed Kelly, the federal government underwrote the costs of the city's coming into compliance with the law over the next nine years. Together, the sewage-treatment plants and the controlling works at the mouth of the Chicago River contributed to cutting the amount of water drained from the Great Lakes for sanitary purposes from more than 6.5 bgd/24.6 bLd to about 1 bgd/3.8 bLd.[76]

The New Deal of Nature Conservation

To illuminate the growth of a conservation ethic during the interwar years, we need to return briefly to 1919 to trace the rise of Chicago's first citywide boss, Anton "Tony" Cermak. Known as the "Mayor of Cook County," he not only perfected the cross-party coalition; he created "a house for all peoples" as well.[77] Born in 1873 in today's Czech Republic, he immigrated with his parents the following year to the coal town of Braidwood, Illinois. Growing up in the mines, he got out soon after turning eighteen. In 1891, he moved to Chicago and worked for a street railway company before starting his own business as a teamster. Joining the Democratic Party as a precinct worker, he earned citywide recognition during the 1900s as the ethnic leader of the antiprohibition, anticharter reform group, the United Societies for Local Self-Government. Cermak personified the Eastern European immigrants who battled against their treatment as inferior subordinates by the Irish ward bosses in control of the party machinery. While his neighbors elected him to the city council, Sullivan's faction refused to slate him in 1919 for the mayor's office. Facing similar opposition from Sullivan's successor, Brennan, he ran a successful campaign three years later to become the president of the Cook County Board of Commissioners.[78]

Cermak realized that the universal desire for outdoor recreation and leisure could become the foundation on which to build a political machine independent of Brennan's domination of the party. On July 1, 1922, he opened his own baseball field, Cermak Park, where he threw the first pitch in the game between the Cermak Indians and the Famous Chicagos. He would continue to sponsor the team throughout the decade, because amateur sports were not only popular but also a source of ethnic identity and pride.[79]

Moreover, he used his official position to seize control of the forest-preserve district despite a Republican majority of eight members to seven Democrats on the board of commissioners. First, he placed all the district's 170 jobs on the civil-service list, which made him the sole source of employment. Then, like Kelly and Crowe, he persuaded one of the Republicans to join his side for a share of the patronage. And, like them, he inflated the payroll to almost fifteen hundred "temporary" workers in the run-up to the November 1923 elections. At the same time, he also oversaw the purchase of nine tracts of land totaling 350 acres/142 ha, which were strategically located to give every area of the metropolitan region additional open space.[80]

From this springboard of power, Cermak catapulted himself up the political ladder over the next eight years to become the first foreign-born mayor of Chicago. Like Kelly, he became a public face of the back-to-nature movement, cutting ribbons at opening ceremonies, awarding trophies at special events, and handing out free tickets to games at his baseball park.[81] He also pursued an ingenious strategy to create a political organization in the residential, commuter suburbs, which tended to be Republican. He went after its second-place runners-up, persuading them to join his side by giving them patronage jobs, contracts, and shared credit for additions to the forest preserves in their areas. But unlike Brennan, Cermak went after the African American vote by converting a key Republican, William Dawson, who would become the boss of the Black Belt. When the Irish party chief died in 1928, Cermak had enough city and suburban loyalists to elevate himself to the top position of the party. After Kelly, Nast, Crowe, and other prominent Irish ward bosses got caught up in the "Whoopee" scandal, he was able to take the last step up to become the Democratic candidate for mayor in 1931, turning Thompson's bid for reelection into his last hurrah.[82]

Cermak's triumph was part of one of the country's periodic, political realignments from one party's majority to the other. Although this historic shift was evident in the 1928 contest between Al Smith and Herbert Hoover, the descending spiral of the economy accelerated it to give Roosevelt and the Democrats a landslide victory in the next presidential election. In a similar way, Chicago's real-estate boom showed signs of losing momentum by 1928 but did not crash until the bottom fell out of the stock market almost two years later.[83] For most, the hardships of the Great Depression and World War II put aspirations of buying a house out of reach. For many, the American dream of homeownership became the nightmare of repossession and downward social mobility. For those who held onto a job, the Roaring Twenties came to an end, replaced by tight budgets with little left to spend on commercial entertainments. For the unemployed, access to open space may have become even more important to relieve the mental stress of forced idleness.

From 1930 to 1945, Chicagoans depended increasingly on the public parks, beaches, and forest preserves to meet their needs for recreation and leisure, because they were free. After the burst of park building during the Progressive Era, however, the city had fallen further and further behind in adding open space compared to its burgeoning population. It had placed fourth best in facilities in 1880, but its 136 parks with 5,400 acres/2,185 ha came in second to last in 1930 among the ten largest metropolitan areas. As the exhaustive, five-volume *Chicago Recreation Survey* stated, "The park acreage and play facilities of Chicago should be multiplied several fold in order to conform to well accepted standards."[84]

The ward bosses' use of its multiple park authorities as patronage plums also resulted in gross inequities of geographic distribution and a lack of comprehensive planning. Such neighborhoods as Bronzeville and South Chicago, with the highest population densities and worst housing conditions, had the least amount of open space, while the all-white bungalow belt gained most of the added facilities. Facing plunging property values and a taxpayers' revolt, policy makers were forced in 1934 to consolidate the city's twenty-two park districts into one, a reform that cut the number of politically appointed commissioners from 114 to 5.[85]

Beyond city borders, the suburban communities and the county governments had been setting aside much larger tracts of green space for public use. During Cermak's administration, the FPDCC increased from 18,000 acres/7,300 ha to 32,800 acres/13,274 ha, or almost its maximum authorization of 35,000 acres/14,200 ha. Despite the economic burden of getting to them during a depression, people kept coming in growing numbers. In 1924, an estimated 7.65 million enjoyed a day in these nature preserves, a number that doubled during the 1930s. Although Cermak made steady progress in purchasing land, he spent most of the operating budget on patronage workers rather than basic facilities, including lavatories. During this period of the commodification of nature, as parks historian Galen Cranz comments, "more and bigger were argument enough."[86] Like the funding of the sanitary district's infrastructure plan, the New Deal underwrote the improvement of public open space. "As a result of this aid," the *Recreational Survey* reported, "it was estimated that in 1934 that the Forest Preserve development program had gone ten years ahead of schedule."[87]

The single-most-important project undertaken by the New Deal's public-works programs was the transformation of the Skokie Swamp on the North Shore from a mosquito-infested and fire-prone nuisance into a model of the conservation ethic and flood control (see Figure 3.2). From the origins of the outer-belt/forest-reserve movement in 1900, Jens Jensen had designated this tract of first-growth trees and wetlands as a top priority. "The Skokie is a

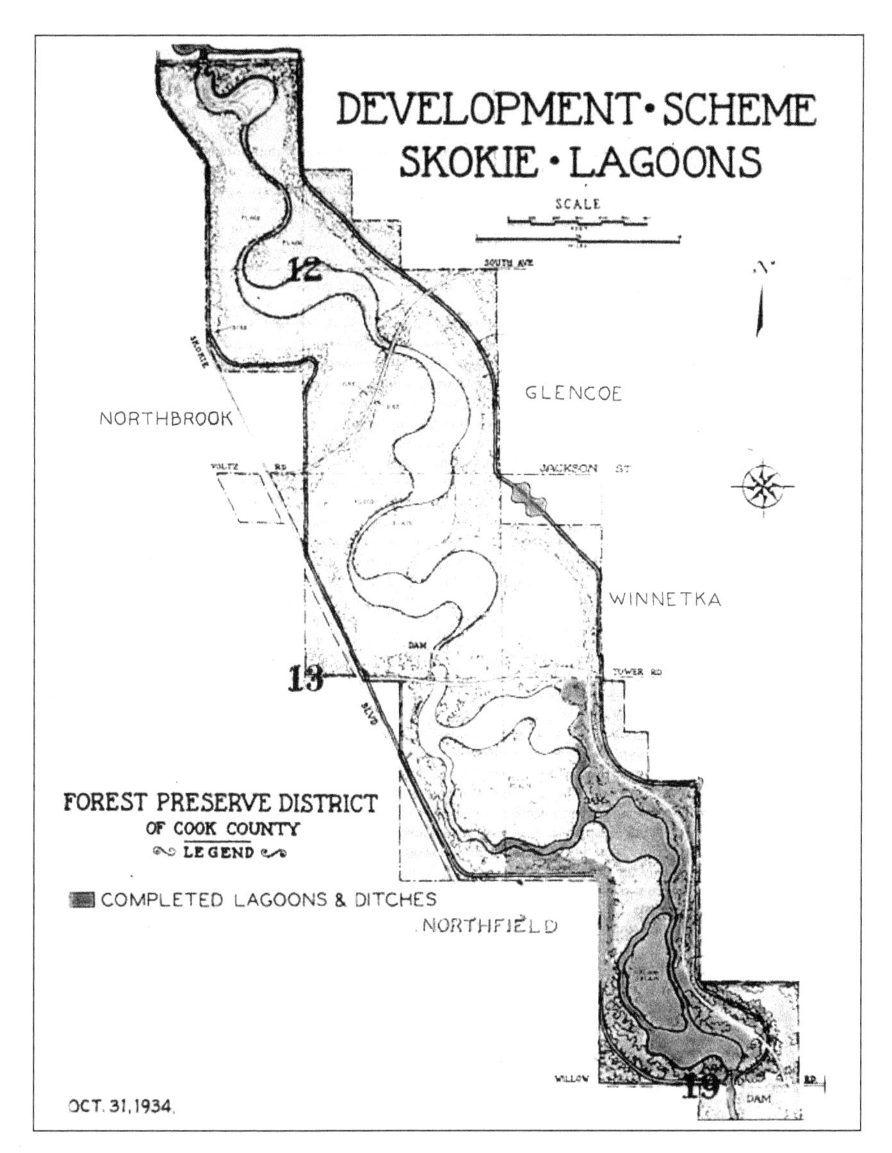

Figure 3.2. Skokie Lagoons Plan, 1934 (Cook County, Illinois, Forest Preserve District of Cook County, *Development Scheme–Skokie Lagoons* [Chicago: Forest Preserve District of Cook County, October 31, 1934])

marsh," he declared, "and next to the lake is considered by many Chicago's most beautiful natural feature."[88] When the FPDCC was finally created sixteen years later, the commissioners, in cooperation with their counterparts in Lake County to the north, purchased 1,300 acres/526 ha in their first wave of acquisitions. Civic leaders on the district's planning group, including Charles Wacker, Dwight Perkins, and Robert McCormick, lobbied the policy makers to act immediately to save what the *Chicago Daily Tribune* agreed was "one of the most beautiful preserves in the country."[89] Their sense of urgency was not hyperbole: real-estate developers were scheming to turn it into residential subdivisions and truck gardens. In 1919 and 1920, its defenders, including Jensen, defeated a special act in the legislature that would have allowed them to drain the marsh. Instead, the commissioners purchased the 2,100 acres/850 ha that the Chicago Audubon Society wanted to preserve as a bird sanctuary for 125 different species.[90]

During Cermak's tenure as the "Mayor of Cook County," the Skokie Swamp remained more of an unimproved nuisance than an attractive amenity. During the real-estate boom, in fact, conditions became more intolerable as new homes were built closer and closer to the approximately 6-mile/9.6 km strip of prairie wetlands lying just to the west of the hydraulic ridge (Green Bay Road) from Evanston to Highland Park. As early as 1922, Chief Forester Ransom Kennicott envisioned that "the little stream flowing through it might be changed to broad lagoons; the islands of trees be reforested; the water held at the proper level by a few dams; and the beautiful reeds and water grasses, which are fast disappearing because of drainage, be restored to their original beauty."[91] Instead, the North Shore suburbs suffered flooded basements, mosquito attacks, and prairie fires that residents had to extinguish with garden hoses to save their homes. Although the state legislature provided some relief by creating mosquito-abatement zones in 1927, the Cermak administration left Kennicott's plan languishing on the drawing board.[92]

Like the SDC's unfulfilled infrastructure plan, his blueprints for the conservation of nature were literally shovel-ready for New Deal, make-work programs. A month after his inauguration in March 1933, President Roosevelt proposed to create an army of 250,000, who would take up the task of reforesting the nation. Moreover, Secretary of the Interior and Director of the Public Works Administration Ickes was a long-time resident of Winnetka and therefore familiar with the problems arising from the Skokie Swamp. Under what became the Civilian Conservation Corps, four thousand of the first wave of recruits were allotted to Chicago. A quarter of them constructed a camp with 115 barracks, mess halls, and a field hospital. In July, they began digging the lagoons and building the dams. By December, Camp Skokie Valley had become the biggest of its kind in the country, with twelve hundred white residents. Eight hundred African Americans were also engaged in the reclamation and flood-control project, but they were organized in segregated "companies" and lived in a separate camp some distance from the site. Nevertheless, the local black-owned newspaper, the *Chicago Defender*, not only was grateful for their inclusion but also took pride in their contributions to "transforming an erstwhile marsh into a beautiful forest and seven lagoons."[93] Working under military command for $1 per day, white and black together brought the $3.5-million project close to completion in seven years, when it opened to the public.[94]

The Skokie Lagoons fulfilled multiple goals of nature conservation, flood control, and mosquito abatement. While a few people may have lamented the loss of its primal state, conservationism meant engineering the environment for the highest use by the greatest number of people. To prepare the area to serve as a public park, the foresters planted 20,000 trees, 12,500 seedlings, and 5,000 shrubs. In addition to landscaping the 1,500-acre/610 ha site, planners ensured the lagoons were large enough for boating and stocked with fifty thousand fish. Less visible, they also were designed to hold 500 million gal./18.9 billion L of floodwater. The earth removed to make these small lakes became 7 to 20 ft./2 to 6 m high embankments around them. The problem of mosquitos' breeding in stagnate water was eliminated by linking the lagoons with channels and controlling their levels with passive, weir-type dams. The area was easily accessed by residents of the North Shore, but city dwellers too could reach this forest preserve's trails, picnic areas, and winter sports facilities on public transportation via steam, electric, or interurban railroads.[95]

Conclusions: The Political Legacy of Conservationism

On February 15, 1933, an assassin shot Mayor Cermak as he was standing next to Roosevelt's car at a victory celebration in Miami Beach, Florida. Three weeks later, he died of complications from his wound. The mystery of whether the bullet was meant for him or the president-elect has never been solved. The more-immediate question for Chicago was the choice of a successor. After becoming mayor two years earlier, Cermak had handed the reins of power over the Cook County Democratic Party Central Committee to "the great harmonizer," Nash.[96] Claiming old age, he anointed his long-time comrade-in-arms, Kelly. Serving out the remainder of Cermak's term of office, Kelly would be elected three times on his own through the Great Depression, World War II, and the postwar period of adjustment. From 1933 to 1947, the political machine of Kelly and Nash tightened its grip on city hall to the point of squeezing the lifeblood out of the Republican Party: it became moribund, never to recover. Complete control of Chicago allowed them to dominate the state government as well as to exert considerable influence at the national level. Mayor Kelly became one of Roosevelt's intimate political advisers, because he could deliver the vote on election day.[97]

During his long term of office, Kelly never forgot the political capital to be gained by promoting outdoor recreation, leisure, and entertainment at public facilities. Inheriting the mantle of the "World's Fair (of 1933) Mayor," he earmarked substantial, if constrained, funding for the park district and mounted mass spectator events, including the first Major League baseball

all-star game. During the war years, moreover, he revived Jane Addams's community-center idea, turning the parks' field houses into beehives of patriotic activity and sports team competition, sponsored by local businesses.

In contrast to the city parks' hyperactive utilitarianism, the FPDCC pursued a more-preservationist approach to conservation. In a revealing 1943 report, its superintendent of maintenance argued that his biggest challenge was saving nature from people. In *Landscape Engineering*, Roberts Mann complained, "The problem is protection, not development. Protection against [accidental] fire[,] . . . the ubiquitous automobile[,] . . . [and] the impact of excessive public use by heedless, trampling hordes." He described his role as a mediator of the paradox between the democratic right of access and the commodification of nature "by city folk who have lost their [country] roots and true perspective."[98] He believed that the goal of his job was to engineer the environment in such a way as to set aside restricted areas for the enjoyment of the public while saving the ecology of nature from its abuses. For example, he called for fully equipped picnic areas but protested cutting down trees to create any more golf courses. In the following years, Mann would become a crusader for public education on the preservation of nature.[99]

The economic downturn of the 1930s aided his cause by causing a commensurate decline in the growth of Chicago's population. The city recorded an increase of a mere 20,000, and suburban Cook County added 60,000 more, for a combined gain of only 2 percent to reach a total of 4,063,000 people. The growth of the five collar counties was equally anemic: their population rose to 506,000 with the addition of 39,000 people, or 8 percent. While the metropolitan region's 4.5 million residents took a toll on its public open spaces, this burden would have been much worse if it had continued the double-digit rates of growth of the past.[100]

Overall, the political legacy of conservationism on Chicago's flood-prone environment during the interwar years was a mixture of significant gains and lost opportunities. To be sure, the construction of sewage-treatment plants on the North and West Sides and the partial completion of the one on the Southwest Side at Stickney represented a major improvement over the overtaxed dilution method. In addition, the Great Lakes benefitted from these infrastructure projects with the reduction of the amount of water drained into the Chicago River.[101] However, the SDC installed equipment that achieved only a 70-percent rate of purification at a time when existing technology could reach more than 90 percent. The equivalent of 1.5 to 2 million people's raw sewage continued to contaminate the six river watersheds, leaving the Chicago River a stinky and unsightly dead zone of pollution.[102] The Des Plaines River, too, was allowed to deteriorate, becoming an open sewer. "The polluted condition of the stream," a public health official reported in 1949, "has . . . curtailed its recreational use greatly."[103]

The public-works programs of the New Deal also extended the sewer lines many miles and enlarged others in the city and the suburbs.[104] At the same time, the SDC continued to permit the combined system rather than begin requiring new developments to install the alternative technology of separate sewerage and drainage pipes. The result was the regular infusion of raw sewage in the Chicago River through hundreds of emergency overflow outlets whenever there was more than 0.5 in./1.3 cm of precipitation. Chief Engineer Ramey admitted that these events occurred at least thirty times a year. An outside study found that even 0.1 in./0.25 cm of rain caused overflows a hundred days each year on average.[105] In paving over the land during the interwar years, city officials allowed sprawl's twin forces of deconcentration and concentration to increase Chicago's risk of flood damages. Suburbanization doubled the capacity of the sewer outlets, while building up the urban environment resulted in a similar, twofold increase in the amount of runoff per square mile. Ramey also kept records of the storms that caused damaging floods. During the "dust bowl" decade of the 1930s, Chicago sank under three of the heaviest downpours in its history. He called the 5.6 in./14.2 cm rainfall of September 11–13, 1936, "the worse since 1885," and he called the 4.1 in./10.4 cm torrent that fell in sixteen hours on July 1, 1938, "the most intense" ever documented. Five years later, on July 6, 1943, another storm dumped 3.2 in./8.1 cm on the city and 4.5 in./11.4 cm on the Calumet District, forcing him to open the Lockport Dam to its maximum rate of flow of 13 bgd/49 bLd.[106]

During the first eight years of the New Deal, the Roosevelt administration supported the conservation of natural resources. But with the outbreak of war in Europe in 1939, it reversed course to engage in a full-throttle exploitation of them. In Chicago, the curtain came down on its long-running political theater of forcing industry to reduce its liquid waste. Its public-works programs were also shut down as its citizens turned their home front into the country's leading arsenal for democracy. By the war's end in 1945, the gains made in improving a flood-prone environment needed to be weighed against the corruption of the public sphere by machine politics and the necessities of national survival.

PART II

The Wet Years

4

The Boom of Suburban Growth, 1945–1965

Baby Booms and Cloudbursts

After 1945, Chicago suffered increasingly damaging floods, which were caused by a combination of suburban sprawl and climate change. When the war ended, the baby boom began. Searching for housing, Americans found that buying dwellings outside the city limits was cheaper than renting homes inside them (see Figure 4.1). Paving over a flood-prone prairie wetland kept raising the speed and volume of storm runoff in an expanding metropolitan area. At the same time, six decades of below-average rainfall began turning into a still-ongoing, extra-wet period. Extreme, flood-damaging events have continued to rise in frequency and severity. A recent study found that "extreme precipitation intensities have increased in all regions of the [c]ontinuous United States and are expected to further increase with warming at scaling rates of about 7% per degree Celsius."[1]

Heavy rains caused Chicago's combined sewer and drainage system to overflow into its six river watersheds and Lake Michigan, the "ultimate sink" of the metropolitan region. Contaminated with raw sewage, the storm runoff also backflowed into people's basements and the city's streets, blocking underpasses and sinking low-lying neighborhoods. In response, sanitary engineers proposed to widen, deepen, and channelize the rivers as the best way to cope with suburban development and its ever-expanding infrastructure of wastewater management.

In Chicago, however, these rivers had gained a special status in the public law and in the popular imagination as cherished, natural landscapes to be saved at any cost from destruction. Since the successful crusades of the Pro-

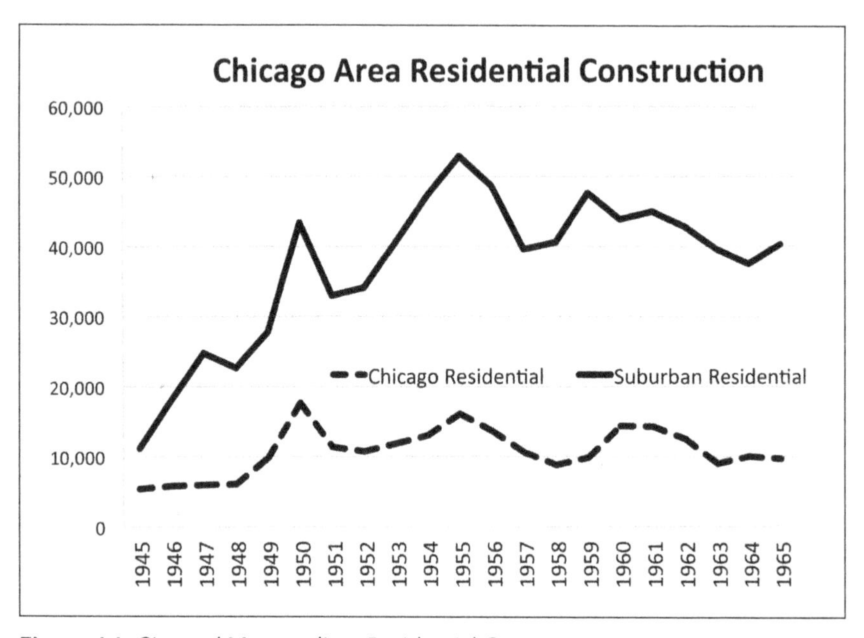

Figure 4.1. City and Metropolitan Residential Construction, 1945–1965 (Data from Carl Condit, *Chicago, 1930–1970: Building, Planning, and Urban Technology* [Chicago: University of Chicago Press, 1974], table 3)

gressive Era, forest-preserve districts have protected the areas along the riverbanks. Forming a 35,000-acre/14,164-hectare (ha) emerald necklace around the city, their picnic grounds, nature trails, and green spaces became immensely popular, especially during the hard times of the Great Depression and the crammed housing of the war years. After 1945, family life, expressways, and outdoor recreation would add millions more visitors. Collectively, they formed a powerful, grassroots constituency of proto-environmentalists. They supported forest-preserve officials, who opposed the engineers' plans to turn the suburban rivers into sanitation canals like the Chicago River.

This chapter interrogates this policy debate, shedding light on human-riverine relationships. Chicagoans' interactions with their rivers during the postwar period helped reshape their ideas about the natural and the built environments. Case studies of the Des Plaines River/Salt Creek on the western side of the metropolitan region and the Calumet/Little Calumet Rivers on its southern side expose the contestation among policy makers, planners, and the grassroots. As the flood-damaging storms grew worse, the politics of Chicago's waters underwent a fundamental transformation. In fighting to save the rivers, nature conservationists became environmentalists, with a modern understanding of ecology. The rising price of damages also forced sanitary engineers to abandon their incremental approaches of the past. In 1965, they would end this debate when they proposed a radical new alternative, the (Deep) Tunnel and Reservoir Plan (TARP). The suburban rivers were saved during this round of conflict over a flood-prone environment.

The Making of a "Crabgrass Frontier"

In 1945, the American experience took a sharp turn away from the past. After a generation of depression and war, the United States embarked on a sustained twenty-year period of domestic prosperity and relative peace abroad. But the Holocaust and the bombing of Hiroshima made hollow the victory of winning the "good war" in the new Age of the Atom Bomb. Americans turned inward and became "homeward bound," according to historian Elaine Tyler May. They got married and had babies in numbers unprecedented before or since. "Both the cold war ideology and the domestic revival," she states, "were two sides of the same coin: postwar Americans' intense need to feel liberated from the past and secure in the future." They sought self-fulfillment through withdrawal into the contained, private realm of family life.[2]

Learning during the interwar years, government and business were prepared to channel the surge in demand for housing these new families in predetermined directions. At the war's end, for instance, Chicago's planners anticipated building at least 350,000 additional units just to relieve immediate, preexisting shortages. Led by the urban economist Homer Hoyt, the city's 1943 *Master Plan of Residential Land Use* reported on the problems of families doubling up in decent housing and squeezing into slum buildings surrounding the central business district (CBD). The planners recommended a Modernist makeover of the commercial core, the clean-sweep demolition of the inner "blighted" ring of racial ghettos and their replacement with apartment towers for white-collar workers, and the conservation of the mostly white-ethnic outer ring of modest bungalows and owner-occupied two-flat apartments. With little vacant land remaining inside city borders, the plan called for the mass production of socially homogenous, "neighborhood units" of detached, single-family dwellings in the suburbs. They would help reduce the real shortages of rental apartments within the city center as well as the perceived excesses of its population densities.[3]

An increasing number of Chicagoans agreed with the planners and policy makers that homeownership represented the American dream. In the 1900s, working- and middle-class city dwellers had become convinced that it generated more than use value; it embodied individual success and projected social status. The result had been a sustained building boom that extended into the late 1920s. Local homebuilders had constructed between eighty thousand and a hundred thousand bungalows: one-and-a-half-story, full-basement brick structures. They formed a crescent-shaped "bungalow belt" of single-family houses and owner-occupied two-flat apartments that stretched from the northwestern corner of the city to the far South Side (see Figure 2.1).[4]

Like raising a family during the Cold War, homeownership after 1945 became something different from the past. It took on a dual meaning as an act of not only personal fulfillment but also civic patriotism in the battle against Communism. In creating a "Consumer's Republic," scholar Lizabeth Cohen confirms that "at the center of Americans' vision of postwar prosperity was the private home, fully equipped with consumer durables." To prove the superiority of the American Way, Chicagoans joined the national crusade to buy more stuff to bury the Soviet Union with Americans' higher standard of living. Over the next two decades, they constructed 674,000 houses in addition to 333,000 apartments. Three out of four of the new single-family dwellings were built in the suburbs.[5]

By the war's end, government and business had completed a fundamental overhaul of residential mortgages that privileged the suburbs over the city and (male) homeowners over renters. Hoyt and other land-use experts working for New Deal agencies, such as the Federal Housing Administration, had created the equations that generated the infamous "redlining" maps of high-risk neighborhoods. Recent analysis reveals that the formulas' purpose was less about calculating land values and more about determining social geography. Any racial/ethnic integration of "homogenous" all-white areas put their homeowners' property values at risk, while nonwhite neighborhoods were completely excluded, or "redlined," from consideration for heavily subsidized mortgages. The equations, moreover, made single-family houses in the suburbs the best buy, because Hoyt and other urban experts from the so-called University of Chicago School believed that overcrowding at the center was the root cause of social disorder, crime, and immorality.[6]

Despite the heavy tilt of the mortgage markets—especially for male veterans—in favor of homeownership at the periphery, many people preferred to live in the city. They resisted the Consumer Republic's siren songs of suburban heaven for a wide variety of reasons. Investigators found that white and black parents alike often put neighborhood amenities above location in describing their ideal place to live. Besides opportunities for homeownership in the city, a study of 250 family records revealed that "welcoming neighbors and many playmates, the availability of multi-unit buildings, the proximity of family and friends, and a sense of connection to a community were also important factors." In addition, the investigation showed that close proximity to the lakefront beaches and free access to the Chicago Park District's (CPD's) recreational sports and leisure activities were major attractions of family life in the center. In contrast, many people, especially from the working class, saw the suburbs as an alien place of exile from their extended social networks of relatives, parishioners, and coworkers.[7]

As this study and others have demonstrated, urban sprawl has never been a one-way flow from the core to the periphery. The lure of the city center produced an unbroken, century-long record of population growth. With 3.4 mil-

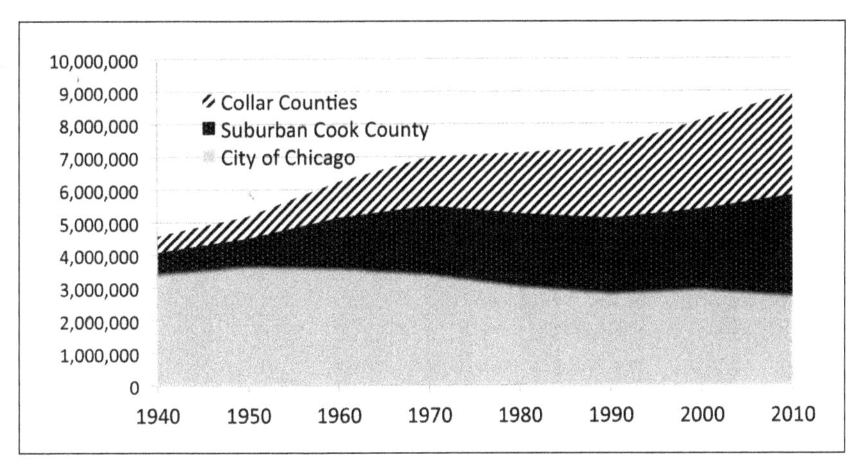

Figure 4.2. Chicago Area Population, 1940–2010 (Data from U.S. Census)

lion people in 1940, Chicago maintained a steady population, despite massive suburban dispersal over the next thirty years. The city's numbers edged up by 6.6 percent during the war decade, but it lost about an equal amount during the following twenty years. The larger metropolitan area continued to spread outward during this period, adding 2.1 million people, a 38 percent increase. It reached 7.6 million people by 1970, when a majority resided outside the city limits (see Figure 4.2).[8]

The birth and metamorphosis of Park Forest from an ideal model of a garden city into a prototype of a Consumer's Republic illustrates how the green fields of suburbia got paved over. Called "America's original G. I. town," it was the brainchild of several New Dealers, including Philip Klutznick, from the Federal Public Housing Authority; public-housing reformer Ferd Kramer; and Elbert Peets, the planner of Greendale, Wisconsin.[9] Significantly, it was unique in an experimental, new-towns program by providing a predominance of single-family houses for private ownership. Forming the American Community Builders in 1946, the three men devised a marketing strategy to jump-start their development to reach the critical mass needed to create a viable "neighborhood unit." Pent-up demand would fill row-house-style rental apartments as fast as they could be put up, creating an instant community and a potential base of customers for the houses planned for the second phase of construction.[10]

The advantages of owning over renting quickly altered Peets's original designs with nature into a mecca for shoppers. Park Forest was located near a commuter railroad station 30 miles/48 kilometers (km) from the CBD on the far South Side. The 3,000-acre/1,214 ha site bordered a public forest preserve and had one of the region's innumerable tiny streams, Sauk Creek, running through it, feeding a little lake. The landscape architect intended to make this low-lying pond the centerpiece of the community. Surrounded by parkland, it would serve the dual purpose of providing aesthetic enjoyment of nature and temporarily retaining flood-damaging storm water. By the groundbreaking in October 1947, however, the final plan had replaced the lake with a shopping mall and the parkland with a parking lot. After completing the three thousand rental units two years later, the developers could not resist eliminating the remaining green space between the houses in Peets's early designs. They built twice as many, or fifty-five hundred houses, over the next several years, when their boomtown reached a white-only population of twenty-five thousand inhabitants. What started as a green vision of the American dream ended up as the gray concrete of an overbuilt landscape.[11]

In part, the gap between dream homes and actual costs accounts for the paving over of a flood-prone environment with high-density suburbs like Park Forest. "Whether one was a small builder or a large developer," architectural historian Clifford E. Clark states, "the need to contain costs was an overwhelming priority." Builders were under tremendous pressure to erect larger homes, because in a Consumer's Republic, the house now represented a "glorification of self-indulgence."[12] Their answer was spilt-level, ranch-style architecture. Like the bungalow, the easiest way to minimize construction costs and lot sizes was to include a basement. Besides the utility and laundry equipment, it contained the recreation room, where the TV became the centerpiece of family life in the mid-1950s. And, like its previous ideal, the ranch style evoked California-like images of living outdoors in nature. Although the split-level version of this model helped keep purchase prices within reach of middle-class consumers, it also put these homebuyers at risk of suffering devastating losses in the future from flood damages.[13]

Assembly-line mass-production homes made Park Forest a mass-production suburb that was itself reproduced many times over throughout the metropolitan region. Along with the community homebuilders came the developers of shopping malls, industrial parks, and corporate-office campuses for white-collar workers. In the next phase in his career, Hoyt became a private consultant to the real-estate industry on the locational analysis of mass consumption. "The scramble to develop these new shopping centers in the past six or seven years," he advised investors in September 1955, "has been like the Klondike gold rush in the Chicago area as well as in other American cities." While he stressed the tsunami of residential construction sweeping outward from the city, he was attentive to the simultaneous, countervailing wave moving toward the center. He reported that customers from as far as 50 miles/80 km farther out in the countryside were driving to the areas' twelve malls, including the one in Park Forest.[14]

Conspicuously missing from Hoyt's reports to the real-estate industry was any mention of the mounting environmental costs of suburban sprawl.

More cars and trucks traveling more miles exacerbated air pollution coming from coal-fired power plants and factories. Adding the partially treated sewage of two million additional people in the periphery further taxed greatly depleted river ecosystems. And the paving over of the land turned open space into urbanized space on a massive scale.

After twenty years of building suburbia, Chicagoans doubled the acreage of the built environment. Other land-use analysts calculated that it lost 16,000 acres/6,475 ha each year to development, or the equivalent of a 450-square-mile/1,165 square km patch of land. In 1945, this was size of the territory served by the Sanitary District of Chicago (SDC). What more than 130 years of city building had taken to accomplish, the Consumer's Republic achieved virtually overnight, but at a mounting cost to the quality of daily life. Professional experts were not the only ones worried about the deterioration of the air, water, and land. At the grassroots too, Chicagoans from city and suburbs alike began sounding the alarm and mobilizing a movement to save the environment from overdevelopment.[15]

For many uprooting apartment dwellers from the city, to be sure, Park Forest and similar developments represented a new and different relationship with the natural world. The significance of suburbia, Christopher Sellers reminds us, "was one not so much of home buying as home owning." He and other environmental historians posit that the seeds of this movement were sown in the transformative experience of living in a suburban nature. In the case of Chicago, at least, the residents of bungalow-belt neighborhoods and other enclaves of single-family houses within the city were not much different from their counterparts living just across the borderline.[16]

The Expansion of Outdoor Recreation

In addition to its diverse economy and world-class cultural attractions, Chicago built an environment that institutionalized outdoor sports and leisure activities. Creating an urban nature, the city followed in the Progressive-Age footsteps of the city beautiful and recreation movements of Daniel Burnham, Jane Addams, and landscape architect and park designer Jens Jensen. The CPD managed more than 2,700 acres/1,093 ha of beaches and green spaces along 17 miles/27 km of the lakefront. It also controlled an equal amount of parkland in the neighborhoods. Many of its 137 parks came equipped with year-round field houses and professional staff. The city government maintained another seventy-eight parks, while most of the schools, public and private, now had supervised playgrounds (see Figure 4.3).[17]

During the war, the CPD pumped up civilian morale and bodies by getting the business community to sponsor "Sports on the Production Line." Companies mobilized their workers and their children into action on the battlegrounds of baseball diamonds, golf courses, and tennis courts. Along

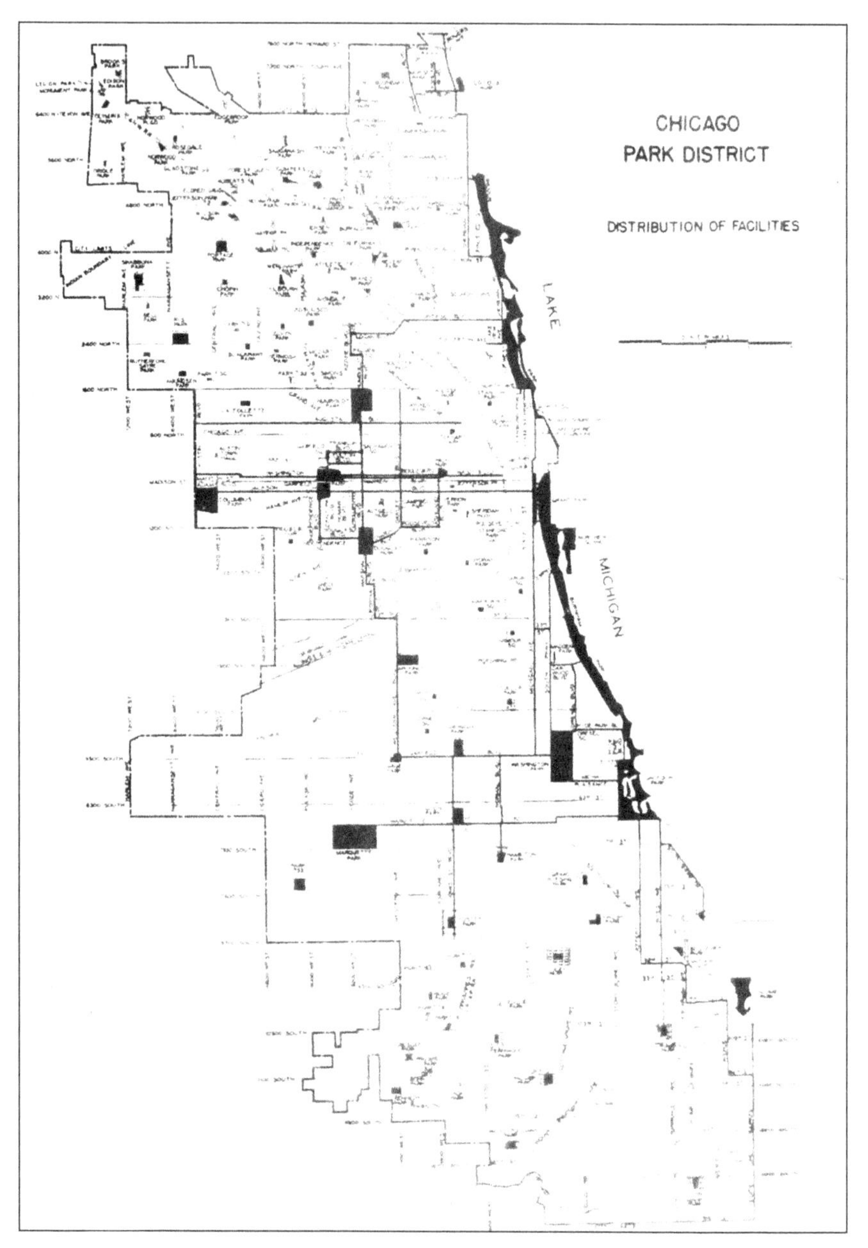

Figure 4.3. Chicago Park District, 1950 (City of Chicago, Chicago Recreation Commission, and Northwestern University, *Chicago Recreation Survey 1937*, vol. 1, ed. Arthur J. Todd et al. [Chicago: Works Progress Administration National Youth Administration, Illinois Emergency Relief Commission, 1937], 98)

the downtown lakefront, for example, the Grant Park Recreation Association organized a softball league that fielded teams from the lawyers, hotel workers, City National Bank, Dole Valve, and Link Belt. The field houses, moreover, were turned into neighborhood centers of patriotic activity seven days a week.[18]

Civic officials had no need to encourage use of the city's outdoor spaces—in fact, just the opposite. As the *Chicago Tribune* announced on the approach of the first postwar Fourth of July in 1946, "Expect record crowds." It predicted at least a million people would flock to the city's beaches and parks, while seats were already sold out on the extra trains and buses added to take them out to the suburbs. In November, 170 public and private organizations took part in the annual Chicago Recreation Conference. In that year alone, the CPD counted twenty-six million participants in supervised recreational activities. It did not attempt to count its less-hyperactive visitors. In any case, Chicagoans' enthusiasm for getting outdoors would only grow stronger in the coming years.[19]

The continuing struggles of African Americans and Mexican Americans for access to a fair share of CPD facilities gave frustrated, albeit powerful, expression to the importance of open space to city dwellers. Although local turf wars among ethnic groups were as old as the city itself, Chicago's great race riot of 1919 was the first to spread from a fight over access to a public space to citywide violence. One interpretation of this tragic turning point in urban race relations casts white and black combatants as defenders of city life, who were fighting against their removal to the suburbs. A commission appointed after the riot found that the city government had institutionalized not only outdoor recreation but also racial/ethnic discrimination. Nonetheless, blacks and Latinos continued to organize protests of inequality, sponsor their own sports leagues, and to fight for use of the parks and beaches through individual acts of defiance. They simply occupied them as a right to the city.[20]

The social and spatial limits of open space within the city led its inhabitants to the "crabgrass frontier" of the suburbs. Again, limiting the focus here to public spaces, the formal opening of the Forest Preserve District of Cook County (FPDCC) to the public a year before the race riots deserves first attention (see Figure 4.4). From the outset, the new agency had encouraged easy access by public transportation to its expanding acreage along the six river watersheds of the metropolitan region. This built on the decade-old tradition of the Prairie Club, which Jensen had formed to help promote the cause of nature conservation. He had led special, weekend trains to a variety of nature sites, where hikers were guided through marshes, ravines, sand dunes, and prairie wetlands. His mission was to save the rivers and the last remaining patches of native trees along their banks. The Prairie Club's excursions became so popular that the CPD formed its own Hiker's Club. After 1945, the

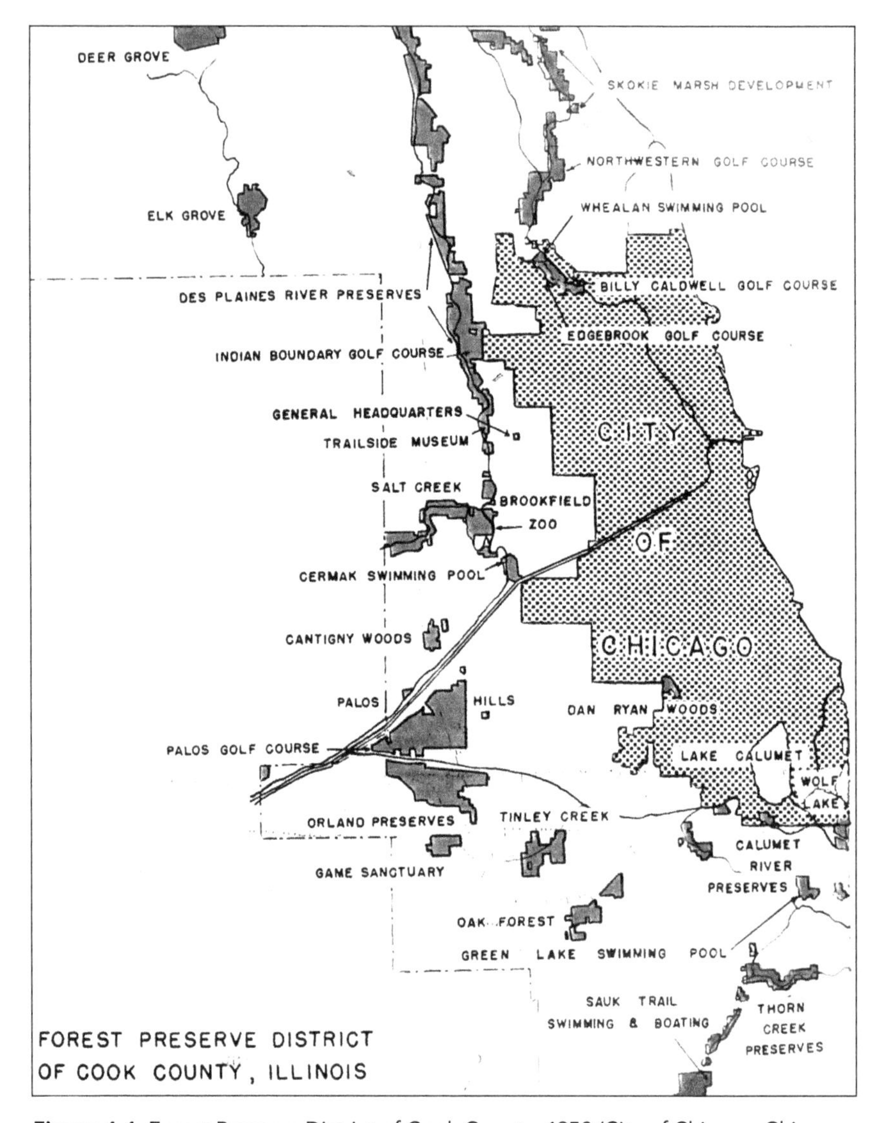

Figure 4.4. Forest Preserve District of Cook County, 1950 (City of Chicago, Chicago Recreation Commission, and Northwestern University, *Chicago Recreation Survey 1937*, vol. 1, ed. Arthur J. Todd et al. [Chicago: Works Progress Administration National Youth Administration, Illinois Emergency Relief Commission, 1937], 164)

newspapers provided a list of the several trips the two groups sponsored each weekend to the forest preserves and beyond.[21]

The easing of gas prices and the opening of new superhighways reinforced preexisting patterns of seeking natural green spaces for recreational and leisure activities. By then, it seemed as though every one of Chicago's ethnic, racial, religious, civic, corporate, and sports groups had marked its favorite spots for gatherings in the forest preserves. The FPDCC had been a major beneficiary of New Deal programs. Because it formed a crescent-shaped arc from the headwaters of the Chicago River on the North Shore to the Calumet River zone of heavy industry at the bottom of Lake Michigan, each neighborhood-based group could find favorite, convenient sites for its respective activities. On a typical weekend in July 1947, several thousand attended the Germania Club's annual picnic in Wolf's Grove, Congress of Industrial Organizations transportation workers headed for athletic contests in Dan Ryan Woods, the Disabled Veteran's Post 849 held an outing in Caldwell Woods, the 26th Ward Democrats met in Kolze's Park, and the Southeast Sportsman's Club sponsored an open fishing contest on its grounds at Wolf Lake.[22]

Well before the suburban housing boom, then, Chicagoans had made the experience of being outdoors in nature an important part of their everyday lives. They formed a large, if largely unorganized, constituency of stakeholders in the environmental protection of the public's existing open spaces. In fact, they supported the plans of local policy makers for the expansion of recreational facilities in the center and watershed/forestland at the periphery. Anticipating growing demands at the end of the war, for instance, the CPD prepared a ten-year development program to add more playgrounds and small parks. In 1945 and 1946, voters approved bond issues to fund the plan. The making of a Consumer's Republic on the crabgrass frontier further reinforced Chicagoans' personal attachments to experiencing nature and their political commitments to saving it.[23]

The Unmaking of a Flood-Prone Environment

In fulfilling the American dream, suburban sprawl had the unintended consequence of turning a prairie wetland into a disaster zone on a regional scale. Of course, Chicagoans were no strangers to damaging floods since the earliest days of city building. But nevertheless, they were caught unprepared, because a sixty-year dry spell had lulled them into forgetting the perils of paving over a marshland. Despite almost twenty years of record-breaking suburban development, the planners and policy makers were unable to reformulate sanitary strategies that fit the new realities on the ground and in the skies. On the contrary, they remained stuck in a mind-set of incrementalism that had worked reasonably well for the previous generation.

Between 1945 and 1963, the region's inhabitants had to suffer through ten "worst-ever" floods and three equally historic corruption scandals before the policy makers at city hall could be persuaded to consider the need for reform. When the rains came, storm surges of bacteria-laden river water had to be released into Lake Michigan, closing the beaches and putting the drinking supply at risk of fueling a deadly epidemic. The growing volumes and speeds of storm runoff into the rivers resulted in different, but no less destructive, consequences for their aquatic ecosystems.[24]

Chicagoans began organizing to protest the failures of the power holders to protect the basements of their homes and businesses from repeatedly filling with contaminated floodwater. They also began mobilizing a proto-environmental movement to save the waters—the lake and the rivers—from ecological disaster. The SDC's treatment plants' partially treated, toxic effluents posed the single-greatest threat to this interconnected, underwater world. The contestation of the grassroots against the sanitary engineers' proposal to turn the suburban rivers into sewage canals like the Chicago River would finally force the politicians and the planners to rethink their strategies of flood control.

After every great inundation of the CBD, Chicagoans have responded by mounting ever-more-heroic, big-technology approaches to turn a marshland into a megacity. In the spring of 1849, the first grand plan to engineer the environment into a profitable, economic resource came after a tidal wave and huge chunks of ice on the river tore through the downtown, causing massive damages. After the next great flood of 1885, municipal reformers put forward an even-more-audacious plan, which promised not only to end the threat of flood damage but also to reverse the flow of the Chicago River's contaminated waters away from the lake's drinking supplies. Opened in 1900, they dreamed of making Chicago the nation's biggest port, linking the Atlantic Ocean and the Great Lakes to the Mississippi River and Gulf Coast.[25]

Although the city boosters' vision of ocean-cruising ships steaming through downtown never materialized, their plan for controlling polluted storm water achieved remarkable success over the course of the next half century. During this period of climate change best known for the dust bowl that developed across the Great Plains, Chicago also experienced subnormal amounts of rainfall. While severe storms occur during years above and below the long-term (1871–2012) annual mean of 34.4 inches (in.)/87.4 centimeters (cm), they tend to occur more frequently during extraordinarily wet years. To contain big downpours and full-sized vessels, the Sanitary and Ship Canal was designed to handle a maximum flow of 10,000 cubic feet per second, or 6.5 billion gallons per day (bgd)/24.6 billion liters per day (bLd). And up to twice as much could be released during a flood-stage event at the Lockport Dam without sinking communities further downstream on the Illinois River. Moreover, the Supreme Court restricted the amount of water Chicago could drain from Lake Michigan to dilute its sewage to just under 1 bgd/3.8 bLd, leaving plenty of capacity to carry away the gradual increase in wastewater coming from the growth of the city and the suburbs.

When the weather changed after 1945, severe storms came more frequently, punctuated by one record-breaking single rainfall and annual accumulation after another. In fact, the city's combined system had been designed to contain only 1 in./2.5 cm of rain in a twenty-four-hour period. Many local building codes throughout the region required that roof downspouts be connected to the combined system of sewers. This water added significantly to the speed and volume of the runoff filling them from paved-over surfaces pouring into the street gutters.

These great surges had to go somewhere. Filling up the underground network of drainpipes leading to the treatment plants, the rainwater became contaminated with raw sewage. It gushed into the rivers through hundreds of emergency overflow valves and backflowed into basements and low-lying areas. If the Chicago River were allowed to rise more than 5 feet (ft.)/1.5 meters (m) above its normal level, the CBD would suffer catastrophic damages. After the Lockport Dam reached its limit, the U.S. Army Corps of Engineers would have to open gates between the rivers and Lake Michigan, releasing billions of gallons of wastewater into the lake. The Corps had been able to avoid this option of last resort until 1944, despite several notable downpours of 3 to 4.5 in./8 to 11 cm within a day's time.[26]

At the war's end, then, Chicago's sanitary authorities were well aware that they had already fallen behind in meeting their current responsibilities, let alone getting prepared for the dawning era of a Consumer's Republic. From the 1920s to the mid-1940s, the SDC engineers had witnessed the volume of runoff pouring into the rivers more than double during heavy rainstorms to reach the maximum amount that could be released at the Lockport Dam. Over the next decade, they would have to grapple with the effects of suburban sprawl, which would increase runoff surges to 18.1 bgd/68.5 bLd during a series of historic rainfalls.

The war itself also set Chicago back in to keeping up with the increased amount of wastewater flowing into the sewer system. From 1900 to 1939, the city had laid an average of 50 miles/80 km of regular sewers and 4.5 miles/7.2 km of much larger and deeper interceptor sewers each year. The rationing of construction materials and manpower during wartime cut off New Deal programs to maintain and extend the sewers into peripheral areas. Equally important was the abrupt halt in the construction of treatment plants to meet the limits on the amount of water Chicago could take from the lake. To add to the resulting amount of untreated and partially treated human waste pouring into the rivers, factories along their banks were allowed to dump their toxic effluents into them. Combined, the 1.2 billion gallons (gal.)/4.5 billion liters (L) of effluents going downstream every day were loaded with the organic equivalent of the raw sewage of a million people.[27]

In November 1944, the city's Engineering Board of Review presented an $80-million improvement plan to the mayor and the aldermen. Proclaiming it "Chicago's No. 1 Post-War Plan," a local newspaper stated that the city "urgently need[ed] adequate sewerage" as well as ready-made jobs for returning veterans and new infrastructure in "blighted areas." The Democratic-ruled administration of Mayor Edward J. Kelly quickly passed the proposal, and the voters followed in June 1945, when they supported a $58-million bond issue to pay the city's share. State and federal grants were expected to cover the remaining costs of the construction program. As if the weather had read the newspaper, damaging floods on the South Side only a month later hammered home the urgency of the problem. While immediate blame could be easily cast on Washington for suspending local public-works programs during wartime, the deeper fault lay in Chicago's political culture among its policy makers and professional experts.[28]

A closer look at the plan and its creators exposes the strong undertow of tradition on the formulation of flood-control strategies that could meet the twofold challenge of climate change and suburban sprawl. The engineers' blueprint of the future was a laundry list of individual projects rather than an overall strategy of water management or a grand design of a technical system. The main author of this incremental approach to keep up with Chicago's development was William Trinkaus, chief engineer of the SDC. He was a second-generation employee, joining his father on the job in 1909 and, like him, rising to its top professional position. In responding to the damaging floods on the South Side between 39th and 87th Streets, he moved its $18-million improvement project to the top of the list. Furthermore, he reminded its residents that the SDC had always fulfilled its duties in the past but admitted that some problems were "almost incurable," such as flooded viaducts. Caught up in a corruption scandal, he would confess to taking fifteen to twenty bribes from contractors over the years, all the while professing he had done nothing wrong.[29]

His successor, Assistant Chief Engineer Horace P. Ramey, had also served a lifetime working for the SDC. He had two years' seniority over Trinkaus, starting in 1907 as a rodman. And he, too, looked to remedies of the past to solve current and future problems. In his 1958 report on flooding in Chicago, he declared his blind faith in the agency's 1924 guidelines for improving the system. In other words, he called for more of the same, except with much bigger pipes, interceptors, and treatment plants. In a similar way, he adhered to equally old ideas in making recommendations for the future of the rivers. His model was a prewar project by the U.S. Army Corps of Engineers to improve navigation on the Calumet/Little Calumet Rivers and expand the capacity of the Cal-Sag sewage channel by enlarging it from 60 to 225 ft./18 to 68.6 m wide. In the late 1940s, the SDC was still seeking the $80 million to pay for it, despite pressure from the federal courts to stop industry from dumping massive amounts of toxic waste into Lake Michigan. Ramey reported that another $120 million would be needed for flood control outside the city limits, including $20 million to deepen and widen the Des Plaines River in the burgeoning western suburbs.[30]

In contrast, one of the city's preeminent planners outlined a much-more-holistic approach to water management on a regional scale. In July 1947, Commissioner of Public Works Oscar Hewitt suggested that "Chicagoland need[ed] a super water authority." Suburbanization, he reasoned, required expanding the jurisdiction of such an agency from a 20-mile/32 km radius to a 40-mile/64 km radius from the center to embrace the collar counties. It would not only replace well water with lake supplies in these peripheral areas but also bring all the sewage systems up to uniform standards of performance. Hewitt admitted that "[the city of] Chicago has lagged behind in building filtration works" and in installing the most up-to-date technologies. Yet even this planner's spatially comprehensive vision was limited to concerns about the pollution of the lake, not the flooding of the land. The overworked treatment plants on the North Shore and the untreated waste of thirty thousand steel workers in the Calumet district were already closing the beaches. The commissioner urged his fellow public officials to enact major administrative and environmental reforms, because "it is man, not the lake, that is the great polluter."[31]

But Hewitt's appeal was drowned under the public outcry arising from the disclosure of massive fraud in the very agencies he wanted to entrust with greater, consolidated power. For more than three years beginning in July 1946, newspaper stories on the sanitary authorities were more about corruption than construction. As the pressures of the war eased, a small, albeit vociferous, group of Republican aldermen was able to pry open the lid of secrecy imposed on public spending by Mayor Kelly and the Democratic ward bosses. In the first three months of following the money, this so-called economy group revealed that the city's Sewer Department had 151 regular jobs listed under civil service rules and 415 hidden elsewhere on the payrolls as temporary "laborers."

Among the many Democratic politicians drawing a second salary as a sewer worker from a nearly $1-million secret budget line was State Senator Frank Ryan. The law allowed public officials to hold two government jobs at once, but they were prohibited from collecting two paychecks on the same day. The payroll records of "special house drain inspector" Ryan showed that he was paid for thirty days' work when his roll-call votes in the legislature proved that he had been in Springfield, the state capital. Assuming the standard bunker mentality of the party bosses, the superintendent of the department and the civil-service commissioners took a defiant stance. Sewer Department Superintendent Thomas D. Garry defended his army of men and women patronage workers, "because civil service workers are lazy." But he denied reporters access to Ryan's work records, because they were a "secret." After the aldermen revealed this information, Garry admitted that there were no records for "special inspectors" like Ryan; they were on the honor system. Under the spotlight of public scrutiny, the lawmaker resigned from his sewer job.[32]

In February 1947, just five months after the city's Sewer Department seemed to be fading from public view, the Republican trustees of the SDC uncovered an even-more-outrageous scandal involving no less than the president of the board, Anton F. "Whitey" Maciejewski. The two reformers on the nine-member board had won their seats in the previous November elections. With officials at the top now willing to listen, the engineers in charge of actually running the system became whistle-blowers. Maciejewski had turned what was becoming the world's largest treatment plant at Stickney into his personal building-supply depot. All the material and labor for his new baseball park had been stolen from ongoing expansion projects, which were trying to catch up with the growth of the western suburbs along the Salt Creek/Des Plaines River watersheds. By far, the worst offense was the destruction of a brand-new, quarter-million-dollar boiler: its pipes had been cut out to make fence poles.

As details of Maciejewski's illegal scheme began leaking, he led a majority of five Democrats in voting to suppress the engineers' report until after election day in April, when voters would choose their municipal officers. At the board meeting following the election, with Maciejewski conveniently taking a vacation in Florida, the superintendents of the Stickney plant complained that patronage appointments had doubled their operational workforce. At least half of their six hundred employees did no work, they reported, but political intimidation had pressured them to put only sixty-eight men on their "unfit" list. Each name had a political sponsor next to it—just about everyone from SDC trustees to city officials and ward committeemen—and each had received final approval from the mayor. Chief Engineer Trinkaus fired fourteen of the three hundred surplus workers; President Maciejewski did not run for reelection.[33]

Just as the boom in public works got underway, then, Chicago's regime of one-party rule undermined widespread popular support for public-works projects to improve the environment. It was replaced by growing skepticism that translated into political demands for structural reforms and budget cuts. The business-minded Civic Federation and the Republican *Chicago Tribune* charged that the SDC was a "hydra-headed monster" that was "rotten to the core" and being led by a "robber." Moreover, many of the double-dipping state officials on the payrolls of the local sanitary authorities were exposed as essentially paid lobbyists, whose real jobs were persuading the legislature to boost the SDC's tax levies and state grants. A *Tribune* editorial opposed giving it any more money, let alone the 50-percent increase in taxes that its henchmen were campaigning for to raise the following year's budget to $12 million. Feeling the blowback from the Maciejewski scandal, the SDC accepted a compromise deal, cutting the tax hike in half. But rather than turn the scandal into an opportunity to take up Hewitt's call for comprehensive reform, the SDC made only a window-dressing gesture by creating a new position, general superintendent. As with Ramey's appointment, an engineer with more than a quarter century of loyal service was promoted from the ranks to take the job.[34]

The Remaking of a Water-Management Policy

A hidebound political culture of machine politics meant fewer economic resources and technical capacities and less popular support to cope with the double-barreled challenge of suburban sprawl and climate change. By 1947, the disclosure of large-scale fraud in virtually every department of city government was causing a crisis for the Democrats. The ward bosses had to force their standard-bearer, Kelly, into retirement. In his place, they installed an honest man, Martin Kennelly, who had helped clean and modernize the park district. Like the new top manager of the SDC, Mayor Kennelly saved the political establishment from having to undertake systemic reforms of the public agencies in charge of environmental protection, including flood control. In the coming years, record-breaking deluges would underscore how each ghost-payroller in the Sewer Department meant one fewer real worker to clean the street gutters' thirty-thousand-plus catch basins and thousands of miles of underground drainpipes to maximize their carrying capacity. In a similar way, each sweetheart deal with private contractors increased the likelihood of system failure as the physical force of runoff surges kept mounting.[35]

Over the next two decades, damaging floods and corruption scandals resulted in the growth of opposition to city hall from the grassroots and shrinking resources from the state and federal governments to fund its public-works projects. For our purposes, a closer look at one of these catastrophic events, the great flood of 1954, serves to illustrate the inability of the power holders to respond adequately to these kinds of disasters. Instead, they motivated people to organize protest movements to save not only their basements but also the environment from their own government. When the planners finally proposed a heroic, big-technology approach in 1965, it was too late for the sanitary authorities. They had lost the public trust, turning the political tide against handing over any more money or power to them.[36]

What Chief Engineer Ramey and others called "the most disastrous flood caused by the heaviest rainfall in the city's history [since 1885]" began on October 2. But unlike that earlier downpour of 6.19 in./15.72 cm in a twenty-four-hour period, this weather pattern of record-breaking heat and thunderstorms persisted for eight days. It started with the thermometer hitting a third-year-in-a-row high of 91 degrees Fahrenheit/33 degrees Celsius, followed overnight by a 4 in./10 cm downpour in a five-hour period. Flash floods caused extensive damages in the southern and western areas of the region, followed by the most rain ever in the CBD—6.72 in./17.07 cm in forty-eight hours—which began cascading down a week later. The month's rain total of

12.06 in./30.63 cm, or four times the average, also surpassed the 1885 figures.[37]

For the first time, hydraulic experts were able to map the great variation in rainwater falling across this vast territory with hard data. After the wet period of climate change attracted the attention of the U.S. Geological Survey in the early 1950s, its hydrologists installed hundreds of stream-flow and rain gauges throughout the metropolitan area. During the deluge of October 2–10, 1954, they recorded a total amount ranging from fewer than 4 in./10 cm. along the North Shore to as many as 12 in./30 cm in the southern Will County and the Calumet District (see Figure 4.5).

Chicago's city dwellers and suburbanites shared widespread property damages, closed businesses and schools, lost wages, and the breakdown of essential services, such as public transportation, telephones, and electricity.[38] In the city center, the effects of the failure of the sanitary authorities to control the storm water were catastrophic, because they let the Chicago River overflow its banks. The managers of the SDC seemed mesmerized into a torpor by their own publicity campaign to regain public support after the Maciejewski scandal. Typical was the boast of Board of Trustees President Anthony A. Olis two years earlier that he was in charge of "the world's largest and most modern sanitation disposal system." Serving a population of four million people within its 450-square-mile/1,165 square km territory, its underground network had grown to almost 3,600 miles/5,800 km of underground pipes. While taking pride in using the 1946 bond money to make sewer hookups throughout 99 percent of its territory, he did not reveal that they also could increase the amount of runoff pouring into the rivers and canals by an additional 2 bgd/7.6 bLd. Moreover, Olis soon had to change the message from self-congratulations to self-defense after Trinkaus and other top managers were exposed in the next corruption scandal. This one uncovered an extortion scheme to force contractors bidding for work to make payoffs to the Democratic Party's coffers. When the storm surge came, the SDC allowed it to raise the river from its normal level of 3 ft./0.9 m below the lake's surface to 3 ft./0.9 m above it. In the downtown area, the floodwater overflowed into the railroad yards, the U.S. post office, the *Daily News* building, and two of Commonwealth Edison's generating stations, knocking out half of the city's power at the height of the emergency.[39]

The sanitary district had to take the option of last resort and do the "unthinkable," opening the gates of the controlling works at the mouth of the Main Branch to reverse the flow of Chicago's river-in-reverse. Approximately 13 billion gal./49 billion L of sewage-contaminated wastewater gushed into the lake. The Health Department increased the amount of chlorine it added to the water supply. Business came to a standstill as tens of thousands of commuters were cut off from their jobs and equal numbers of other workers were forced to stay home until power could be restored to their places of employ-

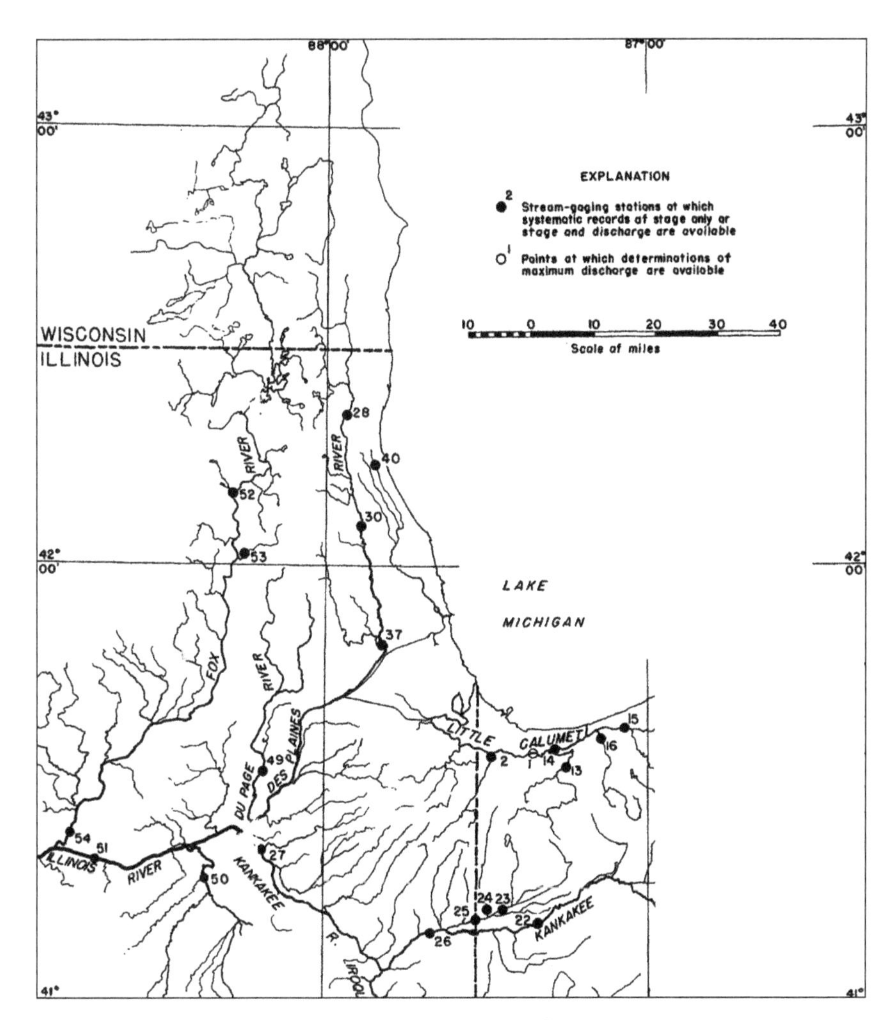

Figure 4.5. Flood Areas, 1954 (U.S. Department of the Interior, Geological Survey, Warren S. Daniels, and Malcolm D. Hale, *Floods of October 1954 in the Chicago Area, Illinois, and Indiana* [Washington, DC: U.S. Department of the Interior, Geological Survey, 1955], fig. 1)

ment. In the city, the flood caused an estimated $10 million in damages. To save face, Olis and the trustees fired Trinkaus, who had already become a political liability for the party bosses at city hall.[40]

The suburbs suffered even greater losses from the maladministration of the public agencies in charge of flood control. Set at $15 million, these damages were the result of sewers backflowing into basements for the most part. Few homeowners carried flood insurance. Many of them living in the Little Calumet watershed on the far South Side experienced the same kind of damages as Mr. and Mrs. Jack Mazzola: the concrete floor of their basement buckled and cracked under the force of the surge, filling it with 4 ft./1.2 m of sewer water. After they threw away all their spoiled goods and cleaned up the mucky residue, a "strong odor remained," according to a reporter. Nearby, in the African American suburb of Robbins, seventy-five families were forced from their homes as large areas of the forty-eight-hundred-member community sank under 15 in./38 cm of water. Its firefighters made forty-seven rescues, while its volunteer organizations sprang into action to care for the refugees. All forty-four hundred families in low-lying Midlothian had to be evacuated.[41]

The scientific instruments of the Geological Survey verified that the extreme flatness of the Little Calumet/Cal-Sag watershed practically eliminated natural drainage. The flow gauges recorded storm water flowing east during the peak of the storm, but then backwashing as the rains stopped. Although dredging the river and digging a ditchlike channel changed the watershed's hydraulics, it remained a marshland. The science pointed to the need for a separate system to deal with such a flood-prone environment, which was being paved over with more and more buildings. The extra initial costs of a two-pipe system would be offset in a long-term wet cycle by forestalling the material and human costs of flooding. The experts began to discount the results of their formulas that calculated the frequency of extreme, "hundred-year storms." The sharp rise in the numbers of these kinds of record-breaking events since the mid-1940s led them to reexamine their assumptions about extreme weather events. In any case, the Geological Survey data left no doubt that the planners would have to redesign the drainpipes under the streets and the overall capacity of the system on a much larger scale to cope with current conditions on the ground and in the skies.[42]

For Chicagoans living in the Salt Creek/Des Plaines River watersheds on the western side of the metropolitan area, all the problems of draining a prairie wetland were compounded by a feverish pace of suburbanization. Typical of these disaster zones was the subdivision built by real-estate developers Smith and Hill just after the war. Their 113 acres/46 ha of green space were located in Forest Park, less than a mile east of the Des Plaines River forest preserve. In May 1947, they asked the Cook County zoning board to permit them to cut the size of the lots in half, doubling the number of houses (and runoff). On their behalf, a consulting engineer also testified that planning was underway to hook up the three hundred new homes to the sewer and storm lines of the SDC. Less than a year later, in March 1948, the "biggest winter rain in history" hit the western suburbs harder than the coming, great flood of 1954. Suffering $1.5 million in damages, four thousand homeowners up and down the Des Plaines River petitioned the state and federal governments for relief. Six years later, they were still waiting.[43]

The difficulties of coping with storm-water surges on the local level were more evident just beyond the SDC's western boundaries, in DuPage County. In the 1950s, this white-collar Consumer's Republic doubled in population to three hundred thousand people. Its public officials put the price tag of the 1954 deluge at $10 million. Nine in./23 cm of rainfall in thirty hours sent the DuPage River gushing over its banks. "Salt Creek," an eyewitness recalled, "normally a gently meandering water course, resembled a mountain stream when spring melts the snow. Smaller creeks and streams contributed to the chaos." The county board quickly came to the conclusion that sprawl was to blame. Its president, Roy C. Blackwell, reasoned that "construction of new homes, streets, driveways, and parking lots all contributed to runoff. Where rainwater would be absorbed into the ground years before, it was now being channeled into miniature rivers."[44]

But even after suffering through the great floods of 1948 and 1954, the residents of DuPage County put protecting their suburban nature above safeguarding their personal property from damages. Like their counterparts in the SDC, the suburban sanitary authorities called for channelizing the rivers and installing bigger drainpipes and treatment plants. The voters defeated this plan under a nature conservatory act, which required that it be approved in a referendum. Two years later, in 1956, the county board sponsored a radical departure from tradition: an eco-friendly, comprehensive land-use plan. Bringing water and land management together, it not only controlled floods but also increased recreational open space along the DuPage River. Basically, the plan called for the purchase and landscaping of 3,000 acres/1,214 ha into six retention basins. After another five years of political stalemate to fund the $1-million proposal, Blackwell gave up on local government as an agent of reform. Instead, he began to advocate changes in the zoning code to shift responsibility for controlling storm runoff to private developers. In new subdivisions, they would be required to set aside 10 acres/4 ha out of every 100 acres/40 ha for retention ponds.[45]

At the state level of government, too, the deluge of 1954 produced political conflict and stalemate rather than concerted, public action. Policy makers in Springfield were unable to decide who should do what. The disaster's inescapable impacts on everyone in the metropolitan area put tremendous pressure on state officials to work out a comprehensive plan. Appointed by the Republican governor, William G. Stratton, a special committee recycled Hewitt's idea for a super authority that would empower the SDC to swallow up the built-up suburbs lying outside its jurisdiction. Already serving seventy-five of them within Cook County, the proposed, metropolitan-wide agency would almost double in territory. The new SDC map added sixty-six communities with seven hundred thousand inhabitants in DuPage County and pieces of the other collar counties. Led by Olis, this special committee was even more ambitious in proposing a budget five times as large as the $80-million improvement plan of 1944, or almost a half billion dollars. But when the legislature answered with a counterproposal to empower the state's Department of Public Works to design the master plan, the governor vetoed it.[46]

Between the great storms of 1954 and 1961, Chicago endured five more failures of the sanitary authorities to prevent damaging flooding or to even come to an agreement on a plan of action to achieve this goal. On the contrary, they remained deadlocked over the political question of who should control the proposed supersized agency. While one of these disasters resulted from the "most intense rain that ever fell," in Ramey's words, the other four required only 2 in./5 cm of precipitation to overwhelm the flood-control systems of the SDC and the suburbs. By then, runoff surges were filling the Sanitary and Ship Canal with four times the maximum amount the engineers could safely release at the Lockport Dam without flooding Joliet and other downstream communities. With more "unthinkable" discharges into Lake Michigan inevitably looming in the future, Chicago was reaching a crisis of public confidence in the ability of government to solve the problem of water management.[47]

If the outlines of a holistic sanitary strategy of pollution and flood control remained vague after 1954, the political picture was becoming perfectly clear to the residents of this expanding metropolitan area. The following year, the Democrats dumped the inept and openly racist Mayor Kennelly in favor of a rising star within the party, Richard J. Daley. Second only to kick-starting the stalled modernization of the CBD, the new mayor's top priority was rebuilding the party's tarnished patronage system. A master politician, the last big-city boss would impose a one-man rule of power and a one-party regime of government on Chicago until his death in 1976, just after winning his sixth term in office. Behind his shield of protection, the sanitary authorities gained immunity from criticism of their shortcomings in day-to-day administration. But even this consummate deal maker could stop neither climate change nor its damaging impacts on the built and natural environments as well as its human toll of emotional trauma and material loss.[48]

Conclusions: The Engendering of an Environmental Movement

The failure of leadership from within the political establishment opened space in the civic arena for professionals and nonexperts alike to offer alternative flood-control strategies. After the defeat of reform in Springfield in 1955, the collar county, state, and national governments began implementing their own programs to deal with this ever-more-frequent and costly problem. Undertaking similar stopgap projects over the next decade, the policy makers at city hall were no longer able to contain the public discourse over water management. On the contrary, debate raged among competing factions of consulting engineers, regional planners, and official experts.

Politicians at all levels of government also had to contend with a groundswell of the grassroots to save the city's waters. Disjointed gestures of eco-protest began to coalesce into an environmental movement between the great flood of 1961 and the great corruption scandal a year later. The ensuing battle of reform to save the lake and the suburban rivers in the forest preserves from the SDC also contributed to the construction of a metropolitan identity. It formed as opposition mounted to the sanitary authorities, the Democratic Party, and, ultimately, Boss Daley.[49]

Ten years into the postwar housing boom, Chicagoans had learned several basic lessons about their flood-prone environment. Of first importance was remembering the insight of the 1885 reformers: a big plan on a metropolitan scale was essential, because water knows no political boundaries. A consensus of opinion, moreover, agreed that the primary cause of the increase in flood damages was suburban sprawl. Paving over a prairie wetland was to blame for larger runoff surges during storms. In contrast, climate change was not considered to be a source of the problem during this period.

Yet the series of record-breaking rainfalls drew an inseparable link in the public imagination between pollution control and flood control. Chicagoans had witnessed the deterioration of the aquatic ecosystems of the rivers due to swelling amounts of sewage and massive discharges of these toxic waters into Lake Michigan. To solve the problem of damaging floods, the land could be engineered in three ways: (1) the watersheds' rivers could be changed to increase the flow of the runoff; the cost of turning them into concrete channels filled with partially treated sewage, of course, would be the forest preserve's ecosystem; (2) storm surges could be stored temporarily until they could be safely released back into the system; or (3) the built environment could be redesigned to retain the rainfall where it fell. The landscape would be transformed into a marshland drained by lagoons, canals, and streams as well as equipped with rooftop gardens, rain gardens, and rain barrels.

A list of projects—one for each level of government—illustrates the continuation of piecemeal, incremental approaches to flood control between 1955 and 1965. Starting at the top, the Supreme Court ordered a scientific study of the SDC's pollution of the lake in its ongoing oversight of Chicago's sanitary strategies. Congress finally approved the U.S. Army Corps of Engineers' plan to enlarge the Cal-Sag Canal to increase its flow and its storage capacities. The Illinois Department of Public Works received funds for sundry projects to channelize the streams and rivers in the collar counties. Its commissioners sponsored land-use master plans as well as bond issues to pay for sewage-treatment plants and the establishment of their own county forest-preserve districts. The SDC embarked on a project below the Lockport Dam near Joliet to dredge a giant storage pool and to cut through the bedrock of the lower Des Plaines River to double its rate of flow. Meanwhile, in the city, the Sewer Department continued to follow the 1944 plan, constructing interceptor sewers under the South Side.[50]

A closer examination of the battle over the fate of the rivers running through the FPDCC sheds light on how the debate over flood control gave birth to an environmental movement to save Chicago's waters. The damages resulting from the 6.24 in./15.85 cm of rain that fell in six hours on July 1, 1957, first drew the media's attention to the links between flooding, water pollution, and the forest preserves. In the western suburbs, Salt Creek filled with raw sewage, forcing DuPage officials to close the Fullersburg (now Salt Creek/Bemis Woods) Forest Preserve on their side of the county border. On the Cook County side, the trustees of the SDC exploited this environmental disaster to make a political end run around their chief engineer, Horace Ramey. They hired an outside consulting firm to design a new master plan without his knowledge. It proposed to widen and deepen the creek as part of a much larger, $2.3-billion construction program to channelize virtually all the waterways within the county to handle a flow of 13 bgd/49.2 bLd, or twice the capacity of the Sanitary and Ship Canal. Hurt by the board's betrayal, Ramey protested that the plan was "extravagant, unrealistic, and unnecessary." In retaliation, the trustees used his half century of loyal service against him; they fired him for old age.[51]

Nonetheless, this expert-versus-expert debate in the public arena broke the party insiders' code of silence, cracking open a political window of opportunity for other voices to offer their perspectives on flood control. In the forefront was the citizens' advisory committee of the FPDCC. Its proposal to hire an outside engineering firm to draw up an alternative, more environmentally friendly plan received the enthusiastic support of the county board. Expressing its determination to save the rivers, the district's superintendent, Charles G. Sauers, stated, "These streams are the backbone and heart of the forest preserves." He objected to the SDC proposal, because it "would destroy their natural beauty and thus damage the forest preserve holdings." For Salt Creek, Sauers suggested that a retention pond could be built within one of the preserves. Governor vetoes, state lawmaker–imposed limits on SDC spending, and the overwhelming defeat of a modest $50-million bond issue held up the plan to engineer the riverine environment of the Chicago metropolitan region.[52]

Even before the next historic thunderstorm hit Chicago in September 1961, city dwellers and suburbanites were already beginning to organize a resistance movement to the plan proposals of the official agencies in charge of flood control. Besides destroying the riverbanks to increase flow rates, all the engineers' designs depended ultimately on using Lake Michigan as the sink of last resort for storm surges. In February 1960, private citizens and regional planners successfully petitioned outgoing Governor Stratton to delay funding of state and SDC plans to channelize Salt Creek and the Des Plaines River. The nonpartisan Northeastern Illinois Metropolitan Area Planning Commission led the lobby campaign to give the FPDCC time to prepare its alternative approach. Despite the preemptive counterattack of the sanitary engineers that any eco-friendly strategy would be "inadequate," the county board presented a $7-million scheme four months later. Its consultants recommended building five impounding dams on the upper Des Plaines River, which would add more parkland at the same time. Sauers also announced that he had worked out a compromise with the state's engineers to use the $5.2 million appropriated to improve Salt Creek in similar ways to minimize environmental damage.[53]

The triumph of the FPDCC was not surprising, given Chicagoans' enduring love affair with its open spaces and facilities. The growth in popularity of spending time in the great outdoors was matched only by the mobilization of political support in favor of increasing open space in the city and suburbs. In the city alone, the park district's count of people engaged in recreational activities reached fifty million a year. Grassroots demands for more forest preserves had already led to an increase of the FPDCC's original charter of 35,000 acres/14,164 ha to 47,000 acres/19,020 ha of land. But this was not enough, according to the *Chicago Tribune* in a May 1961 editorial, because the "forest preserves must grow." The metropolis needed at least 60,000 acres/24,300 ha to keep up with population growth. The newspaper backed two successful bills in the state capital to empower forest-preserve districts to buy nonforested land and to raise more tax money. National lawmakers also granted local governments new authority to spend urban renewal funds to buy open, green spaces, not just "blighted," inner-city land.[54]

Early acts of grassroots activism served the dual purpose of educating the public at large about the perilous state of the rivers and building a political movement to save them. At one end of the social ladder, the affluent women of the city's Garden Club had been finding little patches along the Chicago River to beautify with flowers since the mid-1950s. At the other end, an African American railroad maintenance worker, Jordan Clark, also planted a garden on its banks in the Illinois Central rail yards. By 1961, incoming Democratic Governor Otto Kerner realized that there was enough support to proclaim the inauguration of an annual "Clean Streams Week." Volunteers in the hundreds and then thousands after the September 1961 disaster learned that public authorities had allowed the riverbanks to become garbage dumps. The following year, the publication of Rachel Carson's *Silent Spring* inspired a national epiphany on the meaning of ecology and the interconnectedness of all living things.[55]

Two more years of patchwork grassroots activism led to the organization of the Open Lands Project. The 250 civic organizations belonging to the Welfare Council of Metropolitan Chicago started it to promote more recreational facilities for teenagers. But it would soon take on a life of its own as a lobby for the protection of the environment. By the time of the first Earth Day in 1970, it would be joined by more than thirty other organizations devoted to this cause, ranging from the Wildflower Preservation Society and Prairie Club to the Earth Force and Campaign against Pollution.[56]

In sharp contrast, the sanitary authorities became the objects of ridicule for their incompetence and failure to design a flood-control strategy for Chicago's expanding metropolitan area. After the disaster of September 1961, this criticism gelled into a unified chorus of demand for reform. Producing more than 10 in./25 cm of rain in ten days, this extreme weather event broke the previous monthly record set in October 1954. Given Mayor Daley's success in kick-starting a renaissance of the CBD, the shortcomings of the SDC were no longer tolerable to the city's business community. Its plans to redevelop the downtown area depended on turning the river from an open sewer into an urban amenity, lined with walkways and plazas. Expressing its scorn with biting sarcasm, a newspaper editorial protested the engineers' "basement storage plan" of flood control. Similar complaints from homeowners throughout the metropolitan area poured into the offices of their elected representatives.[57]

The historic downpour of 1961 could not have come at a worse time for Daley, who was facing an insurgency of his base of support in the bungalow belt. Its homeowners were paying the costs of his regime of machine politics. Property taxes kept rising, while neighborhood services, such as garbage pickups, street cleaning, police protection, and park facilities, were all in decline. In what was shaping up to be Daley's toughest reelection contest of April 1963, he would lose this outer ring of wards and a majority of white voters for the only time in his six campaigns. After the deluge, the mayor made a major concession to the business community, agreeing to let it conduct a national search to replace the general superintendent of the scandal-ridden SDC. Its blue-ribbon panel was savvy enough to present only a single candidate, despite the trustees' insistence that it needed at least three names before it could make a decision. Overruling them, the mayor pushed through the appointment of Vinton W. Bacon and promised that he would have complete freedom to manage the agency.[58]

Mayor Daley got far more than he bargained for in permitting an outsider inside this den of thieves. A highly respected civil engineer, Bacon was not only honest and competent; he was also a tireless reformer. His crusade to throw them out showed that the SDC was, in fact, "rotten to the core." From the sabotage of the trustees to the resistance of the union bosses representing patronage workers, Bacon fought a one-man battle. When the trustees rejected his decisions behind closed doors, he took his case into the public sphere, making appeals in newspaper interviews, civic-organization meetings, and state legislative hearings. At first, he made real progress in firing incompetent personnel, canceling sweetheart deals, suing contractors for underperformance, and exposing fundamental flaws in the engineers' plans. A catalog of his disclosures would compose an encyclopedia of corruption.[59]

Bacon's first battle over flood-control strategies represents a case study of a political regime devoted to serving itself alone at the public's expense. In April 1963, the general superintendent made a report to the board of trustees on the two-phase project to prevent the flooding of Joliet. Contractors were already dredging more storage capacity in a pool below the Lockport Dam, and then they would cut through the bedrock of the shallowest stretch of the lower Des Plaines River to increase its flow rate. Reinforced by a team of consulting engineers, Bacon declared that the first $4-million phase was an extravagant "boondoggle" and that the second would be a total waste of $6 million, because it would involve digging in the wrong place. He also stated that he had fired the engineer in charge of flood control for his hopeless plan. This and subsequent meetings turned into table-banging shouting matches, with the trustee in charge of the engineering committee denouncing Bacon's findings while trying to sneak his dismissed minion back into another department. In this contest, the reformer appeared to have won, saving taxpayers at least $7 million.

But after Bacon had served about a year on the job, Daley signaled the trustees to begin rolling back his reforms. In April 1964, they voted 6–2 to fund a restudy of their original plan. The chair of the engineering committee countercharged that Bacon, not the board, was "guilty of concealment, deception, and conspiracy to defraud the taxpayers."[60] He claimed that he had hired at his own expense outside experts, who supported the board's position. However, he could not produce their report when challenged by Bacon to put it in the public record.

Bacon would continue to fight the battle for reform until the board fired him six years later. During this period, almost all of his progress in booting out the thieves would be reversed. Daley had complete control of the civil-service board members, plus the local prosecutors and judges. Although Bacon's political crusade proved ephemeral, his flood-control plan to engineer the environment became a permanent legacy.

The year 1965 marked a major turning point in this unfinished story for Chicago's sanitation strategies and environmental politics. In January, Bacon revealed to a reporter the outlines of a heroic, big-technology plan that "would be one of the great engineering feats in history." But the feasibility study of what would become TARP seems to have been sidetracked by the trustees. In the meantime, Bacon adopted the forest preserves' method of storing storm runoff for temporary periods in retention ponds. Over the next decade, the sanitary district would build two dozen retention ponds, or one-half of the number called for in his plan, along the Chicago, Des Plaines, and Little Calumet Rivers. Their ideal capacity was at least 33 million gal./125 million L each; their actual size ranged from a quarter to ten times as much storage space. Although only two of the retention reservoirs were online by 1977, they helped prevent the rivers from overflowing their banks, causing damages to structures built on their floodplains.[61]

On the eve of the next municipal elections in April 1967, Mayor Daley made the surprise announcement that the city had a bold new plan, "the first of its kind"—and the SDC had received a $1-million grant from the federal government to conduct the technical studies needed to begin construction. Bacon filled in the details of this $1-billion "master water plan." It would permanently end not only damaging floods but also polluted rivers within the territory served by the SDC. The creation of his chief assistant, Frank Dalton, and outside consulting engineers, it was designed to redirect sewer overflows into an underflow system of very large storage tunnels running below the rivers. Later, the contaminated water could be pumped up and properly treated before being released back into the environment. He promised that Chicago's rivers would be turned into natural and recreational amenities clean enough to swim in within ten years.[62]

Bacon's proposal was both old and new, big and small. Chicago was already laying what the SDC called the world's largest interceptor sewers of up to 28 ft./8.5 m in diameter. At the same time, the underflow tunnels would be twice as large and much deeper down at bedrock level. The interceptor sewers were dug using traditional, miner techniques of "drill and blast" with dynamite. The most innovative aspect of TARP was the employment of an experimental, giant boring machine called the "mole," like the one used later to dig the "Chunnel" between Europe and England. The plan was comprehensive in its geographical coverage of Cook County and the rivers running through it, but it was way too small in geographical scale, because it did not embrace the collar counties of the larger metropolitan region.[63]

The engineers failed to take into account not only the past two decades of suburban sprawl but also climate change. At the base of TARP was a simple, fatal flaw in their assumptions about rainfall. Boring down in the technical reports, Judith A. Martin and Sam Bass Warner Jr. uncovered a 1959 study that was used in making all subsequent calculations of storm-surge runoff. This report was limited to a single city block and 0.5 in./1.25 cm of rain in the first hour. Looking back, the resulting size of the storage tunnels turned out to be about half of the size needed to hold runoff surges and several-fold below cost. "This massive construction project proved upon investigation," the historians discovered, "to be the prisoner of professional inertia—an example of the lag and drag imposed by the weight of accepted professional practice. An elected board of commissioners and an expert staff failed to adapt to unfolding environmental criticism and experiments or to citizen initiatives."[64]

The year 1965 was pivotal in a second, related way, because the city's scattered campaigns of environmental protection became a unified movement under the banner "Save Our Lake." On the heels of Carson's book, the *Chicago Tribune* launched this sustained, front-page crusade, which would contribute significantly over the next seven years to the passage of the landmark Clean Water Act. During the transformative year of 1965, the FPDCC reached its fifth anniversary. State lawmakers celebrated by passing a bill to expand its size from 52,000 acres/21,050 ha to 65,000 acres/26,300 ha. Two years earlier, Cook County had joined with its five collar counties to form a pact of mutual cooperation that would add another 5,400 acres/2,185 ha to Chicago's open lands. Taking pride in this project to preserve nature and provide the public with recreational activities, a local newspaper boasted that Chicago was finally entitled to "merit the name of [its motto,] 'The Garden City.'"[65]

5

The Bust of Urban Decline, 1965–1985

Saving the Natural World: Discarding the Built Environment

The half-century battle to reclaim the Chicago River from an open sewer to a recreational waterway clean enough for swimming began as a crusade to save Lake Michigan. In 1966, four years after the publication of Rachel Carson's *Silent Spring*, the city joined her campaign to rid the environment of dangerous chemicals. The staid *Chicago Tribune* sounded the alarm, launching a series of front-page articles under the graphic logo "Save Our Lake." It was "sick" from digesting too much pollution and in imminent danger of suffering the same "death" as Lake Erie.

The newspaper built on Carson's explanation of the ecology of pesticides to help its readers understand how phosphates in their laundry detergent were also poisoning the environment. Pumped into the lake as partially treated sewage, the chemicals were overfeeding its microorganisms. They, in turn, were rising to its surface as massive "algae blooms," depleting its water of oxygen, killing its fish, and making its smelly shorelines off-limits for recreational use.[1]

Over the next two years, Lake Michigan suffered one unnatural catastrophe after another that hammered home Carson's and the *Tribune*'s dire warnings about unintended consequences whenever the balance of nature was upset. First came the revelation that the very government that authorities had entrusted to enforce antipollution laws—the U.S. Army Corps of Engineers and the Metropolitan Sanitary District of Greater Chicago (MSD)—were among their worse violators. Both were using the lake as a free dumping ground for the highly toxic sedimentation they were dredging from the Chicago River. In the past decade, the U.S. Army Corps alone had dumped more than 750 million gallons (gal.)/2.8 billion liters (L) of this cancer- and fecal-laden waste only 6 miles/10 kilometers (km) away from the city's offshore drinking-supply stations. Water samples taken near the dumpsite by the state Board of Health confirmed that disease-causing microbes were spreading in all directions.[2]

Next came the invasion of the alewives. Billions of dead fish washed up on the lakefront beaches after a combination of pollution, overfishing, and invasive species wiped out all of their native predators. The herringlike alewives are a non-native species of 4- to 6-inch (in.)/10- to 15-centimeter (cm) fish. They were left virtually alone to feast on the algae-enriched, aquatic environment of the world's sixth-largest body of fresh water, covering 22,400 square miles/58,000 square km. In 1966, the alewives had stayed offshore but had clogged the city's water-supply intake wells. A revolving screen at one of these island stations had removed 15,000 pounds (lb.)/6,800 kilograms (kg) of them in one hour during the height of the fish run.

The following summer, however, a massive die-off and easterly winds left the city's beaches buried under thick layers of rotting fish. "They keep coming in," exclaimed the man in charge of cleaning up the stinky mess. "In some places they are a foot deep. Look out over the lake and there they are [floating on the surface] as far as the eye can see." After many of his workers gave up and walked off the job, the beached alewives generated a biblical plague of bugs, including swarms of mosquitos that attacked the elite residents of lakefront apartment towers. To complete the catastrophe that kept at least 2.5 million beachgoers away that summer, the lake's depleted alewives allowed the algae to flourish, wash ashore, and bury the beaches under piles of decaying "seaweed."[3]

By the fall of 1967, the campaign to "Save Our Lake" had reached a critical mass of support that turned it into a full-fledged environmental movement. Responding to reader requests, the *Tribune* printed forty-three hundred bumper stickers and sixty-six hundred copies of a forty-eight-page booklet on the subject. Fearing the ghost of Lake Erie, the mobilization of citizen groups forced elected officials at every level of government to give the conservation of Lake Michigan serious attention for the first time since the Progressive Era. In September, Secretary of the Interior Stewart Udall convened a conference of policy makers and engineers from around the Great Lakes to draw plans to save them from pollution. Out of this meeting came a barrage of regulations, litigation, and legislation leading to the passage in 1972 of the comprehensive Water Pollution Control, or "Clean Water," Act.[4]

In the post–*Silent Spring* 1960s, the success of the "Save Our Lake" crusade should come as no surprise. As several historians have demonstrated, a suburbanizing nation of affluent homeowners was predisposed to respond positively to the marine biologist's plea to save the natural world from human abuse. Moreover, an immediate, one-step solution to the problem—remove the phosphates from soap—helped build political momentum for reform at the highest levels of policy making. Perhaps the notion that a small personal cost could result in a great public benefit also helped mobilize support at the grassroots. A willingness to sacrifice a little laundry cleanliness made people feel good about themselves.[5]

A seemingly quick and easy solution to the most immediate problem of the alewives eruption also played a role in reinforcing the belief that the lake should be saved. In 1966, the Michigan Department of Conservation took the desperate step of introducing nonnative species that were their natural predators, starting with Coho and Chinook salmon. "The steadfast faith of Michigan conservation officials that sports fishing was a palliative for urban ills," according to Kristin M. Szylvian, "is evidence of the enduring power of the belief that industrial America could find redemption and virtue in nature." Giving up on reviving the commercial fishing industry, the agency's experts kick-started the moribund economy of lakefront towns and resorts with sports fishing. In this case, at least, it worked as an antidote for the overabundant alewives. Gorging in the equivalent of a fishbowl of super-enriched food, thousands of 4 to 6 in./10 to 15 cm salmon fry took only a year to grow into 18 to 20 lb./8 to 9 kg prize catches. "Coming as a wonderful surprise to all concerned," exuded the *Chicago Tribune* in September 1967, the restocking program set off a sustained boom.[6]

Here was a win-win remedy that stopped the alewives and started a new leisure activity industry. As Chicagoans witnessed the beaches reopening to swimming, they regained their faith in the power of science and technology to effect environmental reform. Moreover, the revival of sports fishing on Lake Michigan created new groups of influential stakeholders in saving it from pollution. Forming a permanent base of political support, they ranged from boat makers, marina operators, and local chambers of commerce to serious anglers from around the world drawn by the lake's supersized sports fish.[7]

In sharp contrast, changing perceptions of the Chicago River as anything other than an industrial corridor filled with wastewater would take a much greater stretch of imagination spanning a fifty-year period. Its unique "envirotechnical" conditions and deeply entrenched attitudes among the local political elite made the process of building a constituency of stakeholders in favor of reclaiming the city's river a much-more-complicated process than saving its lake. In 1970, Mayor Richard J. Daley would confess that he had a dream of being able to fish on the river some day. For the next thirty years, his vision would become the single-most-repeated expression of bitter disappointment in the failure of city hall to reclaim the river for this and other recreational uses.[8]

Behind the mayor's back, so to speak, the mobilization of stakeholders in seeing the river transformed back into a natural state was not so quietly growing in numbers, diversity, and organizational strength. They envisioned the Chicago River as an outdoor playground and tourist attraction of "[wild] nature in the unnatural world." The story of the materialization of their visions of an urban wilderness running through the city center follows in the footsteps of David and Richard Stradling. Their parallel study examines the Rust Belt city of Cleveland, Ohio; its burning Cuyahoga River in 1969; and its "dead" Lake Erie. They illuminate a self-reinforcing process of structural and attitudinal changes that caused a complete metamorphosis of the meaning and even the iconic image of the river on fire.[9]

Like Cleveland, the deindustrialization of Chicago's riverfront and the related loss of jobs brought about a corresponding loss of stakeholders in the river as an economic resource. At the same time, it was gaining new constituencies of users among outdoor and fishing enthusiasts, city-beautification groups, and real-estate developers. As the river showed slow, albeit steady, improvements in environmental quality, its permanent base of stakeholders in its full restoration for recreational use grew in proportion. It would take fifty years before Chicagoans could fully reimagine the river and the lake as a single ecosystem.

This reciprocal process of river reclamation and stakeholder mobilization can be divided into a convenient number of stages. Each incremental improvement in the physical condition of the river fed psychological expectations for further gains as well as enlisted new recruits to the cause of saving it from pollution. In this way, the relationship between the city's people and its aquatic environment underwent change. As we have seen, the period from 1962 to 1972 was dominated by a popular movement for the conservation of the Great Lakes, culminating in the historic Clean Water Act. Riding this wave of popular support, the sanitary district was able to gain approval of its big-technology project, the [Deep] Tunnel and Reservoir Plan (TARP). The sanitary authorities disguised their grandiose scheme of flood control of the river as the final solution to the problem of pollution control of the lake (see Figure 5.1).

From 1975 to 1985, the MSD built the first phase of the Deep Tunnel, while adding chlorine to its sewage-treatment process to forestall its stormwater discharges into Lake Michigan from causing an epidemic. The chemical additive killed not only the dangerous microbes in its partially treated effluents but virtually all forms of life in the Chicago River. Despite becoming a dead zone, the river began its revival from a sewage channel to a nature preserve of the "wildest area within the city limits" in 1979, with the formation of the Friends of the River (FOTR). This lobby group's definition of the waterway as a "working river" reflected the initial weakness of its cause on behalf of urban nature. Its diverse coalition of special-interest groups had to assume a cautious political posture at first, which attempted to include everyone and alienate no one.

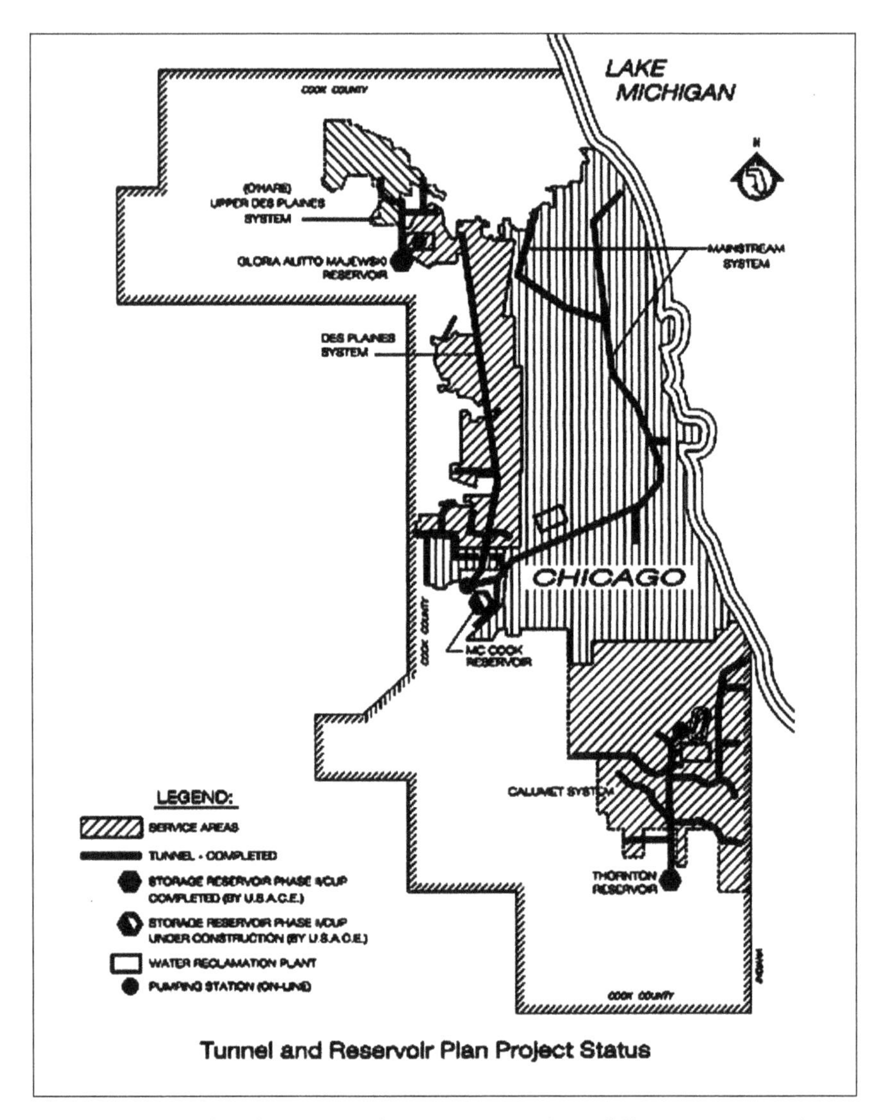

Figure 5.1. Tunnel and Reservoir Plan, 1965–1985 (City of Chicago, Metropolitan Water Reclamation District, "Tunnel and Reservoir Plan Project Status [2012]," available at www.mwrd.org/irj/go/km/docs/documents/MWRD/internet/protecting_the_environment/Tunnel_and_Reservoir_Plan/htm/Tarp_Map.htm)

From 1986 to 1997, the Chicago River would enter a third era of rapid commercial and residential redevelopment. This chapter examines the private investors who created an urban renaissance in an expanding central business district (CBD). During the 1990s, the downtown area would gain twenty thousand residents as condominium apartment towers joined the new office skyscrapers lining the riverbanks. In addition, the renamed Metropolitan Water Reclamation District stopped adding chlorinate to its effluents, because the opening of the Deep Tunnel meant it could hold 1 billion gal./7.6 billion L of storm-water overflow. Portrayed as a cost-saving dividend of the public-works project, the river was the primary beneficiary: it was rewarded with the return of sports fish, nesting birds, native plants, and other wildlife. During the 1990s, the rapid accumulation of stakeholders in the use of the waterway for recreational use would reach a critical mass of support to "Save Our River."

Save Our Communities: The Abandonment of the Central City

But also reaching a boiling point during the 1960s was the resistance of Chicago's African American community against spatial segregation and social exclusion. Fueling black nationalists and civil-rights movements, protests by the grassroots focused attention on the institutionalization of racism in housing, public education, parks, libraries, and sanitation services as well as within the police and the fire departments. In part, Dr. Martin Luther King Jr. shifted his "Freedom Movement" from Selma, Alabama, to the Lawndale neighborhood on the West Side after 1965 because Mayor Daley had crushed the local campaign to integrate the schools. In the three years before his assassination, the civil-rights leader also met a similar outcome of failure in the battle for open housing against city hall and the Chicago Real Estate Board.[10]

As Dr. King had warned when admitting defeat, violence broke out on the West Side. Extremists fought police, looted stores, and burned down neighborhoods. From 1965 to 1968, three uprisings of the grassroots were set off by clashes in the streets between the police and the protesters. The first two were contained, but the third required the U.S. Army to restore order. It was a reaction to Dr. King's death and Mayor Daley's refusal to declare a day of mourning in his honor. In protest, high school students walked out en masse and began marching in the streets toward city hall. They never made it, because angry crowds of spectators became armed combatants engaged in shooting at the police. Firefighters could only watch helplessly from a distance as the West Side's main streets went up in flames.[11]

The civil disorders of March 1968, followed by the "police riot" at the Democratic National Convention in August, set off a panic among white homeowners and businessmen to get out of the city. For shopkeepers and manufacturers burned out of the West Side, serious consideration was not given to reinvesting in the old neighborhoods. While many of their properties became empty lots of despair, some became victims of real-estate vultures, who kept their titles entangled in shady deals and their buildings in ill repair. Combined with the winding down of the Vietnam War, the closing of factories and the stockyards in Chicago eroded political support for maintaining the Chicago River as an industrial corridor. The influence of its lobby further evaporated after the oil embargo and energy crisis of 1973–1974 put many of the city's old-line petro-chemical makers of paints, fuel oils, and plastics out of business.[12]

The neighborhoods bordering the river corridor and the African American communities in the riot zone on the West Side became discarded urban space. Financial and political policy makers turned them into what Peter Marcuse calls fourth-world "ghettos of exclusion."[13] Both the buildings and the people forced to live in them because of poverty and discrimination were treated like throwaway commodities of a culture of consumption. "The cycle of 'blockbusting,' white violence, urban crime, white flight, and community decay that beset Lawndale from the late 1950s on," historian Beryl Satter reports, "was depressingly familiar to anyone who had followed the processes of racial change in Chicago."[14] Only clean-sweep demolitions, such as the urban-renewal projects to redevelop slum districts into model public housing, according to the planners, could lay the groundwork to reverse the bust of urban decline.

But after 1968, the interrelated urban crises of race relations, social disorder, and economic distress fed on each other to produce a mass exodus of whites from the city to the suburbs. White flight caused a sharp jump in deconcentration at the periphery that was driven by the abandonment of the center. Between 1960 and 1980, the city lost 545,000 people, with about three out of four of them white residents moving to newly segregating communities at the edges of the metropolitan region. In contrast to this loss of 15 percent of its urban population, suburban Cook County counted 669,000 newcomers, a 42 percent gain. The greatest rate of sprawl took place in Chicago's five collar counties: they increased by 69 percent, or 758,000 additional residents, to give them a total of 1.85 million people. Together with Cook County, the suburbs' census of 4.1 million was more than a million larger than the city's shrinking numbers.[15]

Along with the flight of white population went the retail businesses and corporate headquarters, turning Chicago into a "multicentered" region of a hybrid, urban/suburban life.[16] Homer Hoyt's shopping malls of the mid-1950s gained doctors' and other professionals' offices, fine restaurants, hotels and convention centers, movie theaters, and eventually office towers filled with white-collar workers. A good example of this process of the "recentering of the metropolitan area" is Schaumburg, a suburb about 25 miles/40 km west of the CBD.[17] Located next to one of the mid-1960s superhighways, it attracted locally bred Motorola and later the world's largest indoor shopping mall to become an instant mini-city. By the mid-1970s, Schaumburg had already become a more-important center of business than Milwaukee, Wisconsin, or Rockford, Illinois. Rush hours on the highways gradually reversed direction as more and more jobs kept gravitating outward toward the crabgrass frontiers of the collar counties. And, just as inevitably, more ring roads were built to link these edge cities. The abandonment of the city intensified the pace of paving over the six watersheds of a flood-prone environment.

"Save Our Lake": The Fulfillment of Conservationism

The transformation of the Chicago River into the Sanitary and Ship Channel had become the city's greatest myth of heroic engineering by the postwar era, despite its serious flaws as an envirotechnical system. It drained billions of gallons a day from the lake through several locks to dilute its chemical and organic wastes as they merged into the Illinois and Mississippi Rivers. Yet the river still became so overloaded with industrial effluents and partially treated sewage that it needed constant dredging to keep its shipping lanes open to shallow-draft barge traffic.[18]

In addition, the MSD had to open the locks during severe thunderstorms, vomiting tons of raw sewage into the lake. By reversing the river's flow, Chicago became the one point of contact between the continents' two largest watershed basins: the Great Lakes and the Mississippi River. The normal flow was "downstream" from Lake Michigan into the river, including a daily average of 2 billion gal./7.6 billion L of partially treated sewage. At the same time, invasive species still moved "upstream" from as far away as the Gulf of Mexico. And during predictable seasonal rainfall of even 0.1 in./0.25 cm, the MSD's underground infrastructure of 5,500 miles/8,850 km of combined sewer and storm pipes overloaded. On average, they poured raw sewage into the river from more than six hundred emergency overflow drains during one out of every three days.[19] The overloaded pipes also backflowed into basements in the city and the almost two hundred suburbs lying within the MSD's boundaries. Costing millions of dollars in damages to mostly uninsured homeowners, the sanitary district's infrastructure put at least 1.5 million people in constant risk of flooding.[20]

Nevertheless, the city's cult of the "technological sublime" drew a sharp, mental dividing line between the riverfront as its back alley for getting rid of its liquid waste and the lakefront as its front yard for recreational use.[21] Chicago had kept this tradition of big technology alive in the contemporary era by building the world's largest sewage-treatment and water-filtration plants. More important here, the engineers of the MSD put forward what would eventually become the biggest urban public-works project in the nation's history, the Deep Tunnel. Reacting to sharp public criticism following particularly damaging thunderstorms for two years in a row in 1957 and 1958, they had proposed to build an "underflow" system of deep tunnels to store the storm-surge overflows until they could be pumped back up and treated. The MSD promised that the multi-billion-dollar TARP would solve all the problems of flood control and lake pollution.

Yet its planners never questioned the status of the river as an industrial corridor, nor its use as the "ultimate sink" for its own partially treated effluents, the daily equivalent of a million people's raw sewage.[22] Local traditions of lakefront beautification stretched equally into the past. The heroic engineering of the artificial waterway took place at the same time as the World's Fair of 1893, the work of Daniel Burnham and Edward Bennett, and their influential *Plan of Chicago*. The *Tribune* tapped into the Progressive Era's spirit of the conservation of nature and the city beautiful in its "Save Our Lake" crusade.[23]

Although funding of the construction phase of the Deep Tunnel was delayed until the oil embargo triggered a recession, the roots of an environmental movement to clean the waters of the Chicago area had been planted in 1966 and 1967 by the "Save Our Lake" crusade. It had conveyed Carson's lesson about the harmful effects of the abuse of nature to broad, receptive audiences around the Great Lakes. The dumping of the rivers' dredging near the drinking-supply stations and the invasion of the alewives had brought her point home in a personal way, not only to the millions of the lake's recreational users but to the entire city.

In the following year, 1968, seasonal cloudbursts again overwhelmed the sewer system. Heavily contaminated rainwater in the underground pipes backflowed into thousands of homeowners' basements, overflowed the Chicago River, and forced the MSD to open the locks and pollute the lake with sewage-laden floodwater. Given the breakneck pace of urban development and suburban sprawl, the area's residents faced a future of more flooding rivers in addition to a "dead" lake killed by the phosphates in laundry detergent. Moreover, the coming decade of extraordinarily wet years would overwhelm the drainage system and force the sanitary authorities to keep polluting the lake with the river's disease-laden sewage.[24]

In this way, the passage of the historic Clean Water Act of 1972 represents the end of one story of environmental reform—the conservation of the Great Lakes—as much as the start of another: the battle to reclaim the Chicago River as a natural ecosystem. Reflecting American federalism, political demands for reform percolated from the local to the state and, ultimate, the federal levels of government. To be sure, lawmakers and bureaucrats in Washington, D.C., had their own self-serving agendas that contributed to the formation of environmental-protection policies.[25]

Nonetheless, the movement to end the pollution of the Chicago River grew out of changing perceptions at the grassroots. During this period, however, only a handful of visionary planners and green activists protested city hall's steadfast defense of the use of the Chicago River as the dumping ground for the city's waste. The new ideas planted by Carson had grown into a popular consensus behind raising the water-quality standards for the country's rivers and lakes to be clean enough for swimming. Congress was now also willing to provide 75 percent of the funding to underwrite this ambitious endeavor. Paradoxically, the need of the MSD to disguise its master plan for flood control of the river as pollution control of the lake was the result of its own success in creating a local, popular culture of big technology.

City hall also used smoke and mirrors to rebrand TARP as an essential part of the national plan to save the Great Lakes to gain funding for what might eventually surpass Boston's "Big Dig" tunnel to become the most expensive urban public-works project in history. Until the insurgency of this grassroots movement, the engineers' master plan had languished on their drawing boards as little more than a fantastic dream. Except for homeowners at risk of basement flooding, it had no constituency of stakeholders willing to bear the property-tax burden necessary to pay for its projected multi-billion-dollar cost. Moreover, the local government's well-deserved reputation for corruption under the regime of "Boss" Daley undercut trust in the MSD as anything other than a cog in his patronage machine of crony capitalism. The state's voters expressed their lack of confidence in local government in the spring of 1967, when they rejected a billion-dollar bond referendum to pay for clean-water infrastructure.[26]

But as the "Save Our Lake" crusade rapidly gathered momentum that summer, the MSD's grand design of the technological sublime captured the public imagination. Significantly, the voters would approve a similar bond act when it came before them a few years later. The engineers promised that TARP would eliminate pollution, flooding, and sewage overflows into the lake. In March, the upsurge of the environmental movement against pollution pried loose a $1-million study grant from the federal government on the plan's feasibility. This seed money spurred the district's trustees to add another $2.5 million to complete the cost-benefit analysis of their big-technology scheme versus other alternatives.[27]

The complete turnaround of Illinois Governor Otto Kerner from defender of industrial polluters to protector-in-chief of Lake Michigan was indicative of the groundswell of support for the "Save Our Lake" crusade. In September 1967, he ignored his own recent vetoes of several clean-water bills and called on his three fellow governors bordering the lake (Indiana, Michigan, and Wisconsin) to join in the common cause of restoring its health. Kerner added his voice to the crescendo of populist demands arising from around the entire Great Lakes, calling for stronger state and national water pollution–control laws.

For the first time, moreover, the attorneys general of the four Lake Michigan states met to coordinate their fight against pollution. In 1967, they settled the states' long-standing federal case dating to the 1920s against Chicago's divergence of too much lake water into its rivers and ship channel. They agreed as well to compile a list of industrial polluters to begin prosecuting the worst offenders under existing antidumping laws. Following Governor Kerner's lead, the Illinois State Board of Health set new, higher standards of water quality: manufacturers now had to remove all oils and floating solids in their wastewater before it entered the sewer system, and no organic or toxic effluents could be dumped directly into the state's rivers and lakes.[28]

Yet the most important policy initiatives to emerge from the "Save Our Lake" crusade came at the local level. Reacting to the newspaper campaign, the trustees of the MSD joined the bandwagon of this popular movement. In June 1966, they announced that the agency was ending its practice of dumping river dredging in the lake. In addition, it was suing the U.S. Army Corps of Engineers to stop the practice on the grounds that it posed a threat to the public health. With funding of the Chicago branch of the federal agency tied directly to the volume of freight moving through the ship canal, the Corps refused to comply. It won the legal dispute of jurisdiction over "navigable waterways" but lost the case in the larger court of public opinion. Political pressure for reform percolated up from the grassroots, local, and state levels into the halls of Congress. By the end of the year, a deal had been worked out in Washington for the Corps to use landfill sites. In exchange, the lawmakers provided "emergency" funding to cover the extra costs of this more-expensive option.[29]

The MSD also raised its image as a guardian of the environment in September, when it issued thirty-day deadlines to violators of the new limits on pollution set by the state. For example, the sanitary authority issued warnings to the Columbia Yacht Club for discharging its tony members' untreated waste into the lake and to the Silver Skillet Food Products Company for dumping its grease, fats, and oils into the river. For the first time, the agency threatened to levy fines and seek court action. District President John Egan echoed the growing choirs of defenders of the environment. "We have long been too tolerant of industrial pollution problems," he declared. "It's time for us to buckle down and fight back."[30]

These initiatives, however, turned out to be a smoke screen of false promises thrown up to hide city hall's complicity in polluting the environment. The failure of the MSD to enforce its own standards of water quality on industrial polluters would make the intervention of the national government essential in the battle to reclaim the Chicago River for recreational use. Soon after announcing its crackdown on a list of three hundred violators in 1968, the sanitary authorities backtracked by refusing to bring any cases to court. On the contrary, it became their guardian shield against prosecution. The MSD's lawyers twisted the law into a convoluted shape that limited the agency to seeking "conciliation" through voluntary negotiations. In questioning its record of "feeble prosecution," a *Chicago Tribune* editorial asked, "Is somebody exerting political influence according to the old tradition of the sanitary district?"[31]

Save Our Basements: The Reprise of Heroic Engineering

Led by Mayor Daley, the MSD became not only the Chicago River's biggest polluter but also the greatest opponent in the political arena of its restoration. His handpicked trustees held the lake cleanup hostage to winning funding for TARP from the state and federal governments. During this eight-year standoff, the agency applied for an exemption from meeting the new standards of water quality for streams and lakes. The sanitary authorities also lobbied to lower these standards as a cost-saving measure, and they refused to enforce their own orders of compliance against industrial polluters.[32]

In May 1966, Daley shifted General Superintendent Bacon from his corps of professional managers to his enemies' list. The mayor had hired him four years earlier to settle a corruption scandal, not to open new ones. Left with a choice between surrender to the trustees or policy deadlock, Bacon cultivated the highest standards of science and technology while continuing his fight to root out corruption. In addition to gaining approval of the retention-pond plan, he built the district's first treatment plant that reached a 99-percent level of purification. Located on the Salt Creek/Des Plaines River in Schaumburg Township, the Hanover plant represented a "milestone," according to the engineer, because it included third-stage treatment facilities that improved the quality of effluents to a level that would do no harm to the aquatic ecosystem.[33]

But Bacon became a political liability when he exposed Daley's patronage machine, setting off a controversy that ultimately cost the reformer his job. He discovered an insider scheme that was rigging the civil-service exams for operating engineers, the men in charge of keeping the machinery running at the MSD's treatment plants and pumping stations. In a criminal conspiracy involving ward bosses and union representatives, the managers of the MSD's personnel department had forged the test results to conform to a prearranged list of winners. After waiting a month for the trustees to act, Bacon seized the fraudulent records that proved to have substitute and original answer sheets with erased and re-marked checkboxes. He also held the examiner, the assistant director of the department, Ronald E. Huston, responsible for changing them.[34]

But in this case of Chicago's three-act public theater of corruption exposed, investigated, and rolled back, a fourth act—attempted assassination—

was added to the script. On Saturday, August 22, the diligent general superintendent drove his car from the North Shore suburbs to work at his downtown office. While he was inside, someone with access to the sanitary district's secure parking lot planted sticks of dynamite on the engine of his car. At the end of the day, Bacon drove back home, stopping at a local gas station, where a surprised attendant discovered the bomb. The assassin had failed to secure one of the wires of the would-be bomb's detonator to a spark plug, precluding an explosion. The newspapers insinuated that links could be drawn between the labor unions, rigged exams, and this murder plot, but no one was ever charged with the crime. Vowing to carry on, Bacon retaliated by broadening his exposures of corruption, seizing the "'secret [patronage] records'" of the personnel department, and fortifying his defenses against the board of trustees in the public sphere. In October, he would face new death threats.[35]

To replace the gangster-style drama of a car bomb, the officials in charge of law enforcement played their roles in the second act of Chicago's classic production of machine politics. The civil-service commission did its bit by suspending the department's director. He duly resigned, leaving Bacon in control of the civil-service system of employment. The state's attorney charged Huston with official misconduct and fraud after stalling as long as possible. The prosecutor then found excuses to put off the beginning of the trial for another year and a half. In the meantime, Bacon got into another fight with the board of trustees over giving the workers in the Fireman's and Oiler's Union Local Seven a raise that was twice as large as needed to secure a labor contract. A month later, in February 1968, the Better Government Association joined Bacon in exposing Daley's patronage machine. The civic-reform group filed a suit in state court that uncovered a scheme to get around Bacon's control of the personnel department. It charged that the MSD was illegally outsourcing its work to private contractors, who made significant contributions to the Democratic Party's coffers. The goal of the case, according to the association's spokesman, was "to close a 'loophole' through which politicians obtain patronage jobs by circumventing the district's civil service system."[36]

When Huston's criminal trial finally reached a jury in May 1968, the judge dismissed the case. The state's attorney had not properly secured the exam answer sheets, allowing the court to dismiss the evidence against Huston. Act three, the rollback, started with an office job as a newly enrolled member of Local Seven. At the closing curtain, the civil-service employee stood center stage, not only vindicated but entitled to back pay and reinstatement in his old job.[37]

Almost a year earlier, Bacon had been effectively pushed off the stage of policy making at the sanitary district by its Democratic majority of the trustees. In January 1969, the trustees had humiliated Bacon with a five-day suspension for a patently phony, non-work-related charge of misrepresenting the agency's public image. A year later, the six Democrats on the board voted to fire him. A Republican trustee, Abe Eiserman, denounced the surprise action as a "'railroad job . . . the most blatant political maneuvering' he'd seen in twenty-five years in politics."[38] Raising the Democrats' majority to eight in the early 1970s, Mayor Daley and his money machine were ready to haul in the contributions of contractors and consultants, who were eager to tap into the multi-million-dollar federal grants following in the wake of the 1972 Clean Water and Environmental Protection Acts.[39]

With the passage of these reforms, the mayor of Chicago became the MSD's top lobbyist in Congress. He pitched its big-technology plan as the only way to meet the national reform goal of saving the Great Lakes, and he tried to persuade lawmakers to sidestep the Supreme Court and give Chicago more lake water to dilute its sewage in the river to avoid the local cost of fully treating it. Mayor Daley's cynical exploitation of his fishing dream was emblematic of the local political regime's Janus-faced posture of environmental degradation of the river. With blue-collar roots in the bungalow belt, "Dick" Daley originally revealed his vision at a press conference. The trustees of the MSD called reporters in to announce another one of their sham crackdowns on river pollution. Appearing with President Egan on June 12, 1970, the mayor said, "I hope we will live to see the day there will be fishing in the river, maybe even a bicycle path along the bank; perhaps swimming."[40] Congress was not convinced.

Three years later, he was still working his river fantasy to break the stalemate over funding the Deep Tunnel. Still deadlocked a year later in 1974, he pleaded before a congressional committee for help in making his dream come true. Mayor Daley took the political theater of deception to a new level of hypocrisy. Desperate, perhaps, he played a thinly veiled race card of the "urban crisis" of civil disorder to present his appeal on behalf of Chicago's big-technology project. "How do you put a value on young people in the inner city being able to fish in the river, which they never had before?" he asked rhetorically. He supplied his own answer with a follow-up question: "Would you say $500 or $1,000? It is very difficult to put a value on that." The Environmental Protection Agency's (EPA's) regional administrator, however, remained skeptical of the value of the super-expensive TARP compared to competing proposals from other sanitary districts bordering the Great Lakes.[41]

The following year, a mix of economic and unnatural disasters combined to tip the political balance of power in favor of the big-technology plan. The shock of the oil-embargo crisis threw the nation into a stubborn, decade-long period of recession combined with rising prices called "stagflation." On the one hand, Congress responded with a stimulus program, including the funding of public works. On the other hand, the Daley administration was desperate for new sources of outside money since the liquidation of Washington's urban-renewal and highway programs. After nearly twenty years of spending far more than the city collected in taxes, Chicago was on the verge of municipal bankruptcy. Suddenly, TARP offered policy makers at all levels of the federal system a "shovel-ready" large-scale proposal that would put a lot of people back to work in the construction industry. It also promised to become a major money machine for Daley's Democratic Party at the time of his death in December 1976 from a heart attack.[42]

Chicago's project remained at the bottom of the EPA's priority list until the state government finally broke the political stalemate over funding. When it allocated $287 million of a bond issue to start the construction of TARP, Chicago's delegation in Congress convinced the lawmakers to override the bureaucrats and provide three-to-one matching funds for 36 miles/58 km of tunnels and pumping stations. Breaking the logjam, they kept federal dollars flowing during the next ten years of building the first so-called pollution-control phase of the Deep Tunnel. The inheritors of Daley's clientele regime turned the money flowing from Washington and Springfield into a fountain of patronage as well as illegal, under-the-table bribes, kickbacks, and reelection contributions.[43]

In addition to the economic crisis of urban decline, the unexpected return of damaging floods in the Chicago area also prompted state and federal policy makers to elevate TARP to the top of their priority list of public-works projects. Between 1944 and 1974, torrential downpours had forced the MSD to open the locks and dump pollution into the lake five times. After 1974, it had to take this course of last resort four years in a row. This cluster of severe weather events cost homeowners between $70 million and $100 million in damages. Over the next seven years, rainstorms forced the sanitary district to contaminate the lakefront and close the beaches four more times.

During this period of construction, a group of academics contested the big-technology approach of TARP with green alternatives. In 1978, the economists offered radically different, small-scale approaches to remaking a flood-prone environment. A professor of urban affairs at Northwestern University, Stanley Hallett, posited that the MSD's big-technology project would fail to save the lake from sewage backflows during severe downpours. The experts were joined by a grassroots coalition of neighborhood and church groups interested in raising the number of public-works jobs and lowering the tax bills of homeowners. Although the Deep Tunnel was a multi-billion-dollar plan, it used giant, automated machines and relatively few workers.

Hallett provided a list of alternative technologies that would do a better job of reducing flood damages for one-third to one-half of the cost of the Deep Tunnel. It included public-works projects, such as universal metering of the water supply to increase sewer-system capacity, Skokie Lagoon–like greenbelts, and porous pavements. Moreover, building owners would be instructed to disconnect their roof downspouts from the sewer system and plug them into rain barrels. All of these eco-friendly measures were designed to store rainwater at its source, preventing it from becoming a runoff surge. The Better Government Association further undercut the rationale behind TARP by giving its phase-two plan of flood control an equally failing grade in the test of costs and benefits. The civic-reform group's watchdogs demanded an immediate halt to this "giant boondoggle" of machine politics.[44]

The sanitary district responded by calling the alternative approaches "good suggestions" but then tearing each one down to conclude that "taken together [they] would not be 'a solution to the total problem.'" The project director of the Deep Tunnel, Frank Dalton, "ridiculed" the whole idea of equipping tens of thousands of homeowners with 50 gal./190 L rain barrels: he quipped that only the barrel makers would benefit. The engineer also rejected the greenbelt proposal by disingenuously dividing the hydrology of the metropolitan region's six watersheds into the city and the suburbs. It was impracticable, Dalton retorted, because "you are talking about built-up areas; you are not talking about the boondocks." In political terms, the eight Democratic members of the MSD's board of trustees never entertained any doubts about the positive value of TARP: they not only reaffirmed their support of the project but also rejected all proposals for an internal review of its economic and environmental values.[45]

From 1965 to 1985, the climate change of more-frequent and more-severe rainstorms became a chronic problem that caused a decline in the quality of daily life throughout the metropolitan region. Although the August 17, 1968, deluge of more than 4 in./10 cm was rare, it was typical of the wet period's impacts on a flood-prone environment: multi-million-dollar damages, transportation disruptions, power blackouts, public-health crises, and beach closings. Pounding the entire area with heavy rains and winds for two hours, the storm hit the Southwest Side hardest, sinking its homes and streets under 2 to 4 feet/0.6 to 1.2 meters of water. "All it takes for a flood around here," a resident of suburban Markham stated, "is a couple of inches of rain and we've got a lot more than that."[46] At least thirty-six thousand homes lost electrical power, and an equal number lost their telephone service. The runoff surge raised the levels of the inland waterways to the danger point, forcing the sanitary authorities to open the locks on the North Shore Channel and the Chicago River. All the beaches had to be closed, because the storm-water backflow contaminated with raw sewage could be clearly seen from helicopters along a 20-mile/32 km stretch of the lakefront.[47]

Sprawl, the MSD warned a year later, in 1969, would make this kind of widespread financial loss and emotional suffering more and more common in the future unless TARP was fully funded. One of the chief architects of the Deep Tunnel, Dalton, made the case for big technology as being the best way to achieve hydraulic control of a flood-prone environment. Over the past fifteen years, he stated, "[two] storms approaching the 100-year intensity" had sunk Chicago. Five others had exceeded the historic, five-year extremes. But this unusual weather was not the cause of damaging floods, according to the engineer. He argued that sprawl best accounted for their mounting costs. "Although [the sewers] were originally designed for what is called a five-year storm," he explained, "the great increase in impervious areas such as streets, expressways, parking areas, and roof tops, frequently coupled with locally improved drainage from such areas, has increased greatly the rate and amount of runoff during any given storm." Within the 300-square-mile/777 square km of the urban core, Dalton calculated, 88 percent had been paved over. He put the price tag at $1.1 billion for TARP as the permanent solution to the hundred-year rainstorm.[48]

Responding in 1977, the Northeastern Illinois Planning Commission forecast that the condition of Chicago's rivers would get worse in the coming years regardless of the benefits of the Deep Tunnel. A 30-percent increase in population in their watersheds by the turn of the twenty-first century would bring "so much new construction that urban runoff is expected to increase despite more cleanup efforts by communities there."[49] The engineers at the MSD started working overtime to complete their plans for the second, multi-billion-dollar "flood-control" phase of TARP. At least three giant, pitlike reservoirs had to be built to hold the overflows from a larger network of deep tunnels twice as long under the Chicago, Des Plaines, and Calumet Rivers.[50]

Like Chicago's Sanitary and Ship Canal, this big-technology project became a wellspring of innovation in mining and excavating techniques. Rather than using the traditional method of blasting a hole through layers of rock, the tunnel makers built a huge, circular boring machine, called the "mole." Its spinning blades ground inexorably forward through the limestone, leaving pulverized rubble behind. And rather than use a traditional, underground railroad to haul it away, the building contractor adopted the grocery-store conveyor belt to produce a significant savings. At times, the moving highway of rubble stretched more than a mile through the tunnels and up the shafts to the surface, where it was sold for use in other construction and landfill projects.[51]

By the time 47 miles/76 km of the Deep Tunnel had been completed in May 1985, Dalton—now general superintendent of an expanded Metropolitan Sanitary District of Greater Chicago—had to recalculate the costs and benefits of his plan. He now put a price tag on flood damages at $43.2 million a year. He also raised construction costs for phase one of TARP to $2.2 billion, or twice his previous estimate. Dalton predicted that this underflow system would eliminate rainfall-induced overflows of raw sewage into the waterways 85 percent of the time. Moreover, it would put 70 percent of the district's homeowners out of harm's way of flood damages from a hundred-year rainstorm. For another $2 billion to complete phase two of the 110-mile/177 km project, he promised, the other 30 percent would also be protected against rivers overflowing and sewers backflowing into their basements.[52]

Belying these optimistic expectations were critical experts and damaging floods. Presenting a different set of figures, Scott Bernstein of the Center for Neighborhood Technology shrunk Dalton's estimated improvement of water quality in the rivers from 85 to 35 percent. In part, this reduction was the result of the practical problem of more than fifty suburbs being unable to connect their sewer systems to the Deep Tunnel. More troubling for the sanitary district was the failure of TARP to stop ruining basements and closing beaches with bacteria-laden floodwaters.

In fact, the problem of storm-runoff surges only got worse during the next two decades of steady expansion of the metropolitan area. Only three months after the Deep Tunnel opened, a thunderstorm forced the sanitary district to dump 58 million gal./220 million L of polluted river water into the lake. Two months later, on November 20, 1985, a 2 in./5 cm downpour caused widespread flood damages. Climate change made that month the wettest November on record, with 7.64 in./19.40 cm of precipitation, causing some homes along Salt Creek in the western suburbs to sink underwater four times. As the MSD's ability to dig more tunnels and reservoirs fell farther and farther behind the pace of sprawl-induced flooding, its big-technology strategy of water management would come into the political arena of civic discourse for the first time.[53]

Save Our Taxpayers: The Contestation of Green Technology

For Chicago's taxpayers, the price of machine politics was fraudulent underestimates and cost overruns. The sheer scale of the MSD's public-works project in terms of mounting multi-billion-dollar expenditures drew the attention of Congress's fiscal watchdog, the General Accounting Office (GAO). And well-founded suspicions of corruption by Illinois's Republican Senator, Charles Percy, set in motion an independent audit of TARP. After only three years of TARP's construction, in 1978, the GAO issued the first of three reports stating that the funding of this sanitary strategy of pollution control and flood control was unsustainable. The experts uncovered a set of cooked books: "We found that this analysis . . . contained numerous inaccuracies, omissions, and unsupported assumptions, which resulted in overstated benefits and questionable costs."[54]

When all the actual costs were added to the MSD's estimates, the price tag for phase one of TARP ballooned from $2.6 billion to $10.2 billion. For example, the sanitary district included neither the interest payments on the bonded debt nor the inflation rates of wages and materials over the long run.

Its estimates also excluded the hundreds of millions of dollars that local taxpayers would have to pay to connect their sewers to the Deep Tunnel. For each household within the MSD, the total bill would add up to $5,750. "The completion of all segments of the program is in doubt," the first report concluded, "because of delays, high and escalating costs, and funding uncertainty. Partial completion will cost a significant amount without realizing all benefits of the installed facilities."[55] In other words, Chicago's sewage-laden storm water would continue to pollute Lake Michigan into the indefinite future.

The two follow-up reports on the flood-control, phase two of TARP not only confirmed the accountants' figures but also laid out an alternative path toward less-expensive, eco-friendly small technologies. A year later, in 1979, the GAO came to the bottom line of the inherent flaw in Chicago's infrastructure design: a combined sewerage and drainage system. The report called this approach to water management an "elusive goal" in a flood-prone environment. It blamed sprawl at the periphery and out-of-date infrastructure in the center for raising the amount of damages to new levels. "However, as we noted," it continued, "flooding in the Chicago area caused an estimated $71 to $102 million in damage to single-family homes [alone] during the last 5 years—and untold inconvenience and disgust." Another similar study estimated that the sanitary district's two hundred suburbs suffered an average of $22 million each year in flood damages, while the city center was at risk of a $100-million loss from a hundred-year rainfall.[56]

The GAO contrasted the Chicago model's soaring costs and completion date stretching into an unforeseeable future with a list of low-cost, green options. Calling their approach "best management practice," the outside experts reasoned that nonstructural flood-control measures should be exhausted before considering whether to employ big technologies. Their report sorted methods of control into three categories: source, collection, and treatment. For example, source-control methods to reduce storm-runoff surges echoed the 1978 list of small-scale technologies outlined by Northwestern University's Hallett. It included turning parking lots into temporary retention ponds, disconnecting roof downspouts, cleaning streets and flushing sewers, and laying porous pavement. Enacting ordinances prohibiting building on flood-prone areas would also decrease the risk of damages. Collection methods involved managing riverine environments as envirotechnical systems of flows. Treatment controls of floodwater ranged from passive settling tanks to various high-speed devices for filtering out solid wastes. Although all of these best-management practices had proven successful, the GAO found that many local governments resisted implementing them.[57]

The GAO's researchers met a wall of resistance from engineers and consultants, such as those working on Chicago's TARP, to their alternative, nonstructural approach to flood control. Bias toward the technological sublime of large-scale projects was a part of our national culture. In contrast, sewer pipes and storm drains were best left out of sight and out of mind. The sanitation bureaucracy was responsible for keeping alive the myth of big technology, according the GAO report. "One of the obstacles to less costly approaches," it carped, "is the influence of architectural and engineering firms. Most communities are not able to design sewage collection projects without assistance from firms that have experience in this area." At the same time, the outside experts conceded that their alternative methods of calculating benefits and costs meant trade-offs between reduced tax bills and greater flood damages.[58]

During the construction of the Deep Tunnel from 1975 to 1985, the MSD's role in the degradation of the urban environment grew in proportion as industry's contribution to the pollution of the river and the lake decreased. Long-term deindustrialization accounted for much of the improvement in the river corridor. While some companies went out of business, others moved away. The state of Illinois lost one out of four of its manufacturing jobs, most within the city of Chicago. An estimated 160,000 positions vanished, affecting 640,000 people. The number of industrial polluters in Chicago's rivers had already plunged since 1965 from 150 to 35 companies.[59]

During this building phase, the chronic postwar loss of industries polluting the waters was accelerated for two independently acting sources of change. The global recession following the oil embargo of 1973–1974 slowed the local economy and threw many of its companies using petro-chemicals into bankruptcy. The federal enforcement of the Clean Water Act also brought about a steady reduction of industrial effluents in the Chicago area. The new law gave the equally new EPA expanded powers of pollution control over the nation's "navigable waterways," including the Chicago River. In 1960, the U.S. Army Corps of Engineers had already won a landmark decision against big business in the Supreme Court case of *U.S. v. Republic Steel Company.*

The nine-year-old case involved the Calumet River. Under the 1899 "Refuge Act," the Corps had been given the authority to remove obstructions to navigation. Its lawyers convinced the justices that the mill's suspended solid wastes were causing underwater shoaling that blocked traffic on the waterway. Their decision meant that the company had to bear the costs of either endlessly paying the Corps to dredge the river or installing equipment to remove the particles from its discharges into the river. "The Refuse Act," according to legal historian Albert Cowdrey, "changed [after the 1960 ruling], becoming, first an antipollution law and, then, a decade later [after the Clean Water Act], a full-scale program of environmental control for the nation's waterways."[60]

The new law made a difference, because it expanded the jurisdiction of the federal and the state governments over the waterways within the boundaries of Chicago's MSD. In a topsy-turvy court case in 1979, the Illinois Department of Conservation sued the U.S. EPA and the Corps for violating the Clean Water Act. In another landmark decision, the presiding federal judge ruled that thirty of the state's rivers qualified for the Clean Water Act's safeguards against pollution, including all six basins in the Chicago area. As Cowdrey indicates, this legislation marked a turning point in these agencies' roles, transforming guardians of economic interests to stewards of natural resources. They were already engaged in fighting the steel mills, oil refineries, and chemical companies in the southern, heavy-industry districts of the Chicago metropolitan region, but the judge's ruling enlarged their jurisdiction for the first time over the MSD and its pollution of the Chicago River. Until this turning point in federal involvement, the agency's long-standing status as a cog in Mayor Daley's patronage machine had gone unchallenged since the removal of Superintendent Bacon.

Although uncovering corruption is difficult, telltale signs of wrongdoing kept rising to the surface of public exposure with painful regularity in the case of Chicago's MSD. Only a month after receiving the single-largest grant ever made by the U.S. EPA for the Deep Tunnel, for example, a scandal exploded in the faces of the agency's trustees. At its regular board meeting in August 1976, one board member accused another, Valentine Janicki, of breaking the rules. Already under criminal indictment for getting kickbacks from a sludge-hauling contractor, Janicki had hired a crony named Napoleon Zbyszewski and paid him $10,000 without board approval to run a commemorative-stamp contest that would celebrate the seventy-fifth anniversary of the big-technology, river-in-reverse project.

In his defense, Janicki asserted that the Chicago Association of Commerce and Industry would co-sponsor the design competition and pay all of its expenses, including $50,000 in prizes. But at the meeting, he admitted that his public-relations scheme was in trouble. On the one hand, the city's most outspoken lobby on behalf of maintaining the river as an economic resource had raised only $4,000. On the other, Zbyszewski had billed the MSD at least that amount for six thousand mailings that had failed to enlist a single additional sponsor. At best, he could claim to have collected a few hundred dollars from about seventy contest entries as compensation for his ongoing consulting fee of $350 per week.[61]

Rather than renounce Janicki's illegal actions, however, the board chose to reenact the three-part play of the city's machine politics: corruption exposed, investigated, and rolled back. It voted seven to one to pay the bills and authorize the stamp contest. This version of the play opened a window on the inner workings of machine politics in Chicago. The trustees gave the responsibility of drafting the authorizing bill to the Law Department, which duly inserted clauses requiring a majority vote by the board for any MSD expenditures and "joint sponsorship" with the businessmen's association. But at the last minute, someone in Janicki's office deleted this latter phrase from the bill, leaving the agency solely responsible for paying for the publicity stunt. The board unanimously, perhaps unwittingly, passed the altered bill. Two years later, in 1978, the stamp contest was still in limbo without a winner. Desperate for extra-local funding of its big-technology projects during a period of economic hardship, the MSD remained mired in a public-relations nightmare of insider graft and crony capitalism.[62]

Local perceptions of the MSD as an opponent of environmental reform were reinforced by its failure to clean up the tons of debris thrown into the river. Especially from the civil disorders of 1968 until the economic recovery of the late 1970s, politicians and policy makers gave the reclamation of the Chicago River low priority. Neither the Corps nor the local authorities would take responsibility for protecting this urban corridor, let alone removing the trash thrown on its banks and lying on its bottom.[63] This semiofficial blind eye, in turn, strengthened perceptions of the Chicago River as the city's garbage dump.

The impressions of a journalist on a boat ride on the North Branch in 1971 are instructive on the river's physical appearance. "Cruising down the river can be plenty of fun," he exclaimed in sardonic disbelief, "if you don't mind dodging picnic tables and seeing some of the worst cases of pollution and neglect in the Midwest." On his journey, he also observed:

> The piling on the east side of the river had a heavy coat of oil and there were oil slicks along the water, trapping branches and logs, which drifted from the north. . . . The cement company was washing out several trucks and the residue was pouring into the river . . . a clear violation of the antipollution ordinance. . . . [Then] we saw two automobiles half in and half out of the water. One had been there for a year.[64]

For the remainder of the decade, scientists in the Chicago metropolitan region studied rivers in decline, the victims of official neglect and abuse. Their condition had deteriorated to a point where fish biologists declared Daley's dream "highly improbable." State Conservation Supervisor Bill Harth found "the Chicago River pollution problem is a tough one. We know it's bad just by looking at it."[65] A year later, in 1974, another newspaper reporter found an "oozing river . . . so badly polluted that its bottom festers with sludge worms and leeches."[66] Raw sewage overflowed into it whenever 0.25 in./0.64 cm or more of rain fell, coating its bottom. In 1981, the U.S. Army Corps of Engineers reported that this sedimentation in a 5-mile/8 km stretch of the Main and North Branches was heavily contaminated with cancer-causing polychlorinated biphenyls. Despite the river's water reaching three times the danger level, a U.S. EPA official proclaimed that this toxic sludge posed "no threat" to the public health.[67] Two years later, Illinois EPA experts announced that the water quality along 2,072 miles/3,335 km of the state's waterways had improved since 1972, but 20 miles/32 km had gotten worse, all in the Chicago area.[68]

Conclusions: Save "Our Friendless River": The Revival of Wild Nature

Public policies of negligence and the degradation of Chicago's rivers meant that any counternarratives of restoration would have to come from private sources. Until the formation of the FOTR in 1979, their activism and demonstration projects were cast in the media as forlorn voices protesting the official vision. In contrast, the new group made an immediate contribution by putting a positive face on the cause of environmental reform. Its members succeeded in lobbying city hall to create a small, demonstration park on the North Branch as a tourist attraction of urban nature.

Without doubt, the Chicago River always had a few "friends" who could envision a recreational corridor of natural beauty flowing through the heart of the city. These early prophets of urban wilderness came from every social class. In 1958, Jordan Clark had started his own, less-formal flower and vegetable garden on the river's edge of the Illinois Central yards, just to the east of Michigan Boulevard. Interviewed in 1965, the African American railroad repairman recalled, "It was so pretty that I planted another patch the next year." He kept expanding, according to the reporter, until the "entire bank along the back of the oiler's shed was blooming with fragrant color."[69]

Another kind of citizen activists—cleanup volunteers—helped highlight the plight of the Chicago River as an orphan of official neglect. In 1973, the "River Rat Society" enlisted packs of Boy and Girl Scouts to reclaim its natural beauty. This children's crusade to "Keep Our Waterways Clean" faced a Herculean challenge. Two years later, for instance, three thousand volunteers hauled away almost 160,000 lb./72,575 kg during a one-day outing along a 7-mile/11 km stretch of the North Branch. These efforts paid off, because wild flowers began taking root in succeeding years: in 1984, a naturalist would find patches of mullein, pineapple weed, motherworts, and daisy fleabanes growing along the banks.[70]

The August 1979 issue of the monthly magazine *Chicago* represents the kickoff of a river-restoration campaign equivalent to the daily newspaper's earlier "Save Our Lake" crusade. Under a lede of "Our Friendless River," this equally conservative publication of an affluent Consumer's Republic made a similar appeal to "reclaim our lost resource." Spurring the organization of the FOTR, the magazine included an "Action Plan." Written by city planner Robert Cassidy, it highlighted the benefits of riverfront improvements in other American cities, such as Denver, San Antonio, and St. Louis. In a follow-up editorial of support, the *Tribune* chided, "The Chicago River is awaiting discovery by a mayor or a county board chairman or a Sanitary District board chairman."[71]

For a brief, fortuitous moment, the river seemed to have found such a champion in the incoming administration of Mayor Jane M. Byrne. The feisty party loyalist had pulled off a "stunning upset" in the crucial primary election of the Democrats, according to Chicago historian Melvin G. Holli. The former department head of Consumer Affairs, Byrne got her revenge on the man who had fired her, the incumbent Michael A. Bilandic. But this self-proclaimed political reformer then executed a complete about-face straight into the arms of the "cabal of evil men" in the city council, whom she had demonized in the campaign. But hamstrung by fiscal restraints in the wake of the second oil-embargo crisis of 1979–1980, the new mayor could do little more than take advantage of her consummate skills in "street theater" to promote her vision of the city beautiful.[72]

In August 1980, Mayor Byrne declared the city's first "River Day." Sponsored by the FOTR, it provided her a platform to present a bold, if undefined, plan to "restore and rejuvenate" the waterway corridor. Like the new lobby group, the mayor accepted its back-alley definition as a "working urban river." Building on these perceptions, she asserted, "We must see that the Chicago River again becomes a thriving part of the local economy."[73] To fulfill these promises, Mayor Byrne spearheaded the passage of an updated riverfront-development ordinance and coordinated the creation of a new North Avenue Turning Basin Park on the North Branch. This was the first site of six FOTR proposals for the beautification of open space along the riverbanks.

Yet like so many public plans for the reclamation of the river, even a mini-park project with the backing of the mayor languished for years inside the MSD. In fact, the agency assiduously avoided drawing up any master plans of land use for its river corridors. Furthermore, it shut the door in the face of Holabird and Root and other highly respected city-planning firms, which offered their services free of charge. The trustees dithered for two years before funding a plan. Mayor Byrne would be out of office for more than a year before the first shovel of dirt would officially initiate the actual building of the park.[74]

Like early green activists, a few architects and planners had always incorporated the Chicago River in their blueprints of the downtown's transformation into a postindustrial showcase of international business and global tourism. Bertrand Goldberg was the most zealous spokesman of these visionary advocates of a river-centered urban renaissance. In the 1960s, he had designed the iconic Marina City (1959–1964), the twin "corncob" apartment towers on the northern edge of the CBD. Despite the stinky smells and floating feces, the architect had included a boat launch and restaurant with a river view. In 1977, he presented a proposal for a much-more-ambitious "River

City," which was also designed as a self-contained "mini-neighborhood." Located on the southwestern edge of the CBD, this multi-billion-dollar megaproject was designed to house eleven thousand people in addition to providing them a full range of services and amenities.[75]

But Goldberg was no dreamer when it came to Chicago's hardball politics of urban development. Ridiculing the current use of the river as a "toilet," he called city hall to account for its semiofficial negligence. "Developers and planners," Goldberg implored at the ground-breaking ceremonies for River City in 1984, "should treat the river like a street front. Instead, they treat it like a back alley. If the city regarded the river's banks the way it regards its beachfront, it would budget money to protect them." In fact, Mayor Byrne's riverfront, land-use ordinance had begun to move city hall in that direction. "If it is going to get the builders to pay attention," he suggested, "it has to zone the river like a street."[76]

With the long-anticipated opening of the Deep Tunnel in sight, perceptions of the river were set to undergo a major change for the better in the public imagination. Poised at the beginning of a sustained building boom, increasing numbers of developers joined Goldberg in looking at the river as a marketable, natural amenity. The two-thousand-strong FOTR represented another growing lobby in favor of its improvement. In the long battle to reclaim the Chicago River, the early green activists and visionary prophets had made a difference. Their example had established a solid foundation of stakeholder support for the emergence of a full-fledged movement of environmental reform. From garden patches and river cleanups, it would gain momentum and move up the federal system to find defenders against its worst enemy: city hall.[77]

Yet the sea change in public perceptions of the Chicago River owes most to its dramatic ecological metamorphosis after May 1985, when the MSD started operating the Deep Tunnel and stopped using chemical disinfectants. The 46-mile/76 km underflow system reduced daily overflows of raw sewage into Chicago's waterways by about 85 percent. As the water became less contaminated with bacteria and no longer poisoned with chlorine, oxygen levels began rising high enough to sustain aquatic life. And with the revival of plants and fish in the river, other forms of wildlife began returning to what had once been natural habitats along its banks.

The restoration of wild nature in the city, in turn, recast the river in a more-positive light and reinforced the ranks of reform activists, who had a stake in reclaiming it for recreational use. Drawn to it were a base constituency of sports fishermen and fisherwomen, bird-watchers, boaters, and other outdoor enthusiasts. It also began winning new groups of stakeholders in the city center, including office workers on their lunch breaks, apartment dwellers on their balconies, and out-of-towners on their tour boats, looking up at the renaissance of architecture in an expanding downtown.[78]

Nonetheless, the local sanitary authority in charge of pollution control of the Chicago River continued to contaminate it with the equivalent amount of raw sewage produced by a city with 150,000 inhabitants. The MSD refused to bear the cost of employing existing technologies at its treatment plants, which could remove 98 to 99 percent of the impurities. Instead, they were operating at a 90-percent level of treatment, constantly pouring dangerous microbes into the river and blanketing its bottom with toxic sludge.

From 1985 to 2011, the remarkable reassembly of the river's biotic community boosted the mobilization of support for further enhancing its natural habitat. Of course, this reciprocal process of interaction between physical appearances and psychological attitudes was also driven by human agency. Impersonal structural shifts in economic and demographic patterns of settlement were causing a centrifugal "multicentering" of jobs and homes throughout the metropolitan region and a refocusing of urban culture and leisure activities in the cosmopolitan downtown.

Individual choices made by investors, developers, tenants, and buyers also played a role in reclaiming the river for recreational use. As this constituency of stakeholders grew rapidly in the 1990s, it came into conflict with the MSD over its current policies and future plans. At the same time, the state and the federal governments were becoming more fully engaged in enforcing their pollution-control laws. With increasing frequency, their EPAs clashed with city hall over local water-management strategies. A quarter century after the passage of the Clean Water Act, the Chicago River finally began to receive its protective battery of defenses against pollution.[79]

6

The Rebirth of Urban Nature, 1985–2011

Regenerating Chicago and Its Aquatic Ecology

Since the opening of the Deep Tunnel in 1985, the pace of climate change and suburban sprawl has outrun efforts to eliminate flood damages and river backflows into the lake. About every five years, clusters of so-called hundred-year rainfalls exceeded the severity of the previous one, creating a "Groundhog Day" scenario of "the most disastrous flood ever."[1] And as in this fantasy film of a revolving time loop, Chicago's sanitary authorities repeated their demand over and over again for more big technology. Crisis-driven decision making in response to these extreme weather events proved to be pivotal in directing the course of public policy for the next quarter century. For our purposes, they are defined as rainfalls great enough to force the sanitary authorities to release river water into Lake Michigan.

The failure of the Tunnel and Reservoir Plan (TARP) to protect Chicagoans from catastrophic damages engendered a political-protest movement based on new ideas about the aquatic ecology of their flood-prone environment. Beginning with the great flood of October 1986, severe weather events repeatedly caused property losses and psychological suffering on a widespread scale throughout the metropolitan region. These disasters created a receptive audience for advocates of alternative, nonstructural approaches to water management. At first, public discourse about the record-breaking rainfalls produced knee-jerk denials by politicians, who engaged in pointing their fingers at some other government agency or upstream community. But over the course of fourteen more hundred-year rainfalls in the next fifteen years, grassroots organizations gained support that framed issues of flood control in terms of the environmental protection of the region's six watersheds.

Without doubt, the success of the Deep Tunnel in improving water quality in the rivers during ordinary rainfalls of less than 1 inch (in.)/2.5 centimeters (cm) accounts for the regeneration of their aquatic ecosystems. And this rebirth of nature, in turn, fostered a growing constituency in favor of their complete reclamation up to the standards of the Clean Water Act of 1972. During the building boom of the 1990s, the Chicago River became the new center of an expanding downtown area. Thousands of riverfront residents, office workers, and tourists used it daily as a recreational amenity. By the end of the twentieth century, the battle to save the rivers came to a showdown against the last major source of intransigence against coming into compliance with the law: city hall.

While the storms of August 1985 raised questions about the capacity of the Deep Tunnel to contain heavy rainfalls, the downpours of September 1986 left no doubts about its shortcomings. During the last two weeks of the month, precipitation totaling 14 to 16 in./36 to 41 cm fell on a prairie wetland, providing the preconditions for flooding by saturating the ground and raising river levels. When cloudbursts struck on October 1 and 4, the entire metropolitan area was inundated. The western suburbs along Salt Creek and the Des Plaines River suffered the worst flooding, sinking four thousand homes and causing $35 million to $40 million in damages. After the Chicago River rose 5 feet (ft.)/1.7 meters (m) above normal, the sanitary authorities had to open the gates three times, discharging 50 million gallons (gal.)/189 million liters (L) of disease-laden wastewater into the lake. The need to close the beaches to swimming to prevent an epidemic further widened the impact of the not-so-unusual-anymore rainy weather.[2]

The flood disaster of 1986 was a model of crisis management by the politicians and policy makers. The president declared Chicago a disaster area; the governor sent in the National Guard to relieve exhausted first responders; and the former Metropolitan Sanitary District, now renamed the Metropolitan Water Reclamation District (MWRD), doubled down on the TARP. For two years preceding the flood-induced crisis, its plan had faced rejection after rejection by the federal government's accountants. But after the flood, the politicians overrode the bureaucrats to begin construction of the second, "flood-control phase" of the big-technology plan. It called for doubling the length of the tunnels to 110 miles/117 kilometers (km) and digging three giant overflow pits at an additional cost of $2.7 billion. By the end of that rain-soaked week in October, a deal had been worked out in Washington for a new funding formula for the first stage of the eight-year plan. In exchange for sweetening the state and the local governments' share of the costs from 25 to

35 percent, Congress authorized $55 million to start building the underground infrastructure to hold another 1 billion gal./3.8 billion L of storm-water overflows.[3]

Like the early grassroots activists on behalf of the riverine environment, the few voices protesting big-technology approaches to water management went largely unheeded, until Chicagoans suffered through two more severe flood clusters. In November 1986, a month after the latest episode of the "most disastrous flood ever," the *Chicago Tribune*'s "environment writer" interviewed several experts on alternative flood-control strategies. Significantly, none was from the MWRD. David Hunter, from the U.S. Army Corps of Engineers, drew a historical dividing line marked by the publication of Rachel Carson's *Silent Spring*. The traditional "structural" approach was to refashion nature into an envirotechnical machine of flows.[4]

Since the publication of Carson's book, however, Hunter had witnessed the alternative, "nonstructural" approach of designing with nature gain support within every level of government. For example, Donald Hey, the head of an environmental-planning firm, was helping suburban Gurnee turn a 3-mile/5 km stretch of the Des Plaines River back into "a marshy wetlands—the kind of swampy stream that settlers encountered in the 1860s." In another suburban initiative, the mayor of Skokie was installing "rain-blockers" on the street drains, which slowed the rate of runoff entering the sewers. The streets themselves served as temporary, storage reservoirs. Without having to add public infrastructure, he was taking a further, eco-friendly step to prevent overflows of the sewer system. He was ordering homeowners to disconnect their downspouts from it and to let their lawns soak up the storm water.[5]

Private philanthropy sponsored the most comprehensive project of environmental protection, a 1,300-acre/526-hectare (ha) patch of prairie wetlands near Grayslake in Lake County. Gaylord Donnelley led a fifteen-year court battle to save it from being built over with thousands of houses. His family's fortunes had risen with Montgomery Ward and Sears, churning out their mail-order catalogs to become one of the nation's largest printers of popular magazines and periodicals. After defeating the homebuilders in 1986, he helped create a master plan "to preserve open land by developing it."[6] Called Prairie Crossing, the project was designed to nestle three hundred homes on 30 percent of the site, which was reengineered into an organic farm and nature preserve. Its developers were dedicated to increasing biodiversity, habitat restoration, and storm-water management. Its Sanctuary Pond, for example, served its namesake for endangered species of fish, while its meandering marshes retained and cleaned storm-water runoff. For a privileged few, Prairie Crossing offered a place to live with nature.[7]

But for the vast majority, the amazing rebirth of the Chicago River following city hall's heralded opening of the main stem of the Deep Tunnel was a convincing reaffirmation of belief in the big-technology fix. Foremost among the converts to the cause of saving the river with TARP were the big investors in downtown commercial and residential real estate. Since the early days of the Richard J. Daley administration, they had envisioned an all-inclusive makeover of the central business district (CBD), ringed by the construction of a high-security, *cordon sanitaire* of luxury apartment towers. In 1974, the topping off of the world's tallest skyscraper, the Sears (now Willis) Tower, marked the symbolic completion of downtown's transformation into a world-class business center.[8]

After a five-year recession lull, Bertrand Goldberg's River City announced the beginning of a new wave of construction frenzy in the city center. His project fit into the master plan by helping fill remaining gaps in the ring of "defensible space" around the CBD. Its southwestern and northwestern flanks on both sides of the Chicago River were lined with obsolete relics of an industrial past. These corridors of light and water became especially attractive sites for redevelopment. By the late 1980s, at least thirteen major projects, including ten thousand residential units, were coming off the drawing boards and taking material form. They ranged from the erection of the mega-scale "City-front Center" at the mouth of the Main Branch to the conversion of the Montgomery Ward warehouses into a mini-neighborhood on the North Branch. And on the South Branch below River City, developers were putting the final touches on a blueprint for a mixed-use "Chinatown Square" of commercial offices, retail shops, premium townhouses, and senior-citizen apartments.[9]

The period from 1998 until 2011 was characterized by the resistance of Mayor Richard M. Daley to bringing Chicago into compliance with the standards set in the Clean Water Act. Pressure mounted from the grassroots as well as the state and the federal Environmental Protection Agencies (EPAs). The mayor stubbornly held on to the old division between the lakefront as a showcase of the city beautiful and the riverfront as a plumbing fixture hidden out of sight and out of mind. Refusing to bear the costs of fully disinfecting its sewage, Chicago became the only big city in the nation holding out against upgrading its waterways to recreational use.[10]

Moreover, the sanitary district became the single worst polluter of Lake Michigan. Suburban sprawl continued its relentless paving over of the land, which caused ever-greater, faster, and more-frequent runoff surges during thunderstorms. The $3.5-billion-and-counting, still incomplete tunnel and reservoir system filled; the overflowing river threatened to flood the city center; and the MWRD was forced to open the locks, discharging billions of gallons of pollution into the lake. The beaches were closed to swimming, leaving frustrated sun worshipers to bake on the sand without relief.

Despite the willing submission to the EPAs by the incoming administration of Mayor Rahm Emanuel in 2011, invasive species on both sides of the barrier between the rivers and the lake have further complicated the interrelated problems of flood control and pollution control in the Chicago area.

Consider the Asian carp, a "monster fish" that eats and out-eats all of its rivals. Set loose in the Mississippi River, it threatens to enter the Great Lakes and destroy its thriving sports-fishing industry. At the same time, the explosive growth of zebra mussels in Lake Michigan and the Chicago River is also upsetting the balance of nature in unpredictable ways, as the alewives did thirty years earlier. These unintended, dire consequences of engineering the environment have finally forced policy makers to start planning in holistic ways that treat the two bodies of water as a single, interdependent ecosystem.

Renewing the Renaissance of the City Center

"August 1987," the *Chicago Tribune* reported on the August 26, "became the wettest month in Chicago's history."[11] In this case, the hundred-year benchmark applied, because the previous maximum amount of precipitation during the month of August—11.3 in./28.7 cm—had occurred during the historic, great flood of 1885. Two days after breaking the record, another 3 in./7.6 cm deluge brought the monthly total up to 17 in./43.2 cm. The historic average for August was 3.5 in./8.9 cm. The previous monthly record had been 14.2 in./36 cm of rain during September 1961. The amount of river water the sanitary authorities had to release into the lake also set a new record, at 1.98 billion gal./7.48 billion L. The beaches were closed. The frequency and severity of recent floods led scientists to connect the dots of climate change. The Illinois Division of Water Resource's French Wetmore explained to residents of Park Ridge how average rainfall amounts had been increasing over the past decade. Attending an "After the Flood Open House" at the local high school, the residents formed a receptive audience, eager to learn how to adapt to living on a prairie wetland during a wet period.[12]

On August 13–14, 1987, Park Ridge and the communities in the western suburbs lying in the watersheds of Salt Creek and the Des Plaines River again suffered the most flood damages. Extreme weather conditions produced "the heaviest rainfall in Chicago history [in a twenty-four-hour period]": 9.5 in./24.1 cm. Hot, humid air from the Gulf of Mexico and the Atlantic collided with a jet stream–driven cold front from the north, setting up a stalled condition of one thunderstorm after the next. At least four deaths from drownings and electrocutions were attributed to this natural disaster. Estimates of the cost of the damages to sixteen thousand homes and businesses in addition to public property ranged from $100 million to $162 million. The flooding also caused widespread turmoil, including power outages to forty-two thousand customers, no phones for another thirty thousand, closed expressways for tens of thousands of commuters, and the sinking of the MWRD's treatment plant on the North Side under 20 ft./6 m of sewage. The storm runoff surge stranded the passengers arriving at O'Hare Airport, which became surrounded by a moat. With no one able to get in or out of the virtually shut-down transportation hub, the marooned travelers were forced to sleep overnight on the floor.[13]

Although emotional distress cannot be quantified, the stories told by flood victims to newspaper reporters help convey a sense of the pain felt from the destruction of their homes and the disruption of their daily lives. The disaster wiped out the savings of thousands of property owners. President Ronald Reagan's immediate designation of Cook and DuPage Counties as a disaster area brought them only limited help in the form of low-interest repair loans. In Riverside Lawn, journalists interviewed residents, including sixty-five-year-old Dorothy Halac. She had just recovered from the disaster of the previous October, when the Des Plaines River overflowed its banks and put her home under 4 ft./1.2 m of water. "If we seem calm," she revealed, "it is because we are cried out, we have no more tears."[14] An upstream neighbor from Wood Dale expressed an equally fatalistic attitude. He told the reporter that his home had been damaged fifteen times from storm-water overflows of the Salt Creek since he had bought it fifteen years earlier. A *Chicago Tribune* editorial used sarcasm to express citizens' feelings of helplessness against the rising waters. "There is not much that can be done about 100-year rains," it stated, "which seem to come along about one a year now."[15] The newspaper seemed to be responding to Governor James Thompson's conflating of a flood-prone environment and flood damages as an act of God that "wouldn't happen again for another 200 years."[16]

The governor's finger-pointing and hyperbole were typical of the defensive reaction of politicians and policy makers to constituents as their frustration turned to anger. In a few cases, neighbors turned on each other. Halac and the other forty residents of Riverside Lawn sued Riverside and Brookfield to force them to pay for a dike to protect their low-lying homesites. But more often, a better-informed public held its elected officials to account for the flood damages. People blamed them for inaction in installing flood-control infrastructure since the October 1985 disaster as well as for their lack of preparation to help people during the emergency. Local officials blamed adjacent suburbs for permitting real-estate developers to pave over their floodplains and for not enforcing state prohibitions on rebuilding structures that had suffered flood damages totaling more than 50 percent of their value.[17]

Although the governor ducked for cover behind vague reassurances of help from the state government, the sanitary district exploited the disaster in its publicity campaign as a reason to support TARP. Just five days before the 9 in. plus/23 cm plus cloudburst, in fact, General Superintendent Frank Dalton had had to defend his big-technology project against new EPA charges of negligence in the on-the-job deaths of ten workers. Unless the project was completed, the engineer threatened, "people are going to see evidence of a severe problem; they are going to see it in their basements." He claimed that the Deep Tunnel was already protecting sixty-three thousand homes; he appealed for funds to safeguard the rest.[18] Two days after the downpour, the sanitary district's commissioner, Aurelia Pucinski, boasted, "The tunnel did what it was suppose[d] to do." She admitted that its success in storing 1 billion gal./3.8 billion L of polluted storm water did not stop sewer backflows into thousands of basements in the city and the suburbs. Yet Pucinski, like Dalton, professed that TARP was the only way to defend Chicago against a wet period of climate change.[19]

At the same time, the failure of the Deep Tunnel to prevent discharges of storm water contaminated with sewage into Lake Michigan on nine occasions during its first five years of operation raised new questions about its costs and benefits. For example, the Citizens for a Better Environment's research director, Robert Ginsburg, contested the sanitary district's big-technology approach. He cited figures of the river reversals into the lake, which would total 3.3 billion gal./12.5 billion L during this cluster of climate-change events. He asked how long it would take a deficit-ridden federal government, if ever, to find $2 billion to pay for its share to complete the flood-control phase of the plan. "Meanwhile," Ginsburg complained, "Chicago remains vulnerable to floods. I think we are going to have to look for other ways to deal with the problem."[20] But even the renewed flooding at the end of the record-breaking month of August did not turn the sanitary authority's single-minded focus of attention away from its giant-scale public-works project.

By the time of the hundred-year flood of 1987, the building boom had turned the riverfront in the CBD into a speculative land rush. A planner from the elite businessmen's Chicago Central Area Committee only a few years earlier had termed city hall's policy of neglect as "laissez-faire"; the same circumstances were now considered to be unacceptable chaos. With support from the Friends of the River (FOTR), the influential group drafted a riverfront protection ordinance, which the incoming administration of Mayor Daley duly ushered through the city council. "By following these guidelines," he trumpeted in March 1990, "we will transform our river into the centerpiece for an expanded downtown . . . a beautiful tourist attraction and recreational development." Yet following local traditions of "talk but don't act," almost a decade would pass before he budgeted any city money to help achieve this goal. In the meantime, the new regulations gave private planners a relatively free hand to design the required public-access spaces.[21]

The main beneficiaries of their work were the inhabitants of the brand-new offices, stores, homes, entertainment venues, and light-manufacturing facilities. These urban pioneers represented a unique wave-in-reverse of white flight from the city. On the contrary, an exodus from the boredom of the suburbs was underway among a bifocal generation of older "empty-nesters" and younger white-collar singles and couples without children. Their attraction to the river covered a broad range of attitudes toward the environment. For some cliff dwellers, a beautiful river view served as a status symbol, a three-dimensional wallpaper for their glamorous lives. For other denizens of these tower blocks, the riverfront walkways and open plazas offered an oasis of healing, an escape outdoors from the surreal interiors of their workaday lives.

Yet others sought to find a hybrid, urban nature in the river's industrial archeology of rust. In the spring of 1985, for example, a sanitary district trustee and founding member of the FOTR, Joanne Alter, stood at the northern tip of Goose Island, on parkland she helped create. "Look at this view," she declared, pointing toward the North Avenue Turning Basin. "Isn't it spectacular? It's amazing how scrap metal can be so interesting." In the coming years, in fact, the avant-garde would portray these abandoned industrial sites as pop-art landscapes of urban doom.[22]

And finally, growing numbers of recreational users looked at the recovering river as a generator of wild nature in the city. Five years after the river's rebirth, the U.S. EPA upgraded its status from "toxic" to "polluted." Biologists found plant life on the bottom of about a quarter of the North Branch, where none had been found a decade earlier. And under the surface of the water, where only sludge worms, leeches, and carp could survive, now there were at least seventeen species of fish, including bass, perch, minnows, and bullheads. Water birds also returned and began nesting along the river's banks, adding to the diversity of its ecosystem. A year later, in the spring of 1991, beavers would be spotted swimming not only in the restored Des Plaines wetlands but also in the South Branch on the edge of the CBD.[23]

During the remainder of the decade, in turn, this self-reinforcing process of biotic regeneration had an accumulative effect in changing more and more people's attitudes toward the urban environment. Their perceptions of the Chicago River underwent a decisive shift from an industrial corridor to a recreational parkway. Unfortunately, this kind of attitudinal change of mind cannot be measured with precision.

Although anecdotal, one of the many stories of grassroots activism helps expose how this reciprocal process of environmental and attitudinal metamorphosis worked to build a political movement to save the river. At the FOTR's fourth annual "Great Chicago River Rescue Day" on June 2, 1996, its organizer, Joyce Peralta, noted, "Every year you have to work harder to find trash." Five hundred "river rats" still hauled away 20,000 pounds (lb.)/91,000 kilograms (kg) of trash on that Sunday, "peel[ing] away layers like an archeologist." As the river corridor improved in appearance, people no longer thought of it as the place to get rid of their (illegal) waste. "People aren't dumping in the clean spots," Peralta observed. "The more we clean our areas, the less people will use them as a garbage can."[24]

Twenty years after the passage of the Clean Water Act, the bar was even rising on the minimum standards of toxic industrial effluents that the U.S.

EPA was willing to tolerate in the Chicago River. By then, it had already forced many of the polluting factories in the metropolitan area to stop dumping their liquid waste into the sanitary district's waterways: instead, the companies were required to pretreat their effluents before they entered the sewer system. When an estimated twenty thousand dead fish floated belly up down the Chicago River in August 1989, the hunt was on to track down the perpetrator of the slaughter. Two years later, an insider tip led to the P & H Plating Company, whose owner admitted guilt. He had ordered that 3,000 lb./1,361 kg of cyanide be poured down the drain rather than pay the $30,000 to $40,000 that proper disposal would cost. The toxic wastewater disabled the sewage-treatment plant and poured into the North Branch.

When the case against the owner got to court in May 1992, U.S. District Judge John F. Grady set a precedent in the enforcement of the pollution-control laws by sentencing him to fifteen months of incarceration behind bars. In sending a violator facing criminal charges to prison for the first time, the court sent a strong message to the entire business community: the national government would no longer tolerate the befouling of the waters in the Chicago area.[25]

For the heavy industries along the Calumet River, Judge Grady's pronouncement was old news. After the massive fish kill-off, the U.S. EPA had cracked down hard on the last remaining holdouts against compliance with the law. This industrial landscape had already undergone the most striking ecological regeneration of the six watersheds in the metropolitan region. By the mid-1980s, the sanitary district could begin promoting the Calumet-Sag Channel subdistrict with a market plan of residential land sales. Its planners envisioned a "River Edge Renaissance" of private development on 1,200 acres/486 ha of its land, with about a third set aside for parks and recreation. Over the next decade, the accumulative results of deindustrialization, TARP, and enforcement transformed this depopulated environment into more of a natural landscape than a suburban community. In 1993, a scientist found "sponges, crayfish, gnats, dragon flies, earthworms and turtles" in the Grand Calumet River. Chinook salmon and steelhead trout could also be found there.[26]

In effect, the MWRD became the last polluter standing as the U.S. EPA tracked down and prosecuted industrial violators of the Clean Water Act. In 1997 alone, it won court orders against five big companies on the Grand Calumet River, which then had to begin paying $5.5 million in fines that went into a special cleanup fund for the river's environmental remediation. After the triumph of the "Save Our Lake" crusade in the early 1970s, in contrast, the Illinois EPA had quickly snapped back into playing its traditional role of political noninterference in the local government of Chicago. Saving its rivers depended on Washington, a spokesperson for the Citizens for a Better Environment confirmed, because "the Illinois EPA is blindly committed to a voluntary pollution prevention program which is obviously not working."[27]

Chicago put Illinois at the very top of the list of states in the amount of toxic chemicals dumped into its waters by local sanitary authorities. In 1992, for instance, they poured a record-high total of 72,200,000 lb./32,750,000 kg of dangerous compounds into the rivers, with the vast majority contributed by the MWRD. In contrast, the amount added to Illinois waterways by private industry was on a steady downward trend, having dropped from the previous year by 11 percent to a record low of 5,700,000 lb./2,585,000 kg. As these numbers underscore, the MWRD stood out as the biggest polluter via toxic chemicals in the Great Lakes/Mississippi River basins. It also drew the attention of the U.S. EPA, because it fell farther and farther behind other sanitary districts in upgrading its treatment technologies to disinfect organic wastes to a desired 99-percent level of purity.

Except for city hall, a self-reinforcing process of environmental restoration and stakeholder mobilization had produced a sea change in public perceptions of the Chicago River. As the Clean Water Act approached its twenty-fifth anniversary, the planning department kept insisting, "It's primarily a working river." It denied the petitions of the swelling community of downtown residents to rezone the remaining patches of industry on the riverfront for redevelopment. Wielding virtually dictatorial power over local government, Mayor Daley had successfully sustained a united front of defiance against the law for a decade. With a Democratic administration in Washington, he had enjoyed an exemption from bringing the MWRD into compliance with its standards of water quality.[28]

Recasting the Politics of Environmental Protest

But even this powerful city boss was not strong enough to hold back the environmental movement when it came to saving the Great Lakes. City hall's refusal to disinfect the 1 billion gal./3.8 billion L of partially treated sewage it poured into the waterways every day became the flash point of conflict in this phase of the battle to reclaim the Chicago River. During a second cluster of hundred-year rainstorms from 1996 to 2002, the MWRD's need to discharge its polluted effluents into Lake Michigan eight times came under the spotlight of public scrutiny. With each beach closing due to its failure to control storm-runoff surges, political pressure mounted for enforcement of the Clean Water Act from the top down of the federal government and the bottom up of the grassroots. In sharp contrast to the earlier "Save Our Lake" crusade, the focus of attention as public enemy no. 1 became a local government rather than a private industry.

A generation of changes in environmental conditions, aquatic sciences, and popular perceptions of Chicago's rivers had accumulated to the point of shattering the city's myth of the technological sublime. In its place, a political consensus began to emerge out of new attitudes toward the urban environment that viewed the rivers and the lake as a single, interdependent ecosystem. The rebirth of the Chicago River after 1985 turned out to have been a mixed blessing. On the one hand, the rebounding, albeit rapidly evolving, biotic communities on both sides of the locks led policy makers and recreational users alike to reenvision them in more-holistic terms. On the other, invasive species could now cross back and forth across what had previously been a toxic dead zone between the Great Lakes and the Mississippi River basins.

The inability of the sanitary authorities to stop the flooding helped tie the river and the lake together in the civic sphere of policy debate. Between 1996 and 2002, the MWRD needed to release 8.3 billion gal./31.4 billion L of polluted storm water into Lake Michigan. More than two-and-a-half times more than during the first cluster, this total amount included a single record-breaking discharge of 4.2 billion gal./12.1 billion L and three other billion-gallon-plus cases of an overloaded Deep Tunnel. By 1996, the cost of TARP had already exceeded $3.2 billion to dig 93 miles/150 km of tunnels that could hold 1.5 billion gal./5.7 billion L of storm water.[29]

But over the next six years, the shortcomings of this never-finished public-works project resulted in more pollution in the lake; more beach closures; and more damaged homes, businesses, and lives. After five years without catastrophic flooding, this cluster of severe weather caused approximately a half billion dollars in damages. The "Groundhog Day" scenario of damaging floods over and over again all along the Salt Creek/Des Plaines River helped forge a consensus around a hydraulic perspective on the metropolitan area's flood-prone environment among local officials and grassroots activists alike. During this phase of crisis management, they successfully lobbied the U.S. Army Corps of Engineers to create a new master plan of storage-retention ponds designed as recreational spaces along a 75-mile/121 km stretch of the waterway.

Beginning in May 1996, the first fifteen months of this period of exceptionally frequent heavy rainfalls can serve as a case study of climate change overwhelming the city's envirotechnical systems of water management. Just a few months earlier, the director of the Midwestern Climate Center, Ken Kunkel, reflected on 1995 as a "goofy year of weather." Looking back a decade, he listed record-breaking droughts, great floods, and killer heat waves to posit that the region was undergoing climate change. "We've seen more extremes than we have in the previous 20 years," Kunkel asserted. Although the scientific expert equivocated on the long-term direction of that change, weather statistics pointed unmistakably toward wetter and hotter.[30]

In May, the first in a series of hundred-year rainfalls sank the northern suburbs under 4 to 5 in./10 to 13 cm of precipitation in one hour. Hundreds of basements were flooded; roads and highways were closed. In Lincolnshire, a retention pond overflowed into a nearby high school, causing $300,000 in damages. A month later, the U.S. Army Corps' hydraulic researchers incorporated the "best available data" in a progress report on creating an updated flood-control plan for the Salt Creek/Des Plaines River. These numbers led to the conclusion that the waterways not only would rise to the hundred-year flood stage more frequently than previous calculations had shown but would also reach a much higher level. Recent extreme weather events were raising them by as much as 2 extra ft./0.6 m to exceed their banks by 5 to 7 ft./1.5 to 2.1 m. As one homeowner confirmed, "There's no place for the water to go."[31] The U.S. Army Corps' Philip Bernstein blamed the greed of private developers and the corruption of public officials for paving over the floodplain. "They don't care," he contended, "if someone else gets flooded downstream." Since beginning the study a few months previously, Bernstein reported, several sites identified to be set aside as nature preserves had been zoned for sale as residential subdivisions.[32]

On July 18, 1996, a downpour set a new record for "the most disastrous rainfall ever" in the state of Illinois: 16.91 in./42.95 cm during a twenty-four-hour period. The slow-moving, low-pressure cell of weather also set a new record, dropping 7.7 in./19.6 cm of rain at Midway Airport on the city's Southwest Side. The governor declared Cook County and all five of its collar counties disaster areas. As during previous hundred-year rainfalls of 3 to 4 in./8 to 10 cm or more, the sewer system filled with 18 billion gal./68 billion L of storm runoff, the Deep Tunnel reached its storage capacity, and an overflow of another 15 billion gal./57 billion L of raw sewage into the Sanitary and Ship Canal raised the Chicago River to its limit of 5 ft./1.5 m above normal. Then the sanitary district needed to release 1.5 billion gal./5.7 billion L of polluted wastewater into Lake Michigan, closing the beaches.[33] The flooding inundated thirteen thousand homes in Will County and three thousand in the city, including the bungalow of Earl and Brenda Hunt. The new owners stood in their basement on a carpet ruined by 7 in./18 cm of sewage-laden water, "breathing dank, moldy air." Mr. Hunt recalled the ordeal of the night before, when he "was in a state of shock" as filthy water filled his basement. Property owners suffered $100 million in damages. An eighty-four-year-old homeowner died of a heart attack while carrying furniture out of his sinking basement.[34]

In the public debate that followed, overbuilding on the floodplains joined the standard explanation of suburban sprawl as the reason for the increase in the frequency of damaging storms. During the 1990s, the rate of deconcentration outpaced the white flight of the 1970s. Chicago ranked tenth in the nation in the rate of settlement at the periphery. While the population of the metropolitan area rose 9 percent, the amount of land paved over increased 40 percent.[35] The president of the board of the MWRD repeated the agency's standard denial of responsibility for sprawl and awesome acts of God. "Ain't a damned thing anybody can do," he prevaricated, "when you get 4 to 5 inches [10 to 13 cm] of rain in one hour." To an extent, environmental activists, such as FOTR Executive Director Laurene Von Klan, had to agree that paving over a prairie wetland was the main source of the problem. "Within a nine year period," the head of the Conservation Foundation of DuPage County added, "we've received the 100-year flood twice."[36]

Nonetheless, environmental activists looked to small-scale approaches to reduce damages across the region rather than the single, big-technology approach of TARP. Scott Bernstein of the Center for Neighborhood Technology, for example, reasoned that "you'd have to redesign the whole area to accommodate a rain like this."[37] He and other experts with an ecological perspective on the metropolitan region had come to the conclusion that structural approaches, such as tunnels and reservoirs, would continue to lag behind the pace of sprawl. Since Vinton Bacon's retention-pond plan of the mid-1960s, the local, state, and federal governments had dug out enough storage capacity on the Salt Creek/Des Plaines River to hold an extra 27 billion gal./102 billion L when they overflowed their banks. A climate changing to produce more-extreme, heavy rainfalls repeatedly overwhelmed all the engineers' structural approaches. Another environmental activist, Dennis Dreher, suggested that more land along the floodplains needed to be reclaimed as wetlands and prairies that would soak up the rain before it became a runoff problem. He approved, for example, the purchase of 4,500 acres/1,821 ha along the Upper Des Plaines River by the Forest Preserve District of Lake County.[38]

In August 1996, the U.S. Army Corp's Philip Bernstein released a blueprint of the long-awaited plan to control the flooding of the Salt Creek/Des Plaines River. It called for a series of Skokie Lagoon–like preserves to contain storm-runoff surges within its 500-square-mile/1,295 square km watershed. And, as with this model conservation project, the $82-million plan of landscape architecture was designed to be used as a space of recreation and leisure. Following eco-planner Donald Hey's lead, Bernstein also incorporated marshlands into his envirotechnical machine of flows. "[It] outlines a grand vision," a newspaper reporter radiated, "of scenic riverfront bounded by deep reservoirs and shallower water storage basins, restored wetlands and sculpted levees and dikes."[39] Moreover, Bernstein's plan presented an alternative to big-technology projects, such as the sanitary district's TARP and DuPage County's Elmhurst Quarry Reservoir. At a cost of $60 million, it had been built after the great floods of 1986 and 1987 to divert floodwater from the Salt Creek/Des Plaines River.[40]

But on February 20, 1997, neither visionary plans nor giant-scale public-works projects could stop nearly 4 in./10 cm of rain and snow from shattering yet another month's daily record. This torrent doubled the previous extreme downpour of 1.98 in./5.03 cm on February 6, 1942. It also doubled the record-breaking 2 billion gal./7.6 billion L of polluted storm water the sanitary authorities had had to release into the lake ten years beforehand. Soaking the Calumet District, the storms put sixty-nine thousand homes under water; nine out of ten of their owners did not have flood insurance. Other Chicagoans, too, suffered in record numbers from blackouts of power and communication services, stalled transportation, and rivers flowing over their banks in the suburbs. The Des Plaines River crested at 9 ft./2.7 m above flood stage, sinking Riverside and River Grove as well as thousands of homes and businesses upstream that had been built on the floodplain. Six months later, another hundred-year rainfall caused another round of widespread human misery, billion-gallon discharges, and multi-million-dollar losses.[41] With TARP chronically outrun by sprawl, political momentum built for alternative, "nonstructural" strategies of water management. City hall's resistance to adopting these eco-friendly approaches, such as rain blockers, that had proved so successful elsewhere began to draw sharp rebuttals from a growing number of groups with a stake in saving the waters for recreational use.[42]

In 1995, an unexpected court case presented a preview of this phase in the fight to bring Chicago into compliance with the Clean Water Act. The suit resurrected the intergovernmental conflict over the amount of lake water Chicago was permitted to divert into the river. The cause of the controversy was not dueling politicians but better technology. Recent advances in the science of hydraulics had been embodied in a new generation of precision instruments. Installed by the U.S. Geological Survey, the "acoustic velocity meters" found that the old measuring devices were undercounting the flow through the three locks from the lake into the waterways by 7.8 percent, or 172.8 million gal./654.1 million L each day. They also showed that an even greater amount was leaking through the locks, which were operated by the U.S. Army Corps of Engineers.

Returning to federal court, the attorney general of Michigan asked the judiciary to revisit its 1967 settlement of the case in light of this new information. Against city hall's lame claim to the extra water as a vested, if illegal, right, Michigan's environmental lobby defended the Great Lakes as a virtually sacred, patriotic symbol. "Protecting the lake is like [protecting] the American flag," a leader of the state's Environmental Council sermonized. The judges sided with Michigan; they ordered the Corps to fix the leaks and the city to begin conserving water by installing meters on the 335,000 homes built without them before 1967.[43]

The case highlighted the interconnections between not only the waters of the lake and the rivers but the drinking supply and the sanitary system as well. The extra water leaking through the navigational locks, supply mains, and plumbing fixtures all fed into the waterways, helping dilute the microbe-infested effluents pouring out of the MWRD's treatment plants. Facing a major reduction in the flow of clean, oxygen-rich water in the rivers, Superintendent Dalton threatened, "The fish will die."[44]

This turn in the nearly century-old legal battle also illuminated the new authority of science in policy making on the environment. The hydraulic instruments were simply the latest high-tech devices to help researchers monitor the lakes and streams of the American Midwest. But just as quickly as the sanitary district had begun measuring indicators of organic life in the Chicago River during the "Save Our Lake" crusade, it had fallen behind in detecting the industrial sources of toxic chemicals. After the plating company's cyanide caused the massive fish kill in 1989, the U.S. EPA had to order the sanitary district to install a warning system to prevent another breakdown of its treatment plants.[45]

A generation of research on the Great Lakes basin had produced a new, more-sophisticated paradigm of ecosystem biology to replace the older model of resource conservation. The instant success of Michigan's sports-fishing experiment had encouraged the other states bordering Lake Michigan to ramp up their own restocking programs: between 1976 and 1994, they put a total of 274 million game fish in the lake. These public agencies adhered to the old mind-set of attitudes toward nature that defined natural resources in economic terms. They had become cogs in the multi-billion-dollar sports-fishing industry. "Even the scientific management of lakes and rivers [under this regime] has traditionally benefited a few popular fish to the great detriment of the rest," the naturalist Joel Greenberg notes. As recently as the 2000s, "fisheries biologists," he bemoans, "have treated these waterways as *farms rather than ecosystems*."[46]

What made them change their minds were the few remaining native and ever-more-disruptive invasive species that kept causing unnatural disasters, such as the algae blooms and the alewives explosions. As the U.S. Army Corps repaired the locks, the fish and other forms of life under the surface of the Chicago River did not die from pollution. On the contrary, this biotic community continued to multiply in numbers and diversity. This growth, in turn, increased the complexity of its interactions with the lake's equally fast-paced evolution. In the 1990s, scientists found 140 nonnative species in Lake Michigan.[47]

By the end of the decade, two of these alien invaders posed the greatest threats to the ecosystems on both sides of the locks linking the two great water basins. The round goby is a little spiny fish that annihilates its competitors by eating their eggs, including those of popular sports fish, such as walleyes and bass. The super-sticky zebra mussel is also small in size but potentially catastrophic in effect. In sufficient numbers, these biotic filter machines can completely change a body of water's chemical and physical qualities, throwing the rest of its life-forms into an unhealthy state of imbalance.

In making the water cleaner, they starve fish dependent on its plankton for food, while super enriching the bottom with nutrients.[48]

In 1998, the decision by the U.S. Army Corps to erect an electric fence on the Sanitary and Ship Canal to stop the goby marked the dawning of a new era of urban ecology. It represented the first milestone in the formation of public policies that gave the recreational uses of Chicago's waterways as a nature preserve equal status with their economic value as an industrial corridor. Its experiment in aquatic-wildlife management perfectly reflected its engineers' reformed attitudes toward them as working, albeit dual-purpose, rivers. They promised that an array of electric pulses under the water would turn back the fish without impeding barge traffic on the surface.

Over the next four years, however, the dithering of the engineers over the design of the fence in the midst of the goby invasion sparked a heated debate over their intentions. Alternatives were proposed to permanently seal the navigational locks. Without knowing whether the U.S. Army Corps' experimental approach would work, recreational stakeholders questioned its compromise solution that put greater value on serving the interests of the shipping industry than the public at large. Protestors used the authority of science to argue that a physical barrier between the lake and the rivers was the only way to safeguard the ecosystems of these two great, fresh-water basins from destroying each other.[49]

The debate exposed new attitudes toward the environment among not only a scientific and policy elite but also a broad range of urban society. The new image of the river corridors as the "wildest area[s] within the city limits" received repeated reinforcement from a string of "first-sightings" reports in the mass media. At the grassroots, the blue-collar dream of the first Mayor Daley had finally come true. Although too contaminated with toxic sediments to eat, the waterways' bass, trout, and other game fish were attracting significant numbers of catch-and-release sportswomen and sportsmen.[50]

Besides telling tales of ever-bigger prize catches, the press covered stories of outdoor enthusiasts, who observed wild newcomers settling in along the riverbanks. In 1998, for example, Canada geese and Mallard ducks were found nesting on the North Branch. Six years later, a bald eagle's nest was spotted on the Little Calumet, the first since 1897. The political symbolism of this historic moment in the restoration of Chicago's rivers was not lost on city hall: it bought the land surrounding this instantly semisacred space to protect it from future development. The flora and fauna of the river added its own voice to the positive reviews the local press was lavishing on its transformation from an open sewer into a natural, if urban, habitat.[51]

For the human settlers in the twenty-thousand new housing units in the downtown area, too, the visionary promise of the river as a front street of natural beauty and urban amenity was beginning to take concrete form.

In 1998, the opening of the Cityfront Center's river-walk promenade and park on the northern side of Main Branch, east of Michigan Avenue, set a high standard of architectural design and landscaping for others to follow. With another two thousand riverfront units in the planning pipeline, a *Sun-Times* reporter exuded, "the central business district is sizzling." The river was attracting homebuyers and apartment dwellers, which in turn led real-estate developers to keep exploiting their seemingly insatiable desire to live next to it.[52]

The expansion of the Montgomery Ward conversion project during the early 2000s illustrates this reciprocal process of the river's reclamation and the grassroots' mobilization of support for its complete restoration. Started a decade earlier, the plan to turn the catalog company's old buildings into a self-contained village had ballooned into a real neighborhood covering the 31-acre/12.5 ha site. Several additional tower blocks and townhouse complexes were bringing the total number of homes up to about three thousand units. According to Sam Persico, the developer of a seven-story, 164-condo building called Park Place, a river walk and the small park planned on adjacent city land were major attractions. His plans called for a landscaped path, which would include several sculptures and a 170 ft./52 m mural of ceramic mosaics depicting the waterway's history.[53]

The Chicago River had become such a hotspot of residential development that patches of homesteaders were sprouting up beyond the expanded boundaries of the CBD. Intermixed with long stretches of industrial landscape, these outposts of urban pioneers were taking root up and down the banks of the waterway. On the South Branch, for example, the privately built Chinatown Square had encouraged the already densely packed community to petition city hall for an adjacent public park. In April 1998, after a decade-long campaign, they got a 12-acre/5 ha site that would include a boat dock in addition to the traditional green spaces and playgrounds.

Even farther away from downtown was the multi-award-winning Bridgeport Village, a 32-acre/13 ha riverfront site with 116 single-family houses. Creating a retro "classical Chicago-style architecture," developer Thomas A. Snitzer "envisioned the neighborhoods that were built 60 or 70 years ago." Plans for more than a mile of landscaped, riverfront trails were helping attract families with young children to what was named the "best overall urban community" of the year by the Home Builders Association of Greater Chicago.[54]

Resisting the Invasion of the Alien Species

The year 1998 was a milestone in the long battle to reclaim the Chicago River in a second way. As the federal government took on the task of building the electric fence, city hall passed the first comprehensive land-use guidelines for the riverfront's entire 17-mile/23 km length. Its planning department had drawn a blueprint that paralleled the U.S. Army Corp's recommendations for the Salt Creek/Des Plaines River. "What we are trying to do," city hall's point man declared, "is reintegrate the river into the neighborhoods that line its banks." Although playing catch-up with the private sector, the River Corridor Redevelopment Plan marked an important turning point in the perceptual metamorphosis of Chicago's waters into a holistic aquatic ecosystem. With Mayor Daley's support, the city ordinance embodied the new consensus of attitudes that cast the river in a positive light as an urban amenity rather than a regrettable, albeit inevitable, by-product of modern life. "It's an awesome plan," a leader of the FOTR rejoiced.[55]

But in finally reaching the goal line of its original mission, the FOTR became the enemies of city hall. This broad-based coalition of stakeholders in the river for recreational use represented the bellwether of mainstream environmental reform. The FOTR continued to follow its middle-of-the-road posturing as a compliant patron of Mayor Daley's clientele regime into the early 2000s. The rewards he granted to help protect the lake offered this nonconfrontational lobby group a promise of putting plans for environmental reform into action sometime in the future. In 2002, for example, Daley vetoed a plan to turn Lake Calumet into a big marina. "A new day calls for a new approach," a spokesman for the newly minted "environmental mayor" crowed. Since the mid-1980s origins of the development plan for this inlet harbor, he asserted, "it's emerging as a refuge and a sanctuary for wildlife now." The following year, the FOTR supported the mayor's proposal to delay making the Chicago River clean enough to swim in for at least another seven years. The local group appeared to accept his standard that it was "good enough" for the tourists to enjoy a boat ride and the outdoor enthusiast to paddle a canoe.[56]

Without making a sharp break, the FOTR position began shifting during the second cluster of severe weather events. Starting from a defensive posture of weakness in a clientele regime, this organization of green activists would come out of the closet, so to speak, to openly declare their opposition to Mayor Daley's vision of Chicago. While applauding the river-corridor plan, it was disappointed in his refusal to pay the city's share. He would only promise to earmark a few million dollars generated by his sleight-of-hand tax-skimming scheme, called Tax Increment Funding (TIF), to the South Branch and the Chinatown river-walk parks; he would not consider adding to taxpayers' bills the costs of installing the tertiary sewage-treatment equipment required to bring the sanitary district into compliance with state and federal water-quality standards.[57]

In contrast, Daley was pouring more than $1.5 billion into completing Daniel Burnham's unfinished plans for the lakefront. Chicago's downtown had been made over into not only an international business center but also a world-class tourist mecca. Between 1990 and 1997, the number of visitors increased by almost two-thirds over the previous decade. They spent 150 percent more, adding $5.4 billion a year to the local economy. The hotel industry responded by building and renovating thirteen thousand rooms. Matching the private sector's investment in the future of Chicago, the mayor remained frozen in place, facing the lake with his back to the river. He was spending $190 million to turn Navy Pier into a tourist attraction and at least equal amounts on the Millennium Park and the Museum Campus/Soldier Field projects. He was also funding an $800-million expansion of the adjacent McCormick Convention Center to turn it into America's biggest facility.[58]

In 2000, open warfare erupted between green activists and local policy makers over their big-technology strategy of water management. Outsiders fired the first salvos of the attack against Mayor Daley's stubborn defense of TARP on his own ground, the city's front yard. Under the banner of "Save Our Lake," the Sierra Club, Lake Michigan Federation, Public Interest Research Group, and other green organizations demanded that the MWRD disinfect its sewage to protect the lakefront beaches from disease-breeding pollution. One of their spokespersons made the argument that "citizens have a right to enjoy their beaches without having to worry about their family's health."[59]

The scientists' ever-more-precise monitoring devices showed that TARP overflows posed a clear-and-present danger to recreational users of the lakefront, allowing them to hammer home their case with indisputable proof. In 2001, three hundred-year rainfalls forced the sanitary authorities to discharge polluted floodwaters through the locks into the lake, closing various city beaches 272 times. The following year, the Sierra Club put an army of volunteers on the beaches to take water samples after another heavy downpour on August 22 forced the sanitary district to release 1.750 billion gal./6.624 billion L of wastewater through the Wilmette Harbor and Chicago River locks. The environmental advocacy group's frightening, eye-opening study showed that the lakefront exceeded the state's standards of safe bacteria levels almost 70 percent of the time. Even when the beaches were closed to swimming, its spokesperson warned, "Children can be at risk if they put bacteria-contaminated sand in their mouths."[60]

Suffering through their eighth hundred-year rainfall in the past six years, Chicagoans trapped in this revolving time loop of damaging floods pushed back to save the lake from their own elected officials. To fend off criticism, the MWRD dismissed the idea of bringing its policies into compliance with the Clean Water Act as "crazy." "What do you expect us to do?" its superintendent exclaimed.[61] Despite its own model treatment plant at Hanover Park, the sanitary district insisted that the costs far outweighed the benefits to upgrade its equipment from a 90-percent to a 99-percent level of purity. The long-running controversy over funding TARP took any suggestion of higher taxes by the district off the table of local politics. By excluding the recreational value of the river as an urban amenity, city hall could condemn the protesters for bad math and bad science. Its counterattack was framed entirely within the old bifocal vision that limited recreational use of the city's waters to the lakefront. It also insisted that the river corridors were the only economical way to get rid of the city's wastes.

The MWRD's top manager offered no alternative to big technology, only more tunnels and bigger reservoirs. At the same time, he conflated the daily operations of the sanitary district with the recent period of climate change. In 2002, however, he had to admit that TARP would not be able to protect the lake for at least another fifteen years. To disinfect storm-water overflows until then, he argued, would be "impractical." He shredded imaginary paper tigers of "just throwing chlorine in the water" and spending a fortune to build giant treatment plants that would be needed only three or four times each year during extreme weather events. For the well-organized eco-guardians of the Great Lakes, the Deep Tunnel's unfailing record of failure to control flooding and pollution made the superintendent's promises ring hollow.[62]

After two successive years of dumping 3 billion gal./11 billion L of storm water contaminated with raw sewage from the rivers into the lake in 2001 and 2002, grassroots pressure percolated up to the state and federal governments to force city hall into compliance with their minimum legal standards of water quality. With a Republican moving into the White House, Mayor Daley lost his exemption from the enforcement of the law by federal agencies and prosecutors. While he still enjoyed the shield of the state government's blind eye of denial, this policy itself came under increasingly critical scrutiny in the public arena. In 1994 and 1996, the U.S. EPA had ordered Illinois to raise its standards of water quality to federal levels. After waiting another four years for the state to meet the deadline, the agency turned to the federal courts. It sought a decree ordering the state government to enforce the 1972 benchmarks against all violators, including local sanitary authorities.

But there was little that the state and the federal governments could do in practical terms when faced with the determined resistance of Chicago's all-powerful boss. On the contrary, Mayor Daley maintained the political advantage while tying up the courts in costly but effective legal maneuvers. He became a master at playing the flood card of homeowner damages and lake pollution. Daley called the state and federal governments' bluff on cutting off the flow of funds into the bottomless pit of TARP.[63] From the cost-benefit analysis of local politics, he stood the most to gain from resisting compliance with the Clean Water Act for as long as possible. Even though the federal government paid for 65 to 75 percent of the Deep Tunnel, the state's and the city's share of such an expensive envirotechnical system imposed a significant burden on local property owners. The mayor cast himself in the role of tax buster, fighting on their behalf against unreasonable demands from outsiders, who would raise their tax bills even higher. In 1999, when the U.S. EPA insisted that the city begin installing rain guards to help prevent river overflows into the lake, Daley got his political revenge: he retaliated against the Washington "establishment" by punishing homeowners with a prominently displayed, special assessment on their tax bills to pay for the alternative, green technology.[64]

The mayor had a second compelling reason for refusing to bring the MWRD up to national standards of sewage treatment. Like the chief executives of all American cities, Daley had to work with fewer and fewer public resources during his long reign of power. Locked into the old dual image of Chicago's waters, he funneled every dollar he could get his hands on into the beautification of the city's front yard. Still seeing the river as its back alley, he could not grasp Goldberg's concept of a parkway of natural wilderness in the city. At the farthest stretch of his imagination, he would be able to picture the river in only mirror-image terms of his bifocal vision as a "second lakefront." Daley could bring environmental reform to a standstill, but he could not stop the round goby and other fish from swimming between the river and the lake.[65]

The looming perception of a "crisis" caused by invasive species on both sides of the locks broke the political stalemate and permanently united the waters of the Chicago area as an interdependent ecosystem. In the midst of two bad years of flooding in 2001 and 2002, the failure of the U.S. Army Corps to construct the electric fence had already allowed the round goby to infest the river waterways. Another unwanted trespasser, the Eurasian ruffe, had followed to further deplete their recovering stocks of smaller, native game fish, such as yellow perch and bluegills. Like the goby, it outcompeted its rivals. A 4 to 6 in./10 to 15 cm fish with spiny fins, the ruffe had a supercharged reproductive system that could lay three hundred to five hundred eggs every twenty days.[66]

But in 2003, an alien invader coming from the other direction, the Asian carp, became the ultimate threat. Imported to clean up commercial fishponds in the South, this aquatic terminator escaped during a flood into the Mississippi River. An insatiable predator, it can consume 40 to 120 percent of its body weight every day, growing to more than 100 lb./45 kg. Each female can lay a million eggs during her lifespan. Like the mutant creatures unleashed in science-fiction flicks by runaway technology, this real-life, 4 ft./1.2 m "monster" would wipe out sport fishing in the Great Lakes if it got through any one of five points of interbasin connection with the riverine waterways of the Chicago area. The Asian carp would eat the baby salmon and trout supplied by the state fisheries before they could grow into the prize catches now expected by fishing enthusiasts.[67]

Driven by this multi-billion-dollar industry's lobby, the environmental coalition to save the Great Lakes shifted into high gear of political advocacy to avert catastrophe. It immediately pushed the district office of the U.S.

Army Corps to reinforce its technological solution to prevent fish from getting across its experimental fence. Finally turning on this shock-impulse barrier in 2002, it announced plans to start constructing a second, more-powerful envirotechnical machine the following year. Reaching Washington, the environmental lobby urged other agencies to take action before a manageable threat spun out of control. The U.S. EPA supported the U.S. Army Corps by underwriting its project to build a fortified electric fence.[68]

The consensus among scientists regarding a solution, however, remained "hydrologic separation." The eco-biologists pointed out that electric shocks would not work against all the creatures and their eggs moving under the waters of the two great basins. According to a report issued by the Alliance for the Great Lakes (AGL), these included "zebra mussels, spiny water flies, and other aquatic invaders that aren't fish." At a conference hosted by the "environmental mayor" on the invasive-species crisis, the assembled group of internationally distinguished experts agreed that damming the interbasin connections should be the top priority of policy makers. They called on them to sacrifice the vested rights of the shipping industry in favor of the greater public interest in the recreational use of the urban environment.

Expressing the fears of fishermen and fisherwomen throughout the Great Lakes, perhaps, scientists from the co-host of the meeting, the U.S. Fish and Wildlife Service, made a plea for far-more-drastic measures. This was an emergency, they believed, that demanded immediate action no matter what the costs. They proposed to nuke Chicago's waterways with deadly chemicals and/or poison gas to turn them back into an aquatic dead zone. "It's for the greater good of the larger whole," a biologist from the University of Wisconsin tried to explain. In other words, they were desperate enough to sacrifice the rivers in the name of saving the lakes.[69]

Without any river reversal of polluted waters into Lake Michigan between 2003 and 2007, Mayor Daley was able to withstand mounting political pressures. This period of high anxiety about the fate of Chicago's waters swelled the ranks of recreational users, who were no longer willing to tolerate any pollution in them. Their attitudes shifted from acceptance of city hall's strategy of water management to resistance. In part, the accumulating weight of scientific evidence tipped the scales of public opinion from support to opposition. The Asian-carp crisis set off a frenzy of studies to find out what was going on below the surface on both sides of the electric fence. They all pointed to the conclusion that contemporary methods of environmental management were not up to the task of keeping a biological separation between these two complex aquatic communities.[70]

In part, the juggernaut of the building boom was responsible for the outbreak of open conflict with Mayor Daley over his refusal to bring the MWRD up to the minimum standards of the Clean Water Act. Just before its historic collapse in 2008, the housing industry was in various stages of building twenty-seven major projects in the downtown area. The mayor was more than willing to help their developers gain passage of an updated riverfront zoning ordinance, but he made sure his promised river walks and parks remained locked up in the planning department.

The restoration of the waterways not only attracted more people; it also drew them closer and closer to the water's edge. A group of kayakers called the Water-Riders were already getting wet with contaminated water that could cause skin infections. Their provocative use of the Chicago River elicited a stern reprimand from the sanitary district. Anyone foolhardy enough to wade or swim in it, its spokesperson warned, was likely to suffer "deadly consequences." It was exactly these dire results from its pollution that just about everyone except city hall had now come to believe were intolerable.[71]

Reversing the Course of City Hall

Beginning in 2007, a string of seemingly endless flood disasters triggered an uprising against city hall's big-technology approach to water management. Like the underwater invaders, the cloudbursts raining down on the Chicago area were beyond the mayor's control. Nine times over the next five years, the MWRD would need to open the locks, releasing 21 billion gal./79.5 billion L of storm-surge runoffs contaminated with disease-causing microbes from the rivers into Lake Michigan. This total amount again exceeded the previous cluster of extreme weather events by two-and-a-half times. This series of thunderstorms also repeated the past by setting a new record as "the most disastrous rainfall ever" in September 2008, when the sanitary district needed to discharge 11 billion gal./42 billion L into the lake. Furthermore, the second- and the fourth-largest releases in the history of the post–Deep Tunnel era took place during this cluster of hundred-year rainfalls.[72]

In July 2007, a *Chicago Sun-Times* editorial expressed the frustrations of lakefront beachgoers banned from getting wet at the very time the summer was starting to get hot. Pointing a finger of blame at "public officials," it complained:

> What, if any, good has the costly Deep Tunnel system been in bringing this 'unavoidable' solution to an end? . . . Let's ask all of them why 40 years after the moon landing we still don't have reliable storm drainage and sewer capacity? By the time we land on Mars, it would be nice for property owners to be able to use their basements as something other than indoor E.-coli collectors.[73]

Joining the revolt of the eco-protesters, the FOTR put the danger to the public health from the MWRD's effluents in the rivers on a par with the lakefront. Reinforcing the Sierra Club and the AGL, the local group charged that kayakers "are being exposed to serious risk right now." Drawing on the authority of science, it issued a study that made the case for disinfecting the sewage. The civic group's calculations showed that existing technologies of chemical treatment and ultraviolet light could be installed at an estimated cost of $250 million to $300 million, spread over a twenty-year period. More-precise monitoring was leading to not only more beach closures but also greater exposure of polluters. Despite denials by the sanitary district, the FOTR accused it of dumping dangerous wastewater into the rivers about every other day on average. "The risk of not disinfecting this effluent," its report concluded, "clearly outweighs the cost of doing it."[74]

After floodwater releases in August 2007, the FOTR raised the stakes in the political battle to replace the entrenched bifocal attitudes toward the urban environment with a holistic, ecosystem-focused point of view. Its executive director, Margaret Frisbie, spoke for recreational users, taxpayers, and business owners alike. "As a society," she implored, "we have a moral obligation to ensure that our shared natural resources are protected from pollution." Anything less was now seen as an unacceptable failure. She took city hall to task for failing to employ every available technology to bring Chicago into compliance with the Clean Water Act. Only half of the rivers and streams in Illinois met national minimum standards, leaving it far behind other states.[75] Over the course of the next two years, the MWRD had to open the locks about every three to four months with disheartening, albeit "'unavoidable,'" predictability. Inspiring a barrage of criticism from the grassroots, the floodwater releases kept the failures of the MRWD's big-technology system to fix the problems of damaged basements and beach closures in the constant glare of public attention.

The state and federal governments also ratcheted up the assault on city hall to persuade Mayor Daley to fulfill his obligation to obey the law. The never-ending "Groundhog Day" of TARP's shortcomings gave extra-local agencies cause to mount a new lawsuit to force Chicago to disinfect the wastewater coming out of its sewage-treatment plants. In 2008, the Illinois EPA fired the first salvo in this phase of the battle to bring city hall into compliance with the Clean Water Act. A five-year state study laid out scientific data and economic statistics that left no doubts about the practical feasibility of achieving the law's minimum standards of water quality. Thirty-six years after the passage of the federal act, the Illinois EPA finally ended the exclusion of the Chicago and the Calumet Rivers from its protection.[76]

In 2008, recreational stakeholders in the rivers' restoration reached a crossroads in their political mobilization of opposition to Mayor Daley. In the biannual elections, they put up reform candidates for seats on the MWRD's board of commissioners. Winning the endorsement of the local press, for example, was Nadine Bopp, a university instructor and environmental planner. Running on the Green Party ticket, she and two Democratic incumbents promised to begin employing the full arsenal of nonstructural approaches to flood control. Among the well-proven alternatives were homeowner rain barrels and subdivision streets and parking lots paved with water-permeable materials. Endorsing their campaigns, the long-time educational director of the FOTR gave expression to the change in attitudes toward the urban environment. "We need to think of the river not as a flowing water," Chris Parson reasoned, "but as an ecosystem with an entire set of plants and animals dependent on it, and on us to protect it."[77]

Firing back in defense of city hall, the MWRD's president of the board, Terrence J. O'Brien, blasted the reformers as "green extremists." Following the official line as a defender of the overburdened taxpayer, the party loyalist promised to stop wasteful spending. Counterattacking on their own turf, moreover, O'Brien challenged the other candidates' scientific authority with his own panel of inside experts. Until they could reach a definitive conclusion from independent research, he declared, the board would hold the line against raising the standards for disinfecting the city's sewage. Warning against a rush to judgment, he predicted that this worthless goal would add at least a $2-billion burden of "unjustified costs" on homeowners, who were already "feeling the pain." Over the next two years, the O'Brien-led, Democratic majority on the governing board would throw up a smokescreen of a work-in-progress by his panel of experts to deflect attention from the mounting opposition to the MWRD's continuing violation of the law.[78]

By 2010, both the state and the federal governments had grown tired of waiting for the agency's scientists to change Mayor Daley's mind. The MWRD held fast to the official line that "disinfecting satisfies only a few and benefits no one." In May, the U.S. EPA issued an ultimatum: agree to bring Chicago into compliance with the law or face a court-directed takeover of its water-management agencies. Claiming to have been blindsided by Washington, the mayor counterpunched, telling the Washington officials to "go swim in the Potomac."[79] While winning some sympathy for the plight of his recession-starved budget, it was not enough to outweigh the value people now placed on having patches of "urban wilderness."

Thrown on the ropes by the blows raining through the federal hierarchy of government, Mayor Daley was caught with his defenses down by the environmental lobby. They got in a knockout blow in this late round in the battle to reclaim the river. A local columnist used the power of irony to claim at least a rhetorical victory. Except in Chicago, he fired back, the rivers running through all the other cities, including the nation's capital, "have been restored enough that swimmers aren't afraid of them."[80]

The state of Michigan joined the attack on Daley's crumbling ramparts of legal defense against holding Chicago responsible for meeting the environmental standards set by the contemporary consensus of scientific opinion. Its attorney general used the battering ram of the federal courts to plead for a "hydrological separation" to protect Lake Michigan and its sport-fishing industry from the Asian carp. Besides the high risk of an inevitable breach of the electric fence, TARP's current failures to contain floodwaters from being discharged into the lake created a clear-and-present danger of irredeemable harm to recreational users, let alone to the larger bionic community of the Great Lakes.[81]

Caught in this crossfire of opposition from the grassroots, the states, and the federal government, Mayor Daley was still able to forestall a capitulation until his retirement from office a year later. The master politician to the end, he narrowly escaped being held accountable for what Henry Henderson of the National Resources Defense Fund labeled a "ridiculous failure of government."[82] Nevertheless, the environmental-reform movement had already penetrated the smokescreen of "junk science" thrown up by Daley's loyalists in charge of the MWRD. Six months earlier, during the November 2010 elections, green candidates for its board of commissioners had pledged to reverse its "good-enough" approach toward the waterways. The election returns ensured that a new majority would vote in favor of the U.S. EPA's plans when its final orders of compliance were issued the following May. By then, a new mayor with long-standing credentials as an eco-guardian of the Great Lakes, Rahm Emanuel, had moved into city hall.[83]

Conclusions: Reimagining a Unified Ecosystem of Urban Nature

The battle for the reclamation of the Chicago River is over. Inspired by Carson's book, the reform movement for its protection from pollution won. Over a half-century period, a reciprocal process of nature restoration and stakeholder mobilization changed public attitudes toward the urban environment. Perceptions of the city's waters shifted from a bifocal vision of front-yard beauty and back-alley eyesore to a holistic point of view of an aquatic ecosystem. Mental maps of downtown moved the river from the edge to the middle of this expanding zone of cosmopolitanism. Coming full circle, the rebirth of wild nature in the Chicago River has inspired environmental reformers to set aside patches of well-manicured beaches on the lakefront to evolve back into their natural habitats.

According to the consent agreement, Chicago will make its rivers clean enough for swimming, and the MWRD will install the required equipment at its sewage-treatment plants to bring the Chicago River into compliance with the Clean Water Act (see Figure 6.1). In return, the state and the federal governments agree to help the city pay for more digging until TARP reaches a storage capacity of nearly 18 billion gal./68 billion L. The plan's completion date has been set back to 2029. Grassroots reform groups have

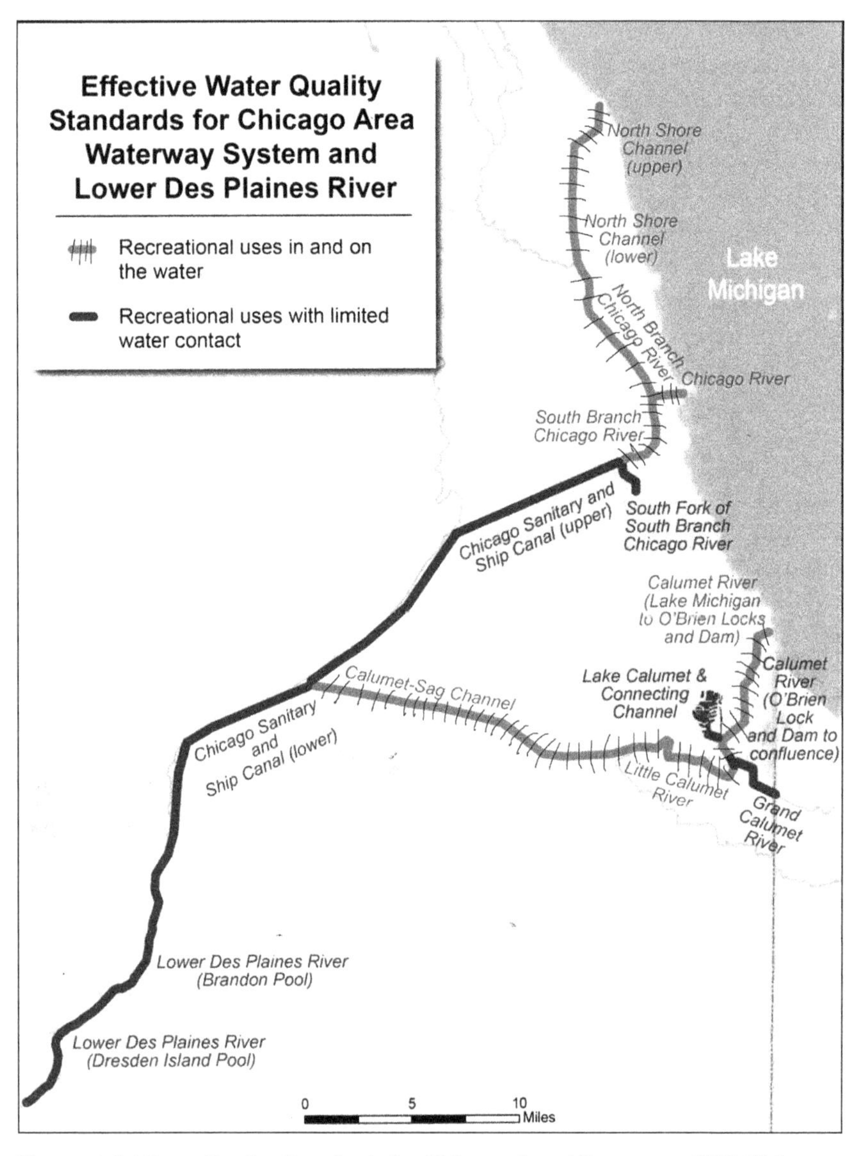

Figure 6.1. Water-Quality Standards for Chicago Area Waterways, 2011 (U.S. Environmental Protection Agency, "Effective Water Quality Standards for Chicago Area Waterway System and Lower Des Plaines River," available at www.epa.gov/sites/production/files/styles/large/public/2015-09/caw-map-20120507-lg.png?itok=tsJ3lzls)

also used the federal courts to force the state and the local governments to implement strategies of flood control that include nonstructural, green technologies.[84]

In executing this U-turn in public policy, Mayor Emanuel also reversed the course of city hall in planning the future development of the waterways. He has taken a leadership role in fulfilling Goldberg's vision of the Chicago River as a front street or parkway of nature in the city. On the heels of his surrender to the U.S. EPA in September 2011, he announced plans to give the river greater priority. And in contrast to his predecessor, he earmarked $16 million to begin building four boathouses in the neighborhoods, complete with bicycle, canoe, and kayak rentals. "The truth is," the mayor reflected, "because of our lake, we've only thought of it as our front yard. I now want to turn our attention to our backyard and have that same vision [for our riverfront] neighborhoods. . . . I want that river to be for all of us what the lake has been." Drawing on "starchitects" like Jeanne Gang, Mayor Emanuel has turned city hall into a full partner with the private sector in charting the course of riverfront development.[85]

Perceptions of the river as an urban wilderness are having their own feedback effects on changing attitudes toward the environment. In the spirit of the city-beautiful movement of the Progressive Era, virtually all of its beaches and harbors were turned into highly ordered recreational and leisure spaces. The dramatic rebirth of the river's native flora and fauna created a powerful impression of the successful regeneration of nature. It inspired environmentalists to also envision the lakefront as a wild place. "Chicago's future," the *Sun-Times* stated, "is that of a post-industrial environmentally friendly 'green' city."[86]

Demonstration patches on the lakefront are duplicating the renaissance of the river's biotic community. Whether on Northerly Island in the downtown area or the Pratt Avenue Beach on the far North Side, these nature preserves have become equally amazing places of dune grasses, wildflowers, and waterfowl not seen in many cases for more than a century. The return of these plants and animals to their shoreline habitats, in turn, is drawing new groups of stakeholders in the environmental protection of Chicago's waters.[87]

Victory in the long battle to reclaim them for recreational use has triggered the start of a new one to save the Great Lakes and Mississippi River basins from destroying each other. In 1998, Congress began a national counterattack against the Asian carp and a rogues' gallery of other alien invaders. In 2010, it put the U.S. Army Corps of Engineers in charge of this mission when it created the Great Lakes and Mississippi River Interbasin Study project. Besides building a fourth-generation electric fence, the U.S. Army Corps has been monitoring life below the surface with a full battery of tests.[88]

While the reams of data its task force has been collecting for scientists to study are first rate, its engineers' continuing pursuit of the myth of the technological sublime is less reassuring. Will their recommendations give ecosystem approaches top priority? "To our knowledge," a U.S. Army Corps brochure on the electric fence boasts, "our barriers are the largest of their kind in the world and the only one on a highly-trafficked, commercially-navigable waterway." It takes special pride, moreover, in its structural approach of building an ever-bigger and more-powerful fail-safe machine. It still maintains a mind-set that envisions Chicago's rivers as economic resources for "treated wastewater, stormwater management, and navigation." In 2011, Congress moved up the deadline for a final report by two years. Its impatience mirrored the nation's anxiety about our ability to protect our waters from the unintended, dire consequences of efforts to turn them into an envirotechnical system of total control.[89]

Conclusion

Cities, Adaptation, and Climate Change

The Rainy City

Five years after Mayor Rahm Emanuel agreed to bring the city into compliance with the Clean Water Act, Chicago remains in a wet period of climate change. On July 23, 2011, two months after signing the consent decree, the "rainiest day . . . since 1871" reminded Chicagoans that reversing the course of public policy was just a first step toward achieving this goal. The account of the storm in the *Chicago Sun-Times* referred to the origins of the systematic record-keeping of weather statistics in the United States. The newspaper reported 6.86 inches (in.)/16.98 centimeters (cm) of precipitation in a twenty-four-hour period.[1]

The torrent was slightly more than what fell during the previous wettest day, which had taken place only three years before. It was the ninth and last "hundred-year rainfall" in a five-year cluster of extreme weather events. It added to the year's totals to make 2011 the second-wettest year, a little behind the historic peak of 2008, when nearly 51 in./130 cm had soaked a prairie wetland. The lightning storm caused all the usual flood damages, submerging basements, knocking out power to 160,000 homes, and forcing the sanitary authorities to release 2.25 billion gallons (gal.)/8.52 billion liters (L) of polluted storm-surge runoff from the Chicago River into Lake Michigan.[2]

Analyzing recent weather patterns, scientists predict a "future [of] intensification of hourly precipitation extremes."[3] Rainstorms and heat waves will continue to set new records. Feeding national data from the 2000–2013 period into a computer, the meteorologists created a range of forecasting scenarios. Their models demonstrate that the damaging effects of climate change are becoming more frequent and severe. They theorize that global warming raises the amount of moisture in the atmosphere, which in turn causes more rain. The risks of flood damages will keep getting greater, because higher humidity levels increase the thermal effects of greenhouse gases. For now, we are trapped in a spiral of intensification.[4]

For a city built on a prairie wetland, the ability of its politicians, policy makers, and citizens to adapt to climate change will require holistic approaches to water management and land regulation as well as to the ecosystems of its lake and rivers. Until then, Chicago will continue to sink and to suffer from flooding. In a pioneering study, the Center for Neighborhood Technology analyzed damage claims during a five-year period between 2007 and 2012 for all of Cook County. Although underrepresentative of all flood victims, the 181,000 claims received by government and private insurance agencies covered 97 percent of the area. The claimants' reports indicated that no one outside the central business district (CBD) was safe from the downside of climate change. The researchers also calculated that 42 percent of the county's land had been paved over. However, they did not find a correlation between the areas with the most impervious surface areas of 60 to 88 percent and flood damages.[5]

Rather than specific hydrological zones, the data showed patterns of a social geography of class. Poor people were disproportionately affected by Chicago's third cluster of heavy rainfalls. Two-thirds of the areas with the most damages were neighborhoods where people earned less than the median income. Beginning on the far West Side, they formed a half crescent of the outlying bungalow belt that arched and widened to the southeast, including South Chicago and the entire Calumet District. During this five-year period, flood damages resulted in a total of $773 million in claims. While their individual amounts ranged widely, they averaged $4,272 each time a basement filled up with sewage and/or a property was covered with water. Even one such disaster subtracted from the resale value of a home by 10 to 25 percent.[6]

Climate change, the study found, raised the costs beyond the loss of material wealth into the realm of significant psychological damages. Four out of five victims of flooding reported feeling stress, and one in ten got sick. For 75 percent of these households, the emotional costs mounted from time spent out of work cleaning up their homes and throwing away their possessions with sentimental value. In response to the failure of government to control storm-water surges, an equal proportion of them had installed their own self-defense devices, such as rain barrels and pumps. But expressing their loss of faith in government, only one in twenty respondents believed these measures would be enough to protect their homes from the next "hundred-year rainfall." The sanitary authorities confirmed their fatalism when they announced that a third, final phase of the Tunnel and Reservoir Plan (TARP) would require another billion dollars of public funding and another quarter century of construction. Until at least 2029, they promised, Chicagoans could expect no relief from flood damages, estimated to cost between $130 million and $188 million each year.[7]

The River City

Closing the beaches, the great flood of July 2011 hammered home the "Groundhog Day" lesson that a combined sewer system would require flood control on a regional scale to bring Chicago up to the federal government's standards of pollution control. By then, the population of the five collar counties had increased to a point of being on a par with the 3.1 million inhabitants of suburban Cook County. Sprawl resumed a half-century trend in settlement patterns toward growth at the periphery at the expense of decline at the center, which shrank to 2.7 million people. On a regional scale, moreover, these simultaneous forces of deconcentration and concentration reproduced the city's social geography of racial/ethnic and class segregation.[8] While the Chicago River downtown reignited a building boom of luxury condominiums on its banks, the city's other rivers continued to overflow theirs in the suburbs of middle- and working-class homeowners. New questions of environmental justice joined the debate in the public sphere over the best ways to adapt to living on a flood-prone marshland during a period of accelerating climate change.[9]

At the same time, the ongoing recovery of the ecosystems of Chicago's six watersheds continued to build political support for the full restoration of their rivers as places of wild nature. In 1997, someone in Northeastern Illinois sighted a white pelican, one of the largest birds in North America. Fast forward nineteen years to the spring of 2016, when the *Chicago Tribune* boasted, "Once-rare white pelicans flood into Lake County."[10] Several flocks numbering in the hundreds each had come to fish in the riverine channels and lakes of this part of the region. They joined other recent, annual visitors, such as loons and swans. Unlike their cousins, brown pelicans, they do not dive into the water to catch fish. Instead, white pelicans fly close to the surface of the water and scoop them up in their large open mouths. "The sky was full of them," a resident of Chain O'Lakes observed. "They just came peeling out of the sky and into the channel." People rushed out of a local harbor bar to see four to five hundred of them flocked together near the piers. "They're beautiful birds, big birds, with four to five feet [1.5 to 1.8 meters] wingspans," a patron exclaimed.[11] They had made another convert to the cause of protecting the environment.

A growing scientific understanding of aquatic ecology and climate change advanced (re)learning how to live on a prairie wetland. The grassroots stakeholders, political supporters, and professional experts in charge of the six county forest-preserve districts took the lead in designing with nature to help control storm-surge runoffs and to expand green space for recreational and leisure activities. They not only followed in the footsteps of the Native Americans and Father Jacques Marquette, who moved his belongings to higher ground from the riverbanks during flood stages of wet weather; they also absorbed the lessons of the Progressive Age reformers Jens Jensen and Frank Lloyd Wright—that the tall grasses and marshlands of the Midwest were inspirations of the sublime, not seas of mud to be paved over. These advocates of open lands and wild rivers did not forget Jane Addams, the playground reformers, and the value that city people placed on getting back to nature.

In the postwar era of the baby boom, the caretakers of the forest preserves came to appreciate the wisdom of Cook County's landscape engineer, Roberts Mann, who believed that the goal of conservation had become a job of preservation. He became a teacher of an environmental ethic that prescribed a stewardship role for those in charge of managing these precious public lands. They needed to be saved from overconsumption and the unintended consequences of thoughtless abuse. Although the origins of the five collar counties' forest preserves spanned a long historical bridge from 1915 to 1971, each had accumulated 20,000 to 30,000 acres/8,000 to 12,000 hectares (ha). For the most part, they are being nurtured back to their presettlement state of nature. The enactment of land-use regulations that require incorporating flood controls in the design of private developments also represents steps toward adapting to a prairie wetland during a period of climate change. The ultimate goal is to turn a weather cycle of more-frequent and more-severe rainfalls from a liability into an asset.[12]

In the conclusions that follow, I lay out three approaches to planning the urban environment, which map different paths to the future. These strategies of water management and land regulation correspond to increasingly larger geographical scales of city, metropolis, and region. They illuminate correspondingly different perspectives on the ways in which Chicagoans have learned how to live on a prairie wetland. The first point of view on the Chicago River reflects city hall's adherence to traditions of adopting private plans as public policies. The Richard M. Daley and Emanuel administrations have benefited downtown developers of postmodernist icons of the elite at the expense of outlying residents in flood-prone neighborhoods. The second casts a spotlight on machine politics by focusing attention on the corruption of the Forest Preserve District of Cook County (FPDCC). The final viewpoint from a satellite looks down on the Great Lakes and the Mississippi River to reveal efforts to solve the problems posed by alien species moving back and forth through the interbasin connections at Chicago. A new consensus informed by hydrology and limnology has been working to formulate plans for the future in holistic, ecological terms.

The perfect place to observe the city center remains Wolf Point, where the Main Channel of the Chicago River splits into the North and the South

Branches. During pioneer days, a tavern and ferry crossing were located on this strategic peninsula, becoming the social crossroads of the community.[13] But today, finding a spot to stand on is difficult, because every square foot has been turned into a construction zone. In what many local pundits consider to be Mayor Daley's parting gift to his real-estate pals in the CBD, the city government rezoned Wolf Point's little park for the benefit of the heirs of fellow Irishman and family friend, John F. Kennedy. In early 2011, they were granted permission to squeeze three oversize skyscrapers on this historic, albeit tiny, 4-acre/1.6 ha plot of land, blocking the view of the lake for the residents living in recently erected luxury condominiums across the river on the western banks of the North Branch.[14]

Mayor Emanuel did not deviate from a political culture of deference to the business community stretching back to the Progressive Era. At least since the days of the Commercial Club's 1909 *Plan of Chicago*, city hall has followed the roadmap of city building laid out by its downtown elites. In his first year in office, he fell into line by giving a major contributor, Hines Interests Limited Partnership, a $30-million grant for public-works improvements. He awarded money earmarked for "low-income blighted areas" to the developer of the Kennedys' skyscrapers for a second plan to build a supersize office tower on the riverfront nearby. Setting off a speculative land grab reminiscent of the Roaring Twenties, other global investors snapped up the few remaining sites on the Chicago River in the downtown area for major construction projects. In April 2016, the Hines Interests spokesman confirmed, "These are some of the last empty spaces along the river [in the CBD]. Filling them is a big event."[15]

The mayor, moreover, did his best to help his friends in the real-estate industry by promoting the improvement of the Chicago River, at least in the CBD. Outdoing the billion-dollar Wolf Point project, the Vista Tower will rise ninety-three stories to become the city's third-tallest skyscraper. Located on the Main Branch across from Cityfront Center and designed by Jeanne Gang, its undulating shapes represent the waves of Lake Michigan. The top prices for its top condominiums may become what real-estate experts call a "game changer." More-reasonable, if upper-end, dwellings in a River City–like megaproject are planned by developers of a property on the east bank of the South Branch near Harrison Street. Their twenty-seven hundred residential units will add to the seventy-eight hundred apartments in the downtown area that are coming on the market by 2018.[16] Following in the tracks of his 1920s predecessor, William Hale "the Builder" Thompson, Mayor Emanuel added a section to the river walk across from Wolf Point to help provide the privileged residents of downtown all the amenities of urban nature. Faced with huge deficits in the city's budgets, however, he had to borrow $100 million to pay for this piece of the waterfront promenade from the federal government.[17]

In a similar way, city hall remained committed to the use of the Chicago River outside the downtown area as an industrial corridor rather than a natural amenity. Fighting against this traditional image of a "working river," even the U.S. Environmental Protection Agency (EPA) retreated in the compliance agreement from fully protecting the South Branch. Beyond Bubbly Creek, water-quality standards were set at a lower level of "recreational uses with limited water contact."[18] City hall's redevelopment plans for a large patch of property 3 miles/4.8 kilometers (km) downstream from Wolf Point revealed a continuing exploitation of the Chicago River as a sacrifice zone. Located on the Southwest Side, the combined 115-acre/46 ha brownfield sites of two electric generating stations lies in the predominately Latino neighborhoods of Pilsen and Little Village. Closing these historic, albeit air-polluting, coal-fired power plants represented a major victory for grassroots protest organizations fighting on behalf of environmental justice. But in 2012, Mayor Emanuel's proposed plan reneged on his promise to give every riverfront neighborhood public access to a reclaimed waterway clean enough for swimming.[19]

On the contrary, city hall's vision of *Chicago Sustainable Industries, Phase One: A Manufacturing Work Plan for the 21st Century* did not give the mostly Mexican American residents living in the shadows of the old smokestacks access to the riverfront.[20] Like the neglect of the South Park trustees in providing African Americans with parks and beaches during the first Great Migration, city hall's map did not show any additions of open space in one of the city's most crowded districts. "Little Village," a local-based report underscored, "is one of the youngest and densest communities in Chicago, and has the least green space per resident. . . . With one of the city's highest obesity rates, and lowest activity rates, access to recreational space and food-producing gardens remains a pressing concern."[21] The mayor's task force did not include any river-walk promenades, boathouse parks, or even a bike path in their planning maps. In fact, the river canal is barely visible, hidden under brightly colored areas zoned for rebuilding. All of these brownfield sites were given over to the private sector. The electric company, for example, will sell its land to the highest bidder, who promises to put up industrial parks of high-tech companies around its remaining equipment. These new tenants were expected to generate few jobs for the people of the community, where unemployment rates stood at 20 to 25 percent for young men.[22]

The focus of recent, debt-ridden administrations on downtown as the path to the future has come at a price of disinvestment in nonwhite neighborhoods throughout the city, such as Pilsen and Little Village. In contrast to the cosmopolitans, who could afford a twenty-four-hour lifestyle at the center, many of their less-fortunate neighbors located on the other side lived in fear for their lives from a deadly combination of unemployment, gangs, drugs, and guns.[23] Such a violence-prone environment—let alone futures without promise of achieving the American dream—helped persuade 180,000 African Americans, or 18 percent of the city's black population, to leave during the opening decade of the new century. Since then, Chicago has continued to lose citizens faster than any other big city and metropolitan region in the nation. From 2013 to 2016, Illinois lost more residents each year than any other state. In the most recent census count, 114,000 people moved away. Analyzing the numbers, demographers found that Chicago was "the root of the problem" because of its violent crime, tax hikes, machine politics, and bad weather.[24] In addition to the poor quality of daily life in some of Chicago's neighborhoods, chronic flooding led their homeowners and shopkeepers to leave.

The Regional City

The Democratic Party's corruption of the FPDCC during the 1990s and 2000s provides a case study of a second option—machine politics—as a way forward in building the urban environment during a wet period of climate change. Under the analytical lens, the outer belt of parks supplies a useful point of view for examining the formation of public policy and planning on the scale of a metropolitan region. This is a story of two generations of Daley sponsorship of two generations of the John Stroger family. In contrast, Emanuel, like Harold Washington, was a product of machine politics, who remained independent from it by staying outside city-hall politics.[25] The African American ward boss and his son, Todd, were consummate party insiders, rising to the top of Chicago's county government as president of its board of commissioners. Following in the footsteps of Ed Kelly, Timothy Crowe, and Anton Cermak, they built an imperious regime of nepotism and graft out of its $3-billion annual budget and twenty-six thousand employees.[26]

In addition to the failure of city hall to save home basements from flooding and lakefront beaches from pollution, the corruption of the FPDCC strengthened belief among Chicagoans that government itself was the problem, not the solution to climate change. Like the save the river protesters, the many friends of the parks and the forest preserves turned from allies to antagonists of their elected officials, because they were harming the environment. In 2001 and 2002, these advocates of outdoor recreation exposed the gutting of the district by its governing board of county commissioners. They had "looted and vandalized" its parklands, turning its 68,000 acres/27,500 ha into stinky garbage dumps of "dilapidated facilities" and its workforce into a useless army of "do-nothing job[holder]s."[27] Yet the backstory of the decline of these riverine open spaces begins with the rise of its long-term president of the FPDCC, John Stroger.

Armed with a college degree in business management, the twenty-four-year-old Stroger had joined the second Great Migration from the South to the South Side of Chicago after World War II. Arriving in 1953, he started a lifelong career as a politician by becoming a precinct captain and getting a city job as an auditor. For the next fifteen years, he worked his way up the patronage ladder for faithful service to the Democratic Party while also earning a law degree to improve his social status. In the midst of the racial and political uprisings of 1968, Mayor Richard J. Daley appointed him the committeeman of the Eighth Ward and slated him to be a commissioner of Cook County two years later. Like several other key African American leaders promoted by Daley, Stroger was a Catholic, which strengthened his personal ties to the Irish faction of the party while limiting them among the mostly Protestant membership of his community. As a supreme mark of loyalty to the Daley family in 1983, Stroger was the only black ward boss to endorse its heir apparent for mayor, Richard M., instead of Washington. Three years after the Irish/Catholic faction of the party finally regained control of city hall in 1989, Mayor Daley returned the favor by elevating Stroger to the presidency of the seventeen-member board of commissioners.[28]

The up-and-coming party functionary learned from a master of boss rule, the first Mayor Daley, how to gain absolute control over government agencies and political organizations. Like Cermak's moniker of respect, "The Mayor of Cook County," Stroger's label among party insiders was "Mr. President." Stroger's tightfisted, one-man reign of power reduced his chief assistant to a demeaning role as the "Ed McMahon of Cook County Government," because his only job—like Johnny Carson's TV sidekick—was to laugh at every joke offered by his boss. In October 2002, however, an investigation by the Friends of the Parks (FoP) and the Friends of the Forest Preserves (FoFP) confirmed that "everything must go through downtown"—that is, through Stroger.[29] As the chief executive of the county government, he also became the president of the forest-preserve district, one of its smaller branches, with a relatively modest $150-million budget and five hundred employees. Given Stroger's much larger and immediate responsibilities for health care and criminal justice, civic reformers did not devote much attention to the forest preserves for almost a decade.[30]

The publication of the environmental reformers' investigation became the closing scene of the first act of another one of Chicago's three-act plays of machine politics: corruption exposed, investigated, and rolled back. At the beginning of the year, another report by reform commissioner Michael Quigley had raised its curtain of public scrutiny on Stroger's neglect of the natural world. The name of his presentation can serve as this version of the play's title: "The Lost Mission." In effect, he accused Stroger of "stealing from its [the forest preserve district's] land acquisition fund and cheating restoration and maintenance efforts while overstaffing recreation [with patronage workers]."[31]

Until the commissioner made this official complaint, the warning alarms set off by the watchdogs of the FPDCC had remained uncoordinated and unheard offstage. But after Quigley put the pieces together into a coherent script of "shabby neglect," the civic melodrama proceeded through its three acts, a sequence of scandals that may be called "The Tree Police," "The Big Mess," and "The Family and Friends Plan."[32]

Because nobody in Chicago is against the forest preserves, the early alarm bells warning of their peril at the hands of elected officials were ignored by civil society. In large part, allegations against the board of commissioners failed to provoke a resounding outcry of public protest, because its members were indicted for acts of mismanagement, not malfeasance. In May 1999, Commissioner Quigley highlighted in his report, the board broke its duty to uphold the now sacred trust of preserving the district's lands forever. Although the piece sold off was fewer than 5 acres/2 ha, the sale of the land for the expansion of a convention center in Rosemont set a dangerous precedent for the future. Over the next two-and-a-half years, this betrayal of Stroger's stewardship role of protecting the county's nature preserves brought to light that he had purchased only 254 acres/103 ha at a cost of $4.5 million, while each of the collar counties had been acquiring thousands of acres of prairie wetlands. For example, DuPage County was spending $75 million to buy 2,000 acres/809 ha.

Stroger had ignored the advice of open-lands advocates early in his administration to spend the $10 million in the land-acquisition fund, because available property was running out. They had urged him to expand the district's holdings to its full authorization of 75,000 acres/30,350 ha. Their land surveys had identified 40,000 acres/16,000 ha that would be suitable additions, but now only half that amount was available. He also had opposed the recommendations of the FoFP to sponsor a $100-million-plus bond referendum to buy more land and to restore the district's recreational facilities. To the contrary, he had driven its operating budget $20 million into the red by adding several high-paying, "do-nothing jobs" to the payroll. Repackaging all of these allegations as "The Lost Mission" in January 2002, Commissioner Quigley's charges of criminal wrongdoing galvanized many grassroots organizations devoted to protecting the environment into action, including the FoP and the FoFP.[33]

His allegation that Stroger had looted the land-acquisition fund to cover growing budget deficits set off alarm bells that raised the curtain of public scrutiny on the first act of this corruption scandal. Repeating the tactics of the Sierra Club in its campaign to save the lake, an army of do-something volunteers from environmental groups focused a spotlight on the decrepit state of the FPDCC. They took to the field, conducting surveys, interviewing workers, collecting data on its two hundred sites, and persuading the press to report its findings. The FoFP reacted to Quigley's bombshell by following the money, the 70-percent increase in the district's operating costs. What they found set the main theme of "The Lost Mission": Stroger was sinking the forest preserves under a top-heavy boatload of cronies, relatives, and neighbors. In August, the state's attorney drew attention to this bloated system of patronage under a regime of machine politics when he indicted at least 10 of the district's police force of 150 men for participating in an illegal scheme to collect overtime pay without doing any work. Perhaps betraying more than he intended, the chief of the force confessed, "I have no idea of what's going on." The finance director, who happened to live just a few doors from Stroger's house, resigned rather than face the coming inquiry on the payroll.[34]

Keeping the attention of civil society two months later, in October 2002, the FoFP issued its scathing report on the "tree police" as a complete waste of $7 million each year, money that was desperately needed to revive the budget-starved maintenance department. The better-trained Cook County Sheriff's Police had an overlapping responsibility to protect the safety of the forty million visitors who came every year to enjoy the forest preserves. The FoFP's volunteers discovered that virtually the entire force comprised politically connected appointees who were completely unqualified for their jobs. The investigators interviewed one of its few professionals, who had resigned a year earlier. He told them that he had quit in disgust after being forced to hire a man who had showed up drunk for his job interview. Stroger filled this former deputy chief's $62,000 position with the son of Alderman Edward Burke, a ward boss and the chairman of the powerful finance committee of the city council. When the state's attorney began to disclose the overtime-padding scandal, Stroger retaliated by hiring a member of his ward organization, who charged an annual retainer of $83,000 a year as his legal adviser. And seemingly as if to punish his antagonists, he cut in half the staff of the maintenance department, the actual caretakers of the forest preserves. Those with the heaviest political clout and lightest workloads were the ones who remained on the district's payroll.[35]

In July 2003, the scandal of "The Big Mess" moved the script of Chicago's political theater of machine politics forward from exposing the causes of corruption to the second act of investigating its effects on the public health and welfare. In this case, the district's recreational facilities were "marred by broken or closed toilets, vandalized picnic shelters, inoperable water pumps, gang graffiti and trash." "In the last few years," a daily visitor lamented, "it's really gone downhill."[36] Although anecdotal, the story of the thwarted efforts of fifty high school students to help clean up the forest preserves gives insight into the reasons why the area had fallen into such a broken-down state of despair. The FoP organized the students' trip to Beaubien Woods in the Calumet District, where they watched two of Stroger's favored maintenance workers sit in their garbage truck during the entire time that they spent collecting several plastic bags full of trash and carrying them to be picked up at the proper collection point. But the garbage collectors purposely ignored the bags, allowing raccoons and other scavengers to tear them apart and the wind to scatter their inedible contents far and wide over the coming week. Frontline workers resented Quigley and the other reformers, blaming them for the cutbacks in the rank and file, while the ward bosses protected the jobs of the best-paid administrators at the top.[37]

Following his mentor's example, Stroger sought to bring the curtain down on the scandal by hiring professionally competent, albeit politically safe, replacements to manage the forest preserves. Because the exposure was contained within a comparatively small division of the county government, he did not have to make the ultimate act of contrition by bringing in an outside expert like Vinton Bacon. In September 2003, he hired a new liaison, Pam Munizzi, to take over daily operations of the district at an annual salary of $104,000. She was a close friend of the Daley family and had been appointed to succeed Richard M.'s brother, John. She took his seat in the state house of representatives as he rose to the senate and the board of the county commissioners as chairman of the finance committee. Munizzi had the support of reformers from her prior service in city government. One of Stroger's critics on the board, for example, admitted that "she always had a solid reputation as a manager at the Chicago Park District." Four years later, the replacements had made significant progress in cleaning up the picnic grounds and making long-deferred repairs.[38]

Coming from offstage, the unscripted intrusion of federal and state prosecutors into the closing scene of the second act kept the curtain of public scrutiny from coming down on the investigative phase of the patronage scandal. In August, Stroger believed that he had begun the rollback of reform. He persuaded his fellow Democrat, Governor Rod Blagojevich, to veto a $100-million bond referendum bill that would authorize the FPDCC to buy land and pay for recreational facilities. He wanted to deliver a political defeat to his antagonists, and he was afraid of facing a vote of no confidence by the taxpayers. But just as he was announcing the hiring of Munizzi a month later, federal agents took down a group of the "tree police" for using the forest preserves as their safe haven for dealing drugs. To satisfy the demands of reformers, Stroger would have to cut the size of the force in half.

Coming on the heels of the drug bust, the overtime-padding case blew up in his face when the state's attorney announced the discovery of a more-sinister plot behind the phony work claims. A hidden mastermind in the finance department set this illegal conspiracy into motion by getting police officers to agree to submit overtime claims in exchange for kickbacks. The criminal mind in charge of the scheme was one of Stroger's precinct captains, who had gained his appointment in 1995 from his elite sponsor. Weakened politically, Stroger could not head off a vote at the state level overriding the veto of the FPDCC bond act. A month later, in December, the county commissioners also handed him a vote of no confidence. The Democrats helped the Republicans defeat the administration's annual budget proposal for the first time in thirty years. Yet the revolt of his fellow ward bosses turned out to be symbolic, because the vote was reversed the next day. Marking a political truce, the investigative stage of the corruption scandal was put into a state of suspended animation until the upcoming primary elections in March 2006. In the meantime, the district's payroll account continued to sink in a sea of red ink.[39]

Despite suffering some humiliating setbacks in his pursuit of power, Stroger was able to reestablish control over the county government, setting the stage for a pushback to restore the prerogatives of the political machine in his next term of office. But a week before the primary election, the seventy-seven-year-old trooper suffered a severe stroke. Left completely incapacitated, "Mr. President" was not seen in public until his funeral two years later. Nonetheless, the Democratic Party's leaders pulled out all the stops to get their constituents to vote for him, promising that he would soon recover and resume his duties on their behalf. His campaign managers also played the sympathy card, appealing for a last hurrah on the basis of a public life devoted to serving the community. His election victory over one of the reform commissioners opens the curtain on act three, "The Family and Friends Plan."

Behind the scenes, Mayor Daley planned to elect the near-dead John Stroger in the primary contest and then put his forty-three-year-old son, Todd, in his place in the November general election. The elder Stroger's condition was kept shrouded in mystery while Daley defeated all the other more-senior and more-experienced African American wannabe candidates. In July 2006, the resignation of one member of the Stroger family was coupled with the announcement of another as his successor in the coming campaign. Like his father, Todd had become a professional politician, starting in 1992 as a state representative. He then took several patronage jobs in the county government until 2001, when the mayor chose him to fill a seat in the city council that had been vacated by the death of its elected officeholder. For faithfully paying his dues to the Daleys, he was rewarded with a promotion to this highly coveted position of power.[40]

The central motif of the tumultuous final act of the corruption scandal, "The Lost Mission," became the exposure of an increasingly vulnerable regime of machine politics. Like the concurrent opposition Mayor Daley was facing from the save the lakes and the rivers movements, a bipartisan coalition of good-government commissioners on the county board joined a phalanx of environmentalists to contest the policy decisions of the party leaders over spending priorities and "do-nothing jobs." The annual battle of the budget pitted a minority of civic-minded reformers against Stroger's majority of loyalists. Reprising the tree-police episode, the fiscal sleuths followed the money trail that led to the unearthing of another raid on the district's funds to cover deficits in the county's payroll account.

Chief Financial Officer Donna Dunnings, Todd Stroger's cousin, defended the district's secret cooking of the books. Although she had earned respect as a sixteen-year veteran of the budget office, Dunnings had come under immediate fire in 2007 for reneging on her promise not to accept a hefty raise above her predecessor's salary. She took the board's grant of a record $175,000 a year. Two years later, the charge of nepotism stuck to Todd Stroger when she suffered a sudden fall from grace and was forced to resign. He had hired one of her close friends to fill a $60,000-per-year job, despite his long record of felony convictions. When Dunnings posted bail for him after another arrest, the mirror images of the Stroger family tree and the organizational chart of the county government came to light in the public sphere. A year after Stroger had to fire her, she was hired again at an annual salary of $79,000 in the county assessor's office.[41]

In contrast to previous engagements in Chicago's political theater of corruption, this play closed with an unexpected final twist in the plot: the rollback got rolled back. For the first time, in February 2010, the grassroots expressed a vote of no confidence in machine politics by dumping Todd Stroger into last place in the primary elections: he got support from a mere 13 percent of the Democrats. To be sure, his constant demands for more budget spending rather than saving undercut his support among taxpayers, but his pillaging of the forest-preserve funds to underwrite his "family and friends plan" cost him backing among an even larger constituency of urban and suburban citizen-guardians of the forest preserves. Prominent in their critique of his administration was his paltry record of adding fewer than 500 acres/200 ha of new lands to the district, while the collar counties kept making substantial additions in the thousands.[42]

The pivotal influence of these advocates of the protection of the environment was made evident in the voters' choice of Toni Preckwinkle over two other candidates representing rival African American and Irish Catholic factions of the Democratic Party. The former history teacher had aligned with the Harold Washington, biracial reform wing of the party twenty years earlier and had been serving Bronzeville as its alderwoman ever since. Under her administration, the FPDCC has been turned into a model of planning ecological approaches to nature preservation.[43] Flood control has been made an integral part of its plans for the adaptation of its recreational facilities to a prairie wetland.[44]

The Ecological City

From the space station orbiting Earth, the drag of machine politics on progress toward the dual goal of pollution control and flood control is not visible.

What can be seen is the contrast between the metropolitan region's natural world of blue and green during the daytime and the artificial lights after nightfall that demark the extent and density of its settlement. Only when a rainstorm has sunk Chicago do the two views overlap by displaying the built environment as well as open space under water.

This global point of view on the Great Lakes and Mississippi River basins is the best way to read the roadmaps of planners with holistic approaches to the problems of a flood-prone environment. For the experts put in charge of providing policy makers with a pathway to the future, an ecological framework of analysis has increasing led to a merging of land and water management. Each of these natural resources affects the uses of the other in significant ways on a prairie wetland. As we have seen, on the one hand, the more it was paved over, the greater the volume and speed of runoff surges of storm water, which in turn damaged human lives on the surface and aquatic ecosystems under it. On the other hand, a still ongoing wet period of intensification resulted in new zoning regulations and best practices. They were designed to turn the rain into an asset for reducing the amount of water drained from the Great Lakes. The new approaches of the eco-planners are rejuvenating aquifers, restoring the health and beauty of the landscape, and reclaiming the region's six watersheds for recreational use.[45]

They have shown, moreover, that building the ecological city involves the adaptation of its economic and social fabric to climate change in holistic ways. In 2010, the Chicago Metropolitan Agency for Planning presented a comprehensive blueprint of the future titled *2040*. In addition to protecting the environment, it included chapters on making investments in human capital, workforce education, economic innovation, and better housing and transportation. To create "livable communities," the plan outlined a wide range of policy objectives, from eliminating so-called food deserts in abandoned neighborhoods by encouraging "sustainable local food" production within them to controlling floodwaters. The plan called for expanding park and open spaces designed with retention ponds, clustering housing, and installing eco-friendly devices, such as rain barrels and green roofs, that would detain storm-water runoff at its source.[46]

In a separate report, *Water 2050*, the planning agency expanded its ecological framework of analysis to incorporate a set of best practices for managing the water supply. Demand management means setting a price on water that fully covers the costs of providing it safely to users and then getting rid of it in a way that does not harm the environment. Putting a price on needless waste and unrepaired leaks in underground pipes and home fixtures works wonders: it results in tremendous savings in the consumption of water and in the energy it takes to pump and process it in and out of the system. In addition to reversing Chicagoans' old-fashioned way of thinking about Lake Michigan as a free commodity—a fountain inexhaustible of everlasting purity—the experts put the greatest stress on enlisting the public in ways that would help absorb and/or slow down storm-water runoff. Unable to predict the future of climate change, they also made provisions for a coming dry period of water shortages. "Climate variability and change," the report predicted, "will very likely influence precipitation patterns, the frequency and severity of droughts, and affect streamflow. For example, since 1970 northeastern Illinois has experienced a 10% increase in precipitation leading to a 35–40% increase in average streamflow."[47]

Like the eco-planners, the engineers and the scientists working for the sanitary authorities have learned to redraw their blueprints of public-works projects to embody more-holistic approaches to water management. Under congressional and executive mandates, the U.S. Army Corps of Engineers was instructed to report on the best ways to save the health of the aquatic ecosystems of the Great Lakes and the Mississippi River from invasions of alien species moving from one basin to the other. Its report of 2014 identified 254 "aquatic nuisance species": fish, crustaceans, mollusks, plants, algae, protozoa, and viruses. From this list of marine disruptors, biologists identified thirty-five as posing the greatest risk of harm from an interbasin transfer. The Asian carp remains the number-one threat if it moves from the rivers into the lakes. While the U.S. Army Corps found dozens of contact points between the two basins, the "Chicago Area Waterways System" was the most likely place for a disastrous interchange. Its envirotechnical rivers in reverse and its use of Lake Michigan as the ultimate sink for storm-runoff surges keep the water flowing in both directions.[48]

In contrast to the Deep Tunnel plan of the late 1960s, the interbasin report presented several multilayered holistic approaches to engineering a flood-prone environment. Rather than offering a single one best way, such as the big-technology TARP, the U.S. Army Corps' experts generated eight alternative models. Leaving aside the no-cost, do-nothing option, the seven others required a combination of chemical treatments, physical barriers, and separation technologies, such as electric fences, screens, and ultraviolet light. Stretched over a quarter century to build, they would cost between $8 billion and $18 billion, with most of the funds earmarked for mitigation of the various plans' impacts on pollution control and flood control. The four models with a physical, hydraulic separation between the basins would permanently end the dumping of billions of gallons of storm water from the rivers into Lake Michigan. To prevent damaging floods, the engineers proposed building a virtually duplicate, parallel system of tunnels and reservoirs to handle all the runoff surges. In effect, these models represent an admission that TARP's original design was only half the size needed for a prairie wetland undergoing rapid urban sprawl.[49]

Of course, the politicians in charge of the Metropolitan Water Reclamation District (MWRD) have supported the U.S. Army Corps' recommendations to spend billions of federal dollars on local public-works projects. The sanitary authorities have negotiated a deal with the federal government to implement the 2011 consent decree that benefits its private building contractors at the expense of its flood-prone constituents. Completion of phase three can take until 2029, because city hall's favored tunnel-and-pit diggers have to work only when the price for bedrock rubble reaches a minimum level of profit. In the meantime, Chicago's combined system dumps raw sewage into the rivers when it rains 0.67 in./1.7 cm or more, or a little more frequently than once a week on average. These overflows keep the rivers polluted with dangerous microbes and in violation of the Clean Water Act. Since the first storm-water reservoir opened in 1998 near O'Hare Airport, environmental reformers estimate that it has saved about $25 million a year in damages. Calling for a speedup in construction, they project additional annual savings of $130 million when the nearby McCook Reservoir in the Des Plaines River watershed and the Thornton Transitional Reservoir in the Calumet District finally open.[50]

Accused by the Friends of the River (FOTR) of crony capitalism for foot dragging on the never-ending TARP, the district also seemed to move at a snail's pace in equipping its sewage-treatment plants with third-stage disinfection machines. Until a year before his retirement in 2011, Mayor Daley's point man on the MWRD's board of trustees had been Terry O'Brien. Serving for twenty-one years, the party loyalist eventually rose to become its president. But his lame-duck sponsor could not carry the day for him in his bid to replace Todd Stroger at the top of Cook County government in the 2010 primary contest against the grassroots insurgency of reformer Preckwinkle. Nonetheless, the sanitary authorities slowed the process of installing long-proven technologies, such as radiation with ultraviolet light. Since the inauguration of Mayor Emanuel, the board's majority of party functionaries put these wastewater purification machines through a pointless, time-consuming gauntlet of cost-benefit studies, laboratory-scale experiments, and test-station analyses. And they repeatedly claimed that making the rivers clean enough to swim in would bankrupt the district's taxpayers. Yet the equipment finally put into operation on the North Branch of the Chicago River and the Calumet River came in at a reasonable cost, according to environmental-protection groups. Progress at the main treatment plant at Stickney on the Southwest Side remained stalled.[51]

Although Chicago's sanitary authorities were among the first to conceptualize the city center and its suburban periphery as a single metropolitan region, they were among the last to see connections between water management and land use as well as between social and physical space. Their plans to continue engineering the environment with big-technology projects proposed several ways to reduce the costs of climate change for its human inhabitants. However, they did not incorporate analogous schemes to restore its wild nature. They also appeared to have been blinded by their cult of the technological sublime to the social links between their giant public-works projects and their constituents' homes and shops. Unlike the eco-planners, they did not grasp that living on a prairie wetland requires not only additional structural reforms to retain runoff surges for temporary periods but also smaller-scale methods to detain the rainwater where it falls to Earth. Appropriately named rain gardens, for instance, are low-lying places that have been planted with native species that thrive in wet conditions but can survive dry periods of drought. Or consider roof gardens, which consume rainwater, reduce runoff surges, grow local food and/or flowers, and save energy to heat and cool buildings by adding an extra layer of insulation. On a public, step-by-step level of adaptation, street, alley, and sidewalk repaving can incorporate permeable materials. Street curbing can also be redesigned to redirect storm water from drain gutters to thirsty varieties of trees planted along the parkways.[52]

The most visionary plan to turn rain into an asset would transform Chicago into a watery landscape, where its aquatic ecosystem would recycle its wastewater and runoff surges. In 2007, architect Martin Felsen, co-founder of UrbanLab and professor at the Illinois Institute of Technology, won a prestigious national competition to imagine the city of the future. His entry, "Growing Water," created a Venice-like terrain of canals cutting through the land from the lake to the west. "The[se] Eco-Boulevards," he explained, "will function as a giant 'Living Machine,' which will treat 100% of Chicago's wastewater and stormwater naturally, using micro-organisms, small invertebrates (such as snails), fish and plants."[53] Building strips of wetland back into the city, Felsen linked them to its public parks and "'emerald necklace'" of forest preserves in the suburbs and collar counties. The planner also clustered high-density housing close to these open recreational spaces. To encourage this settlement pattern, he repurposed the sanitary district's no-longer-needed tunnels deep under the riverbeds into a regional-scale rapid-transit system. Felsen's model of an ecological city foresaw the full integration of the blue and green of the natural world with the bright lights of contemporary society's build environment.[54]

City in a Garden?

From the founders' motto—*Urbs in Horto*—to Daniel Burnham and Edward Bennett's plan and Felsen's dream of growing water, Chicagoans have professed a pastoral ideal of nature. All of these boosters gave expression to this vision of Utopia, which is deeply rooted in Anglo-American culture. Their cities of the future imagined thriving settlements within a natural world under the complete control of its inhabitants. After the Civil War, landscape architects, such as New York's Frederick Law Olmsted and Calvert Vaux, helped Chicago build public parks and private suburbs that were designed to turn a flood-prone marshland into a well-ordered garden.[55]

Although the World's Fair of 1893 remained under the reign of this formal style of the "city beautiful," a rising generation of Progressives led by Dwight Perkins, Jensen, Wright, and Addams ushered in a new fashion of local naturalism that recast the midwestern prairie in a favorable light. Combined with the era's utilitarian goal of conservation, the Arcadian myth of the back-to-nature movement inspired Chicagoans to create more than a citywide network of playgrounds, parks, and beaches. After fighting World War I, they also started constructing an outer belt of forest preserves to protect the wild nature of the metropolitan area's six watersheds from too much progress. And they began building a suburban nature in a bungalow belt at the city's edge and beyond in parklike "neighborhood units" and mini-cities.[56]

But, of course, all this was fantasy for a city built on water. In part, Chicagoans' belief in a pastoral ideal accounts for their collective state of denial. Its settlers shared the view with other pioneers of the Midwest that most of its land was worthless, disease-breeding swamps. In Illinois, the great flood of 1904 triggered a massive effort to channelize the rivers, reclaim their basins, and drain their watersheds. By 2000, 90 percent of the state's presettlement wetlands had been redeveloped into farms, subdivisions, and so on, despite the countermovement of the forest-preserve districts and environmental-protection groups, such as Openlands and Chicago Wilderness.[57] There is only one "biologically significant stream" left in Cook County and its five collar counties, according to *2040—Comprehensive Regional Plan*. In contrast, the same study calculated that a larger, eleven-county area was now covered with 820 square miles/2,124 square km of paved-over, impervious surfaces. A 1 in./2.5 cm rainfall on this area alone—less than 10 percent of a still-expanding metropolitan region—produces almost 1.5 billion gal./5.7 billion L of runoff. Another "hundred-year storm" resulting in historic levels of damages in 2013 and new, record-breaking rainfall totals two years later have not let Chicagoans forget that they live in a period of intensifying climate change.[58]

How, then, can we account for this sorry story of the sinking of Chicago? How can we explain failure after failure of policy makers to save a city erected on a "sea of mud" from constantly suffering damaging floods that cause tremendous harm to its people, aquatic ecosystems, and quality of daily life? Part of the answer to what went wrong lies deep in a popular culture of the technological sublime. Since the explosive growth of Chicago in the 1840s and 1850s, its inhabitants' belief in progress has made it a "city on the make" that clashed with their pastoral ideal of living in a garden.[59] Beginning with the 2-mile/3.2 km water supply tunnel under the lake during the Civil War period and continuing up to the present era of the Deep Tunnel, Chicagoans have been obsessed with building the world's biggest public-works projects. However, this fascination with big technology has come with a heavy price in terms of adapting successfully to living on a prairie wetland. Time and time again, policy makers excluded consideration of less-monumental, albeit effective, alternative sanitary strategies as simply "impracticable."

The rise of Chicago during the Age of Andrew Jackson also made it a private city. Constitutional limitations on city government further reduced the policy options on water management in the name of the public health and welfare. Until the end of the nineteenth century, property ruled over infrastructure planning and land-use regulations.[60] Well into the middle of the twentieth century, moreover, the city's commercial- and industrial-interest groups were given free rein to exploit its natural resources in the name of progress. The policy makers' agreement to designate the Chicago River as a "working river" and Lake Michigan as the ultimate sink of its sewage had the direst consequences of the accumulative toll of victims from water-borne epidemics. Even after the chlorination of the water supply virtually eliminated these fatal outbreaks, the river was allowed to remain an industrial corridor in a delusional obsession to turn it into the world's biggest seaport, linking the Atlantic Ocean and the Gulf of Mexico. Its stigma as the back alley of the city in the popular imagination left the "friendless river . . . the city's most neglected natural resource."[61]

Also planting roots in the Civil War era, class and ethnic conflict became engrained in Chicago's cultural traditions as machine politics. The emergence of political organizations based on patronage and graft can be interpreted as either a reaction to a political economy of privatism or a prideful celebration of ethnic, racial, religious, and/or immigrant-status identities. In any case, the evolution of machine politics from intraparty factionalism to citywide establishments and one-party boss rule helps explain the failure of city hall to remake a prairie wetland into a livable space safe from damaging floods. From the great flood of 1885 until the intervention of the U.S. Supreme Court in the late 1920s, the politicians in charge of the sanitary district adhered strictly to a policy of making its rivers in reverse into a world-class shipping lane rather than an envirotechnical system of pollution and flood control. Huge grants of federal funds to achieve this dual goal during the New Deal period and the postwar Age of Ecology ended up becoming crucial sources of fuel to keep the Democrats' political machine running. Especially after the urban crisis of 1968 set off a panic of white flight, Chicago's record of corruption put Illinois at the very top of the list of criminal convictions of public officials by federal authorities. Between 1976 and 2012, they pursued leads that resulted in sixteen hundred local and state policy makers' facing guilty verdicts, including thirteen judges.[62]

Given the limitations imposed by Chicago's popular and political cultures, what were the approaches to building the urban environment that were dismissed as "impracticable"? What were the roads not taken that might have led from chronic suffering when it rained to harvesting this natural resource as an asset that enhances the quality of urban life? To be sure, what-if or counterfactual history opens its own dangerous roads of speculation, yet it can serve as a useful tool when framed within a context of the policy options offered by contemporaries. After the great flood of 1849 and the cholera epidemic that followed, the city's leaders made the single-most-fateful decision in favor of a combined sewer system. They rejected the alternative approach of separate sanitation pipes and storm drains recommended by their expert engineer, Ellis S. Chesbrough. Their choice was made on the basis of lowest cost, he pointed out, rather than on the maximum protection of the public health.

Creating a technological "path dependency," the policy makers have since blocked entry to crossroads that opened opportunities to switch directions toward a separate system.[63] At several important turning points in policy formation, reformers and experts made recommendations to require a dual system of pipes in undeveloped, peripheral areas of green space. In built-up areas too, the sewer lines needed replacement, because they were obsolete and too small. In 1889, the formation of the Sanitary District of Chicago presented the first chance to review the city's sanitary strategy based on its new metropolitan-scale boundaries. Over the next century, every annexation of additional territory and each reenactment of the district's charter enlarging its borders turned into a lost opportunity to insist on a separate system in rural areas as they became suburbs.

Instead, the sanitary authorities pursed a downtown-centered approach of turning its ship canal into a profitable engine of urban growth. A decade later, in the 1900s, Charles Merriam and his efficiency experts called for retrofitting the CBD to prevent sewer backups from sinking its busy streets. At the end of both world wars, the district's engineers offered long-term plans that included separate storm drains and deep, supersize intercepting sewers to end the backflow of runoff surges into the basements of homeowners and shopkeepers. The question of Chicago's sanitary strategy was also raised in the public sphere after the great floods of the mid-1950s, Bacon's brief era of reform during the 1960s, and again after 1986, when the Deep Tunnel failed to save the lake from the downside of a wet period of climate change. Local reformers and outside consultants repeatedly proposed ways during these policy debates to switch paths from a combined to a separate system in new areas of development as a part of more-holistic, alternative approaches to adapting the built environment to the ecosystems of a prairie wetland.

The inseparable problems of flood control and pollution control posed by a combined sewer system raises a second set of what-if questions about the roads not taken to speed up the reclamation of the rivers from open sewers to public amenities. How much in flood damages could have been avoided, for example, if city hall had followed the advice of its blue-ribbon panels in the 1910s to start installing sewage-treatment plants and a universal system of water meters instead of waiting twenty years to begin these projects? Reductions in the volume of liquid drained from the lake and flowing through the treatment plants would have produced commensurate gains in reducing sewer overflows that forced the sanitary authorities to release contaminated wastewater back into the lake.

While savings to the environment from conservation strategies would have been significant during the dry period before 1945, the human costs in terms of lost property and emotional suffering might have been marginal. In the early 1970s, however, city hall's rejection of the federal government's nonstructural, eco-friendly alternatives to TARP made a difference during an ongoing wet period of intensification of extreme weather events. And Chicago's inhabitants and tourists are still waiting for the sanitary authorities to upgrade their disinfection equipment to standards of purification that the Supreme Court first demanded ninety years ago. If the real past provides any guidance to resolve these what-if speculations, it points to the process of reciprocal interaction between the environmental restoration of the city's waters and the political mobilization to further reclaim them for recreational use.

The history of Chicagoans' back-to-nature movement presents a more-hopeful companion story of adaptation to erecting a city on a prairie wetland. Acting like a counterweight to the exploitation of natural resources has been a constant popular demand for more recreational and leisure open space. From the Civil War's sanitarians to today's deep ecologists, each generation has faced trade-offs between its technological pursuit of material progress and its pastoral ideal of a garden city. The crusade to spend time outdoors took many forms, from individual confrontations over the right of access to public parks and beaches to courtroom duels over the constitutional power of the state to create forest-preserve districts. At the same time, all of these battles were driven by a single-minded pursuit of an Arcadian myth of living in harmony with the natural world. The honor roll of victories on its behalf would start with the Progressives' enshrinement of the region's six watersheds as a greenbelt of sacred space. It would proceed to list the milestones in the campaign to repair their riverine ecosystems from the Skokie Lagoons project in 1933 to 2011, when city hall finally agreed to come into compliance with the Clean Water Act of 1972.

Over the course of this long struggle to save Chicago's waters, the reciprocal interaction between political mobilization and environmental protection has created an enduring legacy of grassroots insurgency. The ascendency of the science of ecology and its popular dissemination strengthened support for the cause of environmental protection. In comparison to the slow spread of conservation values during the first half of the twentieth century, the publication of Rachel Carson's *Silent Spring* in 1962 acted like rocket fuel in launching the modern movement to save Earth from human-made abuses. Growing so large and fast, however, the chorus calling for more open space spun off voices of discord. Consider the clash between advocates of restoring patches of the prairie wetlands to presettlement conditions and local neighbors, who want to preserve the status quo. They want to save the goldenrod and other invasive, albeit beautiful and familiar, species of plants from removal. In a similar way, grassroots demands for environmental justice have broadened and fractured the political base of support in favor of reclaiming the city's waters for recreation and leisure.[64] Nevertheless, the city's massive, if diverse, environmental movement can take credit for redefining the formation of public policy in more-holistic, ecological terms.

Like the weather, the success or failure of Chicago in remaking its flood-prone environment in the future is unpredictable. This historical study suggests that policy makers follow four guidelines in plotting a route to the goal of adapting urban life to a prairie wetland. First, the way forward will take the concerted efforts of everyone from individuals saving tap water to the managers of regional-scale infrastructure. A city built on water needs a million rain barrels and 10,000 miles/16,100 km of permeable pavement in addition to giant overflow reservoirs. Second, the path to pollution and flood control requires the integration of water and land management. Adhering to a regime of best practices, public and private sectors will not only reduce the damages caused by storm-water runoff surges but also regain the use of a valuable natural resource for human recreation and environmental conservation.

Another lesson of the city's history of climate change is the prerequisite of intergovernmental coordination to achieve the first two objectives. Chicago's overlapping and hydra-headed tangle of agencies is a long way from the planners' comprehensive regional perspectives and Felsen's and Gang's visionary dreams of the city of the future. The origins of a final guideline trace back to the beginning of this story, when Father Marquette had to figure out how to live in a flood-prone environment. With the help of Native Americans, his success stemmed from the firsthand experience of using the natural resources of his immediate surroundings to survive. To achieve this primary mission, moreover, he had to observe the changes wrought by the seasonal cycle of the weather. Being forced to flee to higher ground from his riverbank encampment during a spring tidal wave–like freshet taught the explorer that he would literally have to build a city on a hill if he hoped to accomplish his secondary mission of building a city of God on a marsh. Designing with nature today is being put to a supreme test by the Asian carp and other invasive species that threaten irreparable harm to the Great Lakes and the Mississippi River. Setting a course within these four parameters may not result in a city in a garden, but it will help a sprawling metropolitan region reach the goal of learning how to live on a prairie wetland.

Notes

INTRODUCTION

1. Jacques Marquette, "Journal of Father Jacques Marquette, Addressed to the Reverend Father Claude Dablon," *Les Relations des Jesuites* 69 (1674): 181, 172–183, available at http://encyclopedia.chicagohistory.org/pages/10873.html (accessed December 20, 2014). Also see Ann Durkin Keating, *Rising Up from Indian Country: The Battle of Fort Dearborn and the Birth of Chicago* (Chicago: University of Chicago Press, 2012); and Libby Hill, *The Chicago River: A Natural and Unnatural History* (Chicago: Lake Claremont Press, 2000).

2. See Hill, *The Chicago River*, 21–60; and Joel Greenberg, *A Natural History of the Chicago Region* (Chicago: University of Chicago Press, 2002), for more technical explanations of a prairie wetland. Also see Hugh Prince, *Wetland of the American Midwest—A Historical Geography of Changing Attitudes* (Chicago: University of Chicago Press, 1997); Margaret Beattie Bogue, "The Swamp Land Act and Wet Land Utilization in Illinois, 1850–1890," *Agricultural History* 25 (October 1951): 169–180; and Anthony Carlson, "The Other Kind of Reclamation: Wetlands Drainage and National Water Policy, 1902–1912," *Agricultural History* 84 (Fall 2010): 451–478.

3. Marquette, "Journal of Father Jacques Marquette," 172–183.

4. Ibid.; and William Cronon, *Nature's Metropolis—Chicago and the Great West* (New York: Norton, 1991).

5. National Oceanic and Atmospheric Administration's U.S. National Climatic Data Center, *Record of Climatological Observations, 2014*, available at http://www.ncdc.noaa.gov/ (accessed March 1, 2014); and Andreas F. Prein et al., "The Future Intensification of Hourly Precipitation Extremes," *Nature Climate Change*, December 5, 2016, available at http://www.nature.com.flagship.luc.edu/nclimate/index.html?cookies=accepted.

6. Karen Sawislak, *Smoldering City—Chicagoans and the Great Fire, 1871–1874* (Chicago: University of Chicago Press, 1996); Christine Meisner Rosen, *The Limits of Power—Great Fires and the Process of City Growth in America* (New York: Cambridge University Press, 1988); Carl S. Smith, *Urban Disorder and the Shape of Belief: The Chi-*

cago Fire, the Haymarket Bomb, and the Model Town of Pullman (Chicago: University of Chicago Press, 1995); and Donald L. Miller, *City of the Century: The Epic of Chicago and the Making of America* (New York: Simon and Schuster, 1996), 144–160.

7. Robin Einhorn, *Property Rules: Political Economy in Chicago, 1833–1872* (Chicago: University of Chicago Press, 1991).

8. Harold M. Mayer and Richard C. Wade, *Chicago: Growth of a Metropolis* (Chicago: University of Chicago Press, 1969); William Cronon, "To Be the Central City: Chicago, 1848–1857," *Chicago History* 10 (Fall 1981): 130–140; and Harold L. Platt, *Shock Cities: The Environmental Transformation and Reform of Manchester and Chicago* (Chicago: University of Chicago Press, 2005), chaps. 6–7. For broader, comparative perspectives, see Harold L. Platt, "Exploding Cities: Housing the Masses in Paris, Chicago, and Mexico City, 1850–2000," *Journal of Urban History* 36 (September 2010): 575–593.

9. Charles Rosenberg, *The Cholera Years: The United States in 1832, 1849, and 1866* (Chicago: University of Chicago Press, 1962); Edwin O. Gale, *Reminiscences of Early Chicago and Vicinity* (Chicago: Revell, 1902), 319–326; Constance B. Webb, *A History of Contagious Disease Care in Chicago before the Great Fire* (Chicago: University of Chicago Press, 1940); William K. Beatty, "When Cholera Scourged Chicago": *Chicago History* 11 (Spring 1982): 2–13; Thomas Neville Bonner, *Medicine in Chicago 1850–1950—A Chapter in the Social and Scientific Development of a City*, 2nd ed. (Urbana: University of Illinois Press, 1991); and Carl S. Smith, *City Water, City Life: Water and the Infrastructure of Ideas in Urbanizing Philadelphia, Boston, and Chicago* (Chicago: University of Chicago Press, 2013).

10. Harold M. Mayer, "The Launching of Chicago: The Situation and the Site," *Chicago History* 9 (Summer 1980): 68–79.

11. Beatty, "When Cholera Scourged Chicago."

12. James C. O'Connell, "Technology and Pollution: Chicago's Water Policy, 1833–1930" (Ph.D. diss., University of Chicago, 1980); Louis P. Cain, *Sanitation Strategy for a Lakefront Metropolis: The Case of Chicago* (DeKalb: Northern Illinois University Press, 1978); Platt, *Shock Cities*, chaps. 3–4; and Smith, *City Water, City Life*. Also see Chicago Board of Sewerage Commissioners, *Report and Plan of Sewerage for the City of Chicago, Illinois, Adopted by the Board of Sewerage Commissioners December 31, 1855* (Chicago: Scott, 1855).

13. John B. Jentz and Richard Schneirov, *Chicago in the Age of Capital: Class, Politics, and Democracy during the Civil War and Reconstruction* (Urbana: University of Illinois Press, 2013); Louise Carroll Wade, *Chicago's Pride: The Stockyards, Packingtown, and Environs in the Nineteenth Century* (Urbana: University of Illinois Press, 1987); Ross Miller, *American Apocalypse—The Great Fire and the Myth of Chicago* (Chicago: University of Chicago Press, 1990); Sawislak, *Smoldering City*; Richard Schneirov, *Labor and Urban Politics: Class Conflict and the Origins of Modern Liberalism in Chicago, 1864–1897* (Urbana: University of Illinois Press, 1998); Bruce Nelson, *Beyond the Martyrs—A Social History of Chicago's Anarchists, 1870–1910* (New Brunswick: Rutgers University Press, 1988); Stanley Buder, *Pullman: An Experiment in Industrial Order and Community Building, 1880–1930* (New York: Oxford University Press, 1967); and Smith, *Urban Disorder and the Shape of Belief*.

14. Harold L. Platt, "Jane Addams and the Ward Boss Revisited: Class, Politics, and Public Health in Chicago, 1890–1914," *Environmental History* 5 (April 2000): 194–222.

15. Ann Durkin Keating, *Chicagoland—City and Suburbs in the Railroad Age* (Chicago: University of Chicago Press, 2005); Sawislak, *Smoldering City*; and Rosen, *The Limits of Power*.

16. Olmsted, Vaux and Company, *Preliminary Report upon the Proposed Suburban Village at Riverside, Near Chicago* (New York: Sutton, Browne, 1868).

17. Keating, *Chicagoland*, 118, 114–139; Cronon, *Nature's Metropolis*; Carl Abbott, "'Necessary Adjuncts to Its Growth': The Railroad Suburbs of Chicago, 1854–1875," *Journal of the Illinois State Historical Society* 73 (Summer 1980): 117–131; and Z. Eastman, "Ancient Chicago," *Chicago Daily Tribune* [hereafter cited as CT], February 22, 1874, for the quoted phrase.

18. Eastman, "Ancient Chicago." Also see Milo Milton Quaife, *Chicago's Highways, Old and New, from Indian Trail to Motor Road* (Chicago: Keller, 1923).

19. Michael H. Ebner, *Creating Chicago's North Shore—a Suburban History* (Chicago: University of Chicago Press, 1988), 21–25, 91–104; and C. Abbott, "'Necessary Adjuncts to Its Growth.'"

20. Maurice Webster, interviewed in Evanston Township High School and Senior Editors, *Hinky Dinks, Sundaes, and Blind Pigs—an Oral History of Evanston*, ed. Sally Greenwood, Susan Jennett, and Katie Tucker (Evanston, IL, 1964), 4–5; and Ebner, *Creating Chicago's North Shore*, 91, 91–104, for the term "sanctified village."

21. Gwendolyn Wright, *Moralism and the Model Home: Domestic Architecture and Cultural Conflict in Chicago, 1873–1913* (Chicago: University of Chicago Press, 1980); Ann Durkin Keating, *Building Chicago: Suburban Developers and the Creation of a Divided Metropolis* (Columbus: Ohio State University Press, 1988); and Harold L. Platt, *The Electric City: Energy and the Growth of the Chicago Area, 1880–1930* (Chicago: University of Chicago Press, 1991). Also see Christopher C. Sellers, *Crabgrass Crucible: Suburban Nature and the Rise of Environmentalism in Twentieth-Century America* (Chapel Hill: University of North Carolina Press, 2012).

22. CT, July 5, October 25, 1868; April 11, May 9, September 19, 1869; February 6–7, May 1, June 5, August 7, 1870; June 4, 28, 1871. Also see Keating, *Chicagoland*, 92–113; C. Abbott, "'Necessary Adjuncts to Its Growth'"; and Roy Rosenzweig and Elizabeth Blackmar, *The Park and the People: A History of Central Park* (Ithaca, NY: Cornell University Press, 1998).

23. Olmsted, Vaux and Company, *Preliminary Report*, as quoted in Peter J. Schmitt, *Back to Nature: The Arcadian Myth in Urban America, 1900–1930* (New York: Oxford University Press, 1969), 20.

24. Frederick Law Olmsted, as quoted in CT, May 9, 1869.

25. CT, July 23, 1874, October 15, 1876, and September 21, 1878; and Keating, *Chicagoland*, 92–113.

26. CT, September 14, 26, 1871; January 13, 1873; February 1, 1874; Mario Kijewski, David Brush, and Robert Balanda, *The Historic Development of Three Chicago Millgates* (Chicago: Illinois Labor History Society, 1973); Craig E. Colten, "Industrial Wastes in Southeast Chicago: Production and Disposal 1870–1970," *Environmental Review* 10 (Summer 1986): 93–105; James Gilbert, *Perfect Cities—Chicago's Utopias of 1893* (Chicago: University of Chicago Press, 1991), chaps. 6–7; Mary Beth Pudup, "Model City? Industry and Urban Structure in Chicago," in *Manufacturing Suburbs: Building Work and Home on the Metropolitan Fringe*, ed. Robert D. Lewis (Philadelphia: Temple University Press, 2004), 53–75; and Robert D. Lewis, *Chicago Made: Factory Networks in the Industrial Metropolis* (Chicago: University of Chicago Press, 2008), 141–166.

27. Edith Abbott, *The Tenements of Chicago, 1880–1935* (Chicago: University of Chicago Press, 1936); and Dominic A. Pacyga and Ellen Skerrett, *Chicago, City of Neighborhoods: Histories and Tours* (Chicago: Loyola University Press, 1986), 411, 409–450.

28. Blue Wing, "Recollections of Calumet," in *American Field*, April 5, 1884, 322, as quoted in Joel Greenberg, ed., *Of Prairie, Woods, and Water: Two Centuries of Chicago Nature Writing* (Chicago: University of Chicago Press, 2008), 366–367.

29. Emerson Hough, "The Shooting Clubs of Chicago—No. XI: The Grand Calumet Heights Club," *Forest and Stream*, March 21, 1889, 172–174, as quoted in Greenberg, *Of Prairie, Woods, and Water*, 43, 42–48; Greenberg, *Of Prairie, Woods, and Water*, passim, for a useful collection of these writings; and William H. Tishler, ed., *Jens Jensen: Writings Inspired by Nature* (Madison: Wisconsin Historical Society Press, 2012).

30. Elaine Lewinnek, "Mapping Chicago, Imagining Metropolises: Reconsidering the Zonal Model of Urban Growth," *Journal of Urban History* 22, no. 10 (2009): 2.

31. Keating, *Chicagoland*, 14–16, 47, 73, 98, 120; and C. Abbott, "'Necessary Adjuncts to Its Growth.'"

32. Horace P. Ramey, "Floods in the Chicago Area," (unpublished manuscript, February 25, 1958, Harold Washington Library Center, Chicago); Hill, *The Chicago River*, 113–122; CT, February 27–28, 1871; November 22, 1872, May 30, 1874; March 7–15, 1875; April 1–8, 1877; February 28, April 15–21, December 25, 1881; and February 12–18, 1883.

33. Platt, *The Electric City*, chaps. 6–7; Platt, *Shock Cities*, chaps 8–12.

34. Nancy Tomes, *The Gospel of Germs—Men, Women, and the Microbe in American Life* (Cambridge, MA: Harvard University Press, 1998); Bonner, *Medicine in Chicago 1850–1950*; Michael Worboys, *Spreading Germs—Disease Theories and Medical Practice in Britain, 1865–1900* (Cambridge, UK: Cambridge University Press, 2000); Michael P. McCarthy, "Chicago, the Annexation Movement and Progressive Reform," in *The Age of Urban Reform—New Perspectives on the Progressive Era*, ed. Michael H. Ebner and Eugene M. Tobin (Port Washington, NY: Kennikat, 1977), 43–54; O'Connell, "Technology and Pollution"; and Platt, *Shock Cities*, chaps 8–12.

35. Chicago Board of Sewerage Commissioners, *Report*, 5, 6–7, for the two quotations, respectively.

36. Ibid.; and Martin V. Melosi, *The Sanitary City—Urban Infrastructure in America from Colonial Times to the Present* (Baltimore, MD: Johns Hopkins University Press, 2000), on the concept and implications of path dependency.

37. Ellis S. Chesbrough, *Chicago Sewerage: Report of the Results of Examination Made in Relation to Sewerage in Several European Cities, in the Winter of 1856–1857* (Chicago: Board of Sewage Commissioners, 1858), 55, passim; Chicago Board of Sewerage Commissioners, *Report*; and Platt, *Shock Cities*, chaps. 3–4.

38. CT, March 26, 1867; Jack Wing, *The Great Chicago Lake Tunnel* (Chicago: Jack Wing, 1867); Platt, *Shock Cities*, chaps. 3–4; and Smith, *City Water, City Life*.

39. National Oceanic and Atmospheric Administration, *Record of Climatological Observations*.

40. Center for Neighborhood Technology, *People, Water, and the Great Lakes: Ready for Change?* (Chicago: Center for Neighborhood Technology, August 2012). Also see Center for Neighborhood Technology, ed., *The Prevalence and Cost of Urban Flooding—a Case Study of Cook County, Ill.* (Chicago: Center for Neighborhood Technology, updated May 2014).

41. Judith A. Martin and Sam Bass Warner Jr., "Local Initiative and Metropolitan Repetition: Chicago, 1972–1990," in *The American Planning Tradition: Culture and Policy*, ed. Robert Fishman (Washington, DC: Woodrow Wilson Centre Press and Johns Hopkins University Press, 2000), 263–296.

42. Center for Neighborhood Technology, "Urban Flooding"; and City of Chicago and Chicago Metropolitan Agency for Planning, *2040—Comprehensive Regional*

Plan (Chicago: Chicago Metropolitan Agency for Planning, October 2010), available at http://www.cmap.illinois.gov/documents/10180/17842/long_plan_FINAL_100610_web.pdf/1e1ff482-7013-4f5f-90d5-90d395087a53 (accessed May 5, 2016); CT, April 17–20, 2013; and CBS, "Stormy Weather: Wettest June on Record for Illinois: Second Wettest Month Ever, June 30, 2015," available at http://chicago.cbslocal.com/2015/06/30/stormy-weather-wettest-june-on-record-for-illinois-second-wettest-month-ever/ (accessed April 14, 2016).

CHAPTER 1

1. *Chicago Daily Tribune* [hereafter cited as CT], February 16, 1883. Also see Lisa Krissoff Boehm, *Popular Culture and the Enduring Myth of Chicago, 1871–1968* (New York: Routledge, 2004), chap. 1; and Libby Hill, *The Chicago River: A Natural and Unnatural History* (Chicago: Lake Claremont Press, 2000).

2. CT, February 12, 1883.

3. Ibid., February 17, 1883, for the first quotation and quoted phrase; and ibid., February 18, 1883, for the second quotation.

4. Peter J. Schmitt, *Back to Nature: The Arcadian Myth in Urban America, 1900–1930* (New York: Oxford University Press, 1969). For the World's Fair, see R. Reid Badger, *The Great American Fair: The World's Columbian Exposition and American Culture* (Chicago: Nelson Hall, 1980); Madeline Weimann, *The Fair Women: The Story of the Woman's Building, World's Columbian Exposition, Chicago, 1893* (Chicago: Academy Chicago, 1981); James Gilbert, *Perfect Cities—Chicago's Utopias of 1893* (Chicago: University of Chicago Press, 1991); Arnold Lewis, *An Early Encounter with Tomorrow—Europeans, Chicago's Loop, and the World's Columbian Exposition* (Urbana: University of Illinois Press, 1997); and Joseph Alan Gustaitis, *Chicago's Greatest Year, 1893: The White City and the Birth of a Modern Metropolis* (Carbondale: Southern Illinois University Press, 2013).

5. Robin Faith Bachin, *Building the South Side: Urban Space and Civic Culture in Chicago, 1890–1919* (Chicago: University of Chicago Press, 2004), 129. Also see Daniel Bluestone, *Constructing Chicago* (New Haven: Yale University Press, 1991), chap 2; William H. Wilson, *The City Beautiful Movement* (Baltimore, MD: Johns Hopkins University Press, 1989); Allen F. Davis, *Spearheads for Reform—The Social Settlements and the Progressive Movement, 1890–1914* (New York: Oxford University Press, 1967); Robert A. Slayton, *Back of the Yards—The Making of a Local Democracy* (Chicago: University of Chicago Press, 1986); and Maureen A. Flanagan, *Seeing with Their Hearts: Chicago Women and the Vision of the Good City, 1871–1933* (Princeton, NJ: Princeton University Press, 2002).

6. Carl S. Smith, *The Plan of Chicago: Daniel Burnham and the Remaking of the American City* (Chicago: University of Chicago Press, 2006); Thomas S. Hines, *Burnham of Chicago: Architect and Planner* (Chicago: University of Chicago Press, 1978); Ross Miller, *American Apocalypse—The Great Fire and the Myth of Chicago* (Chicago: University of Chicago Press, 1990); and Carl S. Smith, *Urban Disorder and the Shape of Belief: The Chicago Fire, the Haymarket Bomb, and the Model Town of Pullman* (Chicago: University of Chicago Press, 1995).

7. Clifford Putney, *Muscular Christianity: Manhood and Sports in Protestant America, 1880–1920* (Cambridge, MA: Harvard University Press, 2001).

8. Ibid.; Gijs Mom, *The Electric Vehicle: Technology and Expectations in the Automobile Age* (Baltimore, MD: Johns Hopkins University Press, 2003), for the quoted phrase "dream machine"; and Kenneth T. Jackson, *Crabgrass Frontier—The Suburbanization of*

the United States (New York: Oxford University Press, 1985), for the final quoted phrase. Also see Gunter Barth, *City People: The Rise of Modern City Culture in Nineteenth-Century America* (New York: Oxford University Press, 1980); Steven A. Riess, *Touching Base: Professional Baseball and American Culture in the Progressive Era* (Westport, CT: Greenwood Press, 1980); Steven A. Riess, *City Games—The Evolution of American Urban Society and the Rise of Sports* (Urbana: University of Illinois Press, 1989); Sam Bass Warner, *Streetcar Suburbs: The Process of Growth in Boston, 1870–1900* (Cambridge, MA: Harvard University Press, 1962); Robert C. Twombly, "Saving the Family: Middle Class Attraction to Wright's Prairie House, 1901–1909," *American Quarterly* 27 (March 1975): 57–72; and Margaret Garb, *City of American Dreams: A History of Home Ownership and Housing Reform in Chicago, 1871–1919* (Chicago: University of Chicago Press, 2005).

9. Joan E. Draper, "The Art and Science of Park Planning in the United States: Chicago's Small Parks, 1902–1905," in *Planning the Twentieth-Century American City*, ed. Mary Corbin Sies and Christoper Silver (Baltimore, MD: Johns Hopkins University Press, 1966), 98–119; Galen Cranz, "Models for Park Usage Ideology and the Development of Chicago's Public Parks" (Ph.D. diss., University of Chicago, 1971); Galen Cranz, *The Politics of Park Design: A History of Urban Parks in America* (Cambridge: Massachusetts Institute of Technology Press, 1982); Joseph P. Schwieterman, Dana M. Caspall, and Jane Heron, *The Politics of Place: A History of Zoning in Chicago* (Chicago: Lake Claremont Press, 2003); Rebecca C. Retzlaff, "The Illinois Forest Preserves District Act of 1913 and the Emergence of Metropolitan Park System Planning in the USA," *Planning Perspectives* 25 (October 2010): 433–455; and Robert Lewis, "Modern Industrial Policy and Zoning: Chicago, 1910–1930," *Urban History* 40 (February 2013): 97–113.

10. Benjamin Heber Johnson, *Escaping the Dark, Gray City: Fear and Hope in Progressive-Era Conservation* (New Haven, CT: Yale University Press, 2017), 11, chaps. 1–4, 7. Also see Samuel P. Hays, *Conservation and the Gospel of Efficiency: The Progressive Conservation Movement 1890–1920* (Cambridge, MA: Harvard University Press, 1959).

11. Johnson, *Escaping the Dark, Gray City*, chaps. 5–6.

12. See Smith, *Urban Disorder and the Shape of Belief*; Bruce Nelson, *Beyond the Martyrs—A Social History of Chicago's Anarchists, 1870–1910* (New Brunswick: Rutgers University Press, 1988); Eric L. Hirsch, *Urban Revolt—Ethnic Politics in the Nineteenth-Century Chicago Labor Movement* (Berkeley: University of California Press, 1990); and Richard Schneirov, *Labor and Urban Politics: Class Conflict and the Origins of Modern Liberalism in Chicago, 1864–1897* (Urbana: University of Illinois Press, 1998), on urban society. See Hill, *The Chicago River*, 61–138; and Harold L. Platt, *Shock Cities: The Environmental Transformation and Reform of Manchester and Chicago* (Chicago: University of Chicago Press, 2005), chaps. 11–12, on the urban environment.

13. Harold L. Platt, *The Electric City: Energy and the Growth of the Chicago Area, 1880–1930* (Chicago: University of Chicago Press, 1991); Harold L. Platt, "Creative Necessity: Municipal Reform in Gilded Age Chicago," in *The Constitution, Law, and American Life—Critical Aspects of the Nineteenth-Century Experience*, ed. Donald G. Nieman (Athens: University of Georgia Press, 1992), 162–190; Margaret Marsh, "From Separation to Togetherness: The Social Construction of Domestic Space in American Suburbs, 1840–1915," *Journal of American History* 76 (September 1989): 506–527; and Dolores Hayden, *Building Suburbia: Green Fields and Urban Growth, 1820–2000* (New York: Vintage, 2003).

14. Michael P. McCarthy, "Chicago, the Annexation Movement and Progressive Reform," in *The Age of Urban Reform—New Perspectives on the Progressive Era*, ed. Michael H. Ebner and Eugene M. Tobin (Port Washington, NY: Kennikat, 1977), 43–54;

and Michael H. Ebner, "The Result of Honest Hard Work: Creating a Suburban Ethos for Evanston," *Chicago History* 13 (Fall 1984): 48–65. Also see Morton White and Lucia White, *The Intellectual Versus the City, from Thomas Jefferson to Frank Lloyd Wright* (Cambridge, MA: Harvard University Press, 1962); Raymond Williams, *The Country and the City* (New York: Oxford University Press, 1973); and Christopher C. Sellers, *Crabgrass Crucible: Suburban Nature and the Rise of Environmentalism in Twentieth-Century America* (Chapel Hill: University of North Carolina Press, 2012).

15. M. McCarthy, "Chicago, the Annexation Movement and Progressive Reform"; Miller, *American Apocalypse*; Carl W. Condit, *The Chicago School of Architecture* (Chicago: University of Chicago Press, 1964); and Ann Durkin Keating, *Chicagoland—City and Suburbs in the Railroad Age* (Chicago: University of Chicago Press, 2005). Also see John R. Stilgoe, *Metropolitan Corridor—Railroads and the American Scene* (New Haven, CT: Yale University Press, 1983).

16. M. McCarthy, "Chicago, the Annexation Movement and Progressive Reform"; Ebner, "The Result of Honest Hard Work"; Platt, "Creative Necessity"; and Michael H. Ebner, *Creating Chicago's North Shore—a Suburban History* (Chicago: University of Chicago Press, 1988).

17. CT, May 26, 1889.

18. Platt, *Shock Cities*, chaps. 11–12; and James C. O'Connell, "Technology and Pollution: Chicago's Water Policy, 1833-1930" (Ph.D. diss., University of Chicago, 1980).

19. See G. P. Brown, *Drainage Channel and Waterway* (Chicago: Donnelly, 1894), for an account of the origins of the Chicago Sanitary District and a copy of the enabling act of 1889. Also see Christopher Hamlin, *A Science of Impurity—Water Analysis in Nineteenth Century Britain* (Bristol: Hilger, 1990); and Nancy Tomes, *The Gospel of Germs—Men, Women, and the Microbe in American Life* (Cambridge, MA: Harvard University Press, 1998).

20. Platt, *Shock Cities*, 366–377.

21. CT, May 26, June 30, 1889; Brown, *Drainage Channel and Waterway*; and M. McCarthy, "Chicago, the Annexation Movement and Progressive Reform."

22. Mayor Cregier, as quoted in CT, July 3, 1889. Also see ibid., July 3, 11–13, 18, 25, 27, 31, August 2, 18–19, September 4, 19, 27, October 5, 10, 13, 1889.

23. H. C. Miller, as quoted in ibid., July 3, 1889.

24. Judge Prendergast, as quoted in ibid., October 13, 1889.

25. Ibid.

26. Ibid., December 13, 1889, for the quotation; ibid., December 3–14, 1889, for the election.

27. Christopher Lasch, *The New Radicalism in America, 1889-1963—The Intellectual as a Social Type* (New York: Knopf, 1965); and Robert H. Wiebe, *The Search for Order, 1877-1920* (New York: Hill and Wang, 1967), respectively, for the quoted phrases. Also see John Higham, "The Reorientation of American Culture in the 1890s," in *Writing American History: Essays on Modern Scholarship*, ed. John Higham (Bloomington: Indiana University Press, 1970), 73–102; T. J. Jackson Lears, *No Place for Grace: Anti-modernism and the Transformation of American Culture, 1880–1920* (New York: Pantheon, 1981); Riess, *City Games*; and Paula Fass, "How Americans Raise Their Children: Generational Relations over Two Hundred Years," *Bulletin of the German Historical Institute* 54 (Spring 2014): 7–20.

28. Richard Hofstadter, *The Age of Reform: From Bryan to F. D. R.* (New York: Vintage Books, 1955).

29. Ibid. Also see Richard Sennett, *Families against the City: Middle Class Homes of Industrial Chicago, 1872–1890* (Cambridge, MA: Harvard University Press, 1970); Flan-

agan, *Seeing with Their Hearts*; John B. Jentz and Richard Schneirov, *Chicago in the Age of Capital: Class, Politics, and Democracy during the Civil War and Reconstruction* (Urbana: University of Illinois Press, 2013); Margaret Garb, *Freedom's Ballot: African American Political Struggles in Chicago from Abolition to the Great Migration* (Chicago: University of Chicago Press, 2014); and Colin Fisher, *Urban Green: Nature, Recreation, and the Working Class in Industrial Chicago* (Chapel Hill: University of North Carolina Press, 2015).

30. Higham, "The Reorientation of American Culture," 79. Also see Putney, *Muscular Christianity*, 1–10; William L. O'Neill, "Divorce in the Progressive Era," *American Quarterly* 17 (Summer 1965): 203–217; and Ann Douglas, *Feminization of American Culture* (New York: Knopf, 1997).

31. Wiebe E. Bijker and T. J. Pinch, "The Social Construction of Facts and Artifacts: Or How the Sociology of Science and the Sociology of Technology Might Benefit Each Other," in *The Social Construction of Technological Systems: New Directions in the Sociology and History of Technology*, ed. Wiebe E. Bijker, Thomas P. Hughes, and Trevor J. Pinch (Cambridge: Massachusetts Institute of Technology Press, 1987), 17–50; David A. Hounshell, *From the American System to Mass Production, 1800–1932: The Development of Manufacturing Technology in the United States* (Baltimore, MD: Johns Hopkins University Press, 1984), 189–216; and Gustaitis, *Chicago's Greatest Year*, 76–94.

32. Schmitt, *Back to Nature*, xvii.

33. Gustaitis, *Chicago's Greatest Year*. Also see Stephan J. Schmidt, "The Evolving Relationship between Open Space Preservation and Local Planning Practice," *Journal of Planning History* 7 (May 2008): 91–112.

34. Roy Rosenzweig, *Eight Hours for What We Will: Workers and Leisure in an Industrial City, 1870–1920* (Cambridge, UK: Cambridge University Press, 1983).

35. Smith, *Urban Disorder*, 101–176; Jentz and Schneirov, *Chicago in the Age of Capital*; and Nelson, *Beyond the Martyrs*.

36. Smith, *Urban Disorder*, 125.

37. Bachin, *Building the South Side*, 127; and Lasch, *The New Radicalism in America*, 3–37. Also see Davis, *Spearheads for Reform*; Helen Lefkowitz Horowitz, *Culture and the City: Cultural Philanthropy in Chicago from the 1880s to 1917* (Lexington: University of Kentucky Press, 1976); Louise C. Wade, *Graham Taylor—Pioneer for Social Justice, 1851–1938* (Chicago: University of Chicago Press, 1964); Kathleen D. McCarthy, *Noblesse Oblige: Charity and Cultural Philanthropy in Chicago* (Chicago: University of Chicago Press, 1982); Robert Gottlieb, *Forcing the Spring: The Transformation of the American Environmental Movement* (Washington, DC: Island Press, 1993), chap. 2; and Emily Talen, "Do-It-Yourself Urbanism: A History," *Journal of Planning History* 14 (September 2014): 135–148.

38. Paul Boyer, *Urban Masses and Moral Order in America, 1820–1920* (Cambridge, MA: Harvard University Press, 1978).

39. James F. Findlay, *Dwight L. Moody, American Evangelist, 1837–1899* (Chicago: University of Chicago Press, 1969); Darrel M. Robertson, *The Chicago Revival, 1876: Society and Revivalism in a Nineteenth-Century City* (Metuchen, NJ: Scarecrow, 1989); Justin H. Pettegrew, "Onward Christian Soldiers: The Transformation of Religion, Masculinity, and Class in the Chicago YMCA, 1857–1933" (Ph.D. diss., Loyola University Chicago, 2006); and William J. Baker, *Playing with God: Religion and Modern Sport* (Cambridge, MA: Harvard University Press, 2007).

40. Boyer, *Urban Masses and Moral Order*; Harold L. Platt, "Jane Addams and the Ward Boss Revisited: Class, Politics, and Public Health in Chicago, 1890–1914," *Environ-*

mental History 5 (April 2000): 194–222; Davis, *Spearheads for Reform*; Flanagan, *Seeing with Their Hearts*; and Schneirov, *Labor and Urban Politics*. See also Christopher C. Sellers, *Hazards of the Job: From Industrial Disease to Environmental Health Science* (Chapel Hill: University of North Carolina Press, 1997); and Tomes, *The Gospel of Germs*.

41. Pettegrew, "Onward Christian Soldiers," 177; and Dwight L. Moody, as quoted in Findlay, *Dwight L. Moody*, 56, for the second quoted phrase. Also see Baker, *Playing with God*, chap. 2.

42. Pettegrew, "Onward Christian Soldiers," 23.

43. Putney, *Muscular Christianity*, 11–44; and Pettegrew, "Onward Christian Soldiers," 111–127.

44. Moody, as quoted in Findlay, *Dwight L. Moody*, 327.

45. Findlay, *Dwight L. Moody*, 326–421; and Pettegrew, "Onward Christian Soldiers," 111–203.

46. Pettegrew, "Onward Christian Soldiers," 111–203.

47. Theodore Roosevelt, as quoted in Higham, "The Reorientation of American Culture," 78; and Putney, *Muscular Christianity*, 12.

48. Putney, *Muscular Christianity*, 66; and Thomas Winter, "The YMCA and the Construction of Manhood and Class, 1877–1920," *Men and Masculinities* 2 (January 2000): 272–285.

49. Putney, *Muscular Christianity*, 29.

50. Ibid., chap. 2.

51. Frances E. Willard, as quoted in Michael Taylor, "Rapid Transit to Salvation: American Protestants and the Bicycle in the Era of the Cycling Craze," *Journal of the Gilded Age and Progressive Era* 9 (July 2010): 349; and Gustaitis, *Chicago's Greatest Year*, 78, respectively, for the quotations; and Frances E. Willard, *A Wheel within a Wheel—How I Learned to Ride the Bicycle with Some Reflections by the Way* (New York: Fleming H. Revell, 1895). Also see M. Taylor, "Rapid Transit to Salvation," 337–364; Gustaitis, *Chicago's Greatest Year*, 76–94; Ellen Gruber Garvey, "Reframing the Bicycle: Advertising-Supported Magazines and Scorching Women," *American Quarterly* 47 (March 1995): 66–101; and Baker, *Playing with God*.

52. Garvey, "Reframing the Bicycle," 66.

53. Ibid.; Richard Hammond, "Progress and Flight: An Interpretation of the American Cycle Craze of the 1890s," *Journal of Social History* 5 (Winter 1971–1972): 235–257; and Putney, *Muscular Christianity*, chap. 6.

54. Rev. Jenkin Lloyd Jones, as quoted in M. Taylor, "Rapid Transit to Salvation," 347; and Boehm, *Popular Culture and the Enduring Myth of Chicago*, chap. 2.

55. Dominick Cavallo, *Muscles and Morals: Organized Playgrounds and Urban Reform, 1880–1920* (Philadelphia: University of Pennsylvania Press, 1981), 22. Also see Putney, *Muscular Christianity*, 4.

56. Davis, *Spearheads for Reform*; Cranz, *The Politics of Park Design*; and Bachin, *Building the South Side*, chap. 3.

57. Charles Zueblin, "Municipal Playgrounds in Chicago," *American Journal of Sociology* 4 (September 1898): 145–158; Graham R. Taylor, "Recreation Developments in Chicago Parks," *Annals of the American Academy of Political and Social Science* 35 (March 1910): 88–105; Marian Lorena Osborn, "The Development of Recreation in the South Park System of Chicago" (master's thesis, University of Chicago, August 1928), chaps. 1–3; Karen M. Mason, "Mary McDowell and Municipal Housekeeping: Women's Political Activism in Chicago, 1890–1920," in *Midwestern Women—Work, Community and Leadership at the Crossroads*, ed. Lucy Eldersveld Murphy and Wendy Hamand Venet

(Bloomington: Indiana University Press, 1997), 60–75; Michael McCarthy, "Politics and the Parks: Chicago Businessmen and the Recreation Movement," *Journal of the Illinois State Historical Society* 65 (Summer 1972): 158–172; and Robert E. Grese, *Jens Jensen—Maker of Natural Parks and Gardens* (Baltimore, MD: Johns Hopkins University Press, 1992), 28–45, 120.

58. Bachin, *Building the South Side*, 129. Also see Maureen A. Flanagan, "Gender and Urban Political Reform: The City Club and the Women's City Club of Chicago in the Progressive Era," *American Historical Review* 95 (October 1990): 1032–1050.

59. Osborn, "The Development of Recreation," chaps. 2–3; Fisher, *Urban Green*, chaps. 2–3; and Dwight Heald Perkins, compiler, *Report of the Special Park Commission to the City Council of Chicago on the Subject of a Metropolitan Park System* (Chicago: City of Chicago, 1904).

60. Fisher, *Urban Green*, chap. 4; and Chicago Commission on Race Relations, *The Negro in Chicago: A Study of Race Relations and a Race Riot* (Chicago: University of Chicago Press, 1922). Also see Allan H. Spear, *Black Chicago: The Making of a Negro Ghetto, 1880–1920* (Chicago: University of Chicago Press, 1967).

61. Badger, *The Great American Fair*; Weimann, *The Fair Women*; and Russell Lewis, "Everything under One Roof: World's Fairs and Department Stores in Paris and Chicago," *Chicago History* 12 (Fall 1983): 28–47.

62. Wilson, *The City Beautiful Movement*, 75.

63. Ibid., chap. 3; Gilbert, *Perfect Cities*; A. Lewis, *An Early Encounter with Tomorrow*; and Platt, *The Electric City*, chap. 4.

64. Wilson, *The City Beautiful Movement*, chap. 3; Garb, *City of American Dreams*, 136–139; Homer Hoyt, *One Hundred Years of Land Values in Chicago—The Relationship of the Growth of Chicago to the Rise in Its Land Values, 1830–1933* (Chicago: University of Chicago Press, 1933), chap. 4; M. McCarthy, "Politics and the Parks"; Christine Meisner Rosen, "Businessmen against Pollution in Late Nineteenth Century Chicago," *Business History Review* 69 (Fall 1995): 351–397; Ocean Howell, "Play Pays—Urban Land Politics and Playgrounds in the United States, 1900–1930," *Journal of Urban History* 34 (September 2008): 961–994; Smith, *The Plan of Chicago*; and Robert Fishman, *Urban Utopias in the Twentieth Century—Ebenezer Howard, Frank Lloyd Wright, Le Corbusier* (New York: Basic, 1977).

65. U.S. Census Bureau (1880–1940), *Population*; Edith Abbott, *The Tenements of Chicago, 1880–1935* (Chicago: University of Chicago Press, 1936); Flanagan, *Seeing with Their Hearts*, chap. 5; Thomas Lee Philpott, *The Slum and the Ghetto: Neighborhood Deterioration and Middle-Class Reform, Chicago, 1880–1930* (New York: Oxford University Press, 1978); and Gwendolyn Wright, *Moralism and the Model Home: Domestic Architecture and Cultural Conflict in Chicago, 1873–1913* (Chicago: University of Chicago Press, 1980). Also see Daniel T. Rodgers, *Atlantic Crossings: Social Politics in the Progressive Age* (Cambridge, MA: Harvard University Press, 1998).

66. Platt, *Shock Cities*, 377–387; Gustaitis, *Chicago's Greatest Year*, chap. 14. Also see Alan M. Kraut, *Silent Travelers—Germs, Genes, and the "Immigrant Menace"* (Baltimore, MD: Johns Hopkins University Press, 1994); and Howard Markel, *Quarantine! East European Jewish Immigrants and the New York City Epidemics of 1892* (Baltimore, MD: Johns Hopkins University Press, 1997).

67. Warner, *Streetcar Suburbs*. Also see Paul Barrett, *The Automobile and Urban Transit—The Formation of Public Policy in Chicago 1900–1930* (Philadelphia: Temple University Press, 1983).

68. Frank Lloyd Wright, *The Natural House* (New York: Horizon Press, 1954), 23, for the quoted phrase; Gilbert, *Perfect Cities*, chap. 6; Twombly, "Saving the Family"; and Marsh, "From Separation to Togetherness."

69. Gilbert, *Perfect Cities*, chap. 6; Alec C. Kerr, ed., *History: The City of Harvey, 1890–1962* (Chicago: First National Bank in Harvey, 1962); Garb, *City of American Dreams*; Sigfried Giedion, *Mechanization Takes Command—A Contribution to Anonymous History* (New York: Oxford University Press, 1948), 512–627; Joseph C. Bigott, "Bungalows and the Complex Origins of the Modern House," in *The Chicago Bungalow*, ed. Dominic A. Pacyga and Charles Shanabruch (Chicago: Arcadia Publishing, 2003), 31–52; and Joseph C. Bigott, *From Cottage to Bungalow: Houses and the Working Class in Metropolitan Chicago, 1869–1929* (Chicago: University of Chicago Press, 2001).

70. Gilbert, *Perfect Cities*, 195.

71. Ibid., 196, for the quoted phrase; ibid., chaps. 2, 6; Smith, *Urban Disorder*; and Stanley Buder, *Pullman: An Experiment in Industrial Order and Community Building, 1880–1930* (New York: Oxford University Press, 1967).

72. Gilbert, *Perfect Cities*, chaps. 2, 6–7; and Pettegrew, "Onward Christian Soldiers," chap. 2.

73. Platt, *Shock Cities*, chap. 11; and Carolyn G. Shapiro-Shapin, "'A Really Excellent Scientific Contribution': Scientific Creativity, Scientific Professionalism, and the Chicago Drainage Case," *Bulletin of the History of Medicine* 71 (Fall 1997): 385–411. Also see Tomes, *The Gospel of Germs*; Kraut, *Silent Travelers*; and Judith Walzer Leavitt, "'Typhoid Mary' Strikes Back: Bacteriological Theory and Practice in Early 20th-Century Public Health," *Isis* 83 (December 1992): 608–629.

74. Frank Lloyd Wright, as quoted in Twombly, "Saving the Family," 62.

75. Ibid., 59, 57–72; and Neil Levine, *The Architecture of Frank Lloyd Wright* (Princeton, NJ: Princeton University Press, 1996), chaps. 1–2.

76. Marsh, "From Separation to Togetherness," 527. Also see Margaret Marsh, "Suburban Men and Masculine Domesticity, 1870–1915," *American Quarterly* 40 (June 1988): 165–186; and G. Wright, *Moralism and the Model Home.*

77. Marsh, "Suburban Men and Masculine Domesticity," 174. See also Twombly, "Saving the Family"; Wiebe, *The Search for Order*; and Olivier Zunz, *Making America Corporate, 1870–1920* (Chicago: University of Chicago Press, 1990).

78. Frank Lloyd Wright, "The Art and Craft of the Machine," in *Frank Lloyd Wright: Writings and Buildings*, ed. Edgar Kaufman and Ben Raeburn (1901; repr., New York: Horizon, 1960), 53–73; David A. Hanks, "Chicago and the Midwest," in *The Arts and Crafts Movement in America, 1876–1916*, ed. Robert Judson Clark (Princeton, NJ: Princeton University Press, 1972), 63–94; Tomes, *The Gospel of Germs*; Garb, *City of American Dreams*, chaps. 3–4; and Suellen Hoy, *Chasing Dirt—The American Pursuit of Cleanliness* (New York: Oxford University Press, 1995).

79. Clifford Edward Clark, *The American Family Home, 1800–1960* (Chapel Hill: University of North Carolina Press, 1986), 180.

80. Scott Sonoc, "Defining the Chicago Bungalow," in *The Chicago Bungalow*, 9–30; Bigott, "Bungalows and the Complex Origins of the Modern House"; and Bigott, *From Cottage to Bungalow.*

81. Twombly, "Saving the Family." The house is the Avery Coonley estate, which was located on a 10-acre/4-hectare site with the Des Plaines River bordering three sides.

82. Bigott, *From Cottage to Bungalow*; and Harvey Levenstein, *Revolution at the Table: The Transformation of the American Diet* (New York: Oxford University Press, 1988).

83. Clark, *The American Family Home*, 171, for the quoted phrase.

84. Garb, *City of American Dreams*, chaps. 1–2, 5; Elaine Lewinnek, "Inventing Suburbia: Working-Class Suburbanization in Chicago, 1865–1919" (Ph.D. diss., Yale University, 2005); Will Cooley, "Moving Up, Moving Out: Race and Social Mobility in Chicago, 1914–1972" (Ph.D. diss., University of Illinois, Urbana, 2008), chap. 1. Also see Stephan Thernstrom, *Poverty and Progress—Social Mobility in a Nineteenth Century City* (Cambridge, MA: Harvard University Press, 1964); and Marc A. Weiss, *The Rise of the Community Builders—The American Real Estate Industry and Urban Land Planning* (New York: Columbia University Press, 1987).

85. Garb, *City of American Dreams*, 118; and Clark, *The American Family Home*, 183, for the first and the second quotations, respectively. Also see Garb, *City of American Dreams*, chaps. 5–6; and Lewinnek, "Inventing Suburbia," 160–167.

86. U.S. Census Bureau (1900–1910), *Population*.

87. Hoyt, *One Hundred Years of Land Values*, 201, chap. 5.

88. Ibid., 203. Also see Spear, *Black Chicago*; Robert Ezra Parks, Ernest Watson Burgess, Roderick Duncan McKenzie, and Louis Wirth, *The City* (Chicago: University of Chicago Press, 1925); and Andrew Wiese, *Places of Their Own: African American Suburbanization in the Twentieth Century* (Chicago: University of Chicago Press, 2004).

89. Garb, *City of American Dreams*, chaps. 5–6; and Ebner, *Creating Chicago's North Shore*.

90. Retzlaff, "The Illinois Forest Preserves District Act"; Theodore S. Eisenman, "Frederick Law Olmsted, Green Infrastructure, and the Evolving City," *Journal of Planning History* 12 (November 2013): 287–311; Perkins, *Report of the Special Park Commission*; Daniel H. Burnham and Edward H. Bennett, *Plan of Chicago* (Chicago: Commercial Club, 1909); and Cook County of Illinois, Forest Preserve District, *The Forest Preserves of Cook County* (Chicago: Forest Preserve District of Cook County, 1918).

91. Of course, political contestation over open space began in the city with the first proposals for public parks. See Roy Rosenzweig and Elizabeth Blackmar, *The Park and the People: A History of Central Park* (Ithaca, NY: Cornell University Press, 1998).

92. City of Chicago, Department of Health, *Annual Reports*, 1895–1896; and Platt, *Shock Cities*, chap. 11.

93. Platt, *Shock Cities*, chap. 11.

94. Julia Sniderman Bachrach, "River Views," in *Midstream: The Chicago River, 1999/2010*, ed. Richard Wasserman and Julia Sniderman Bachrach (Chicago: Columbia College Chicago Press, 2012), 121, 117–131, for the quoted phrase; Platt, *Shock Cities*, chap. 11; and Harold L. Platt, "Chicago, the Great Lakes, and the Origins of Federal Urban Environmental Policy," *Journal of the Gilded Age and Progressive Era* 1 (April 2002): 122–153.

95. G. Taylor, "Recreation Developments in Chicago Parks," 96. Also see Jane Addams, *The Spirit of Youth and the City Streets* (New York: Macmillan, 1909); Cranz, *The Politics of Park Design*; and Osborn, "The Development of Recreation."

96. Jennifer Gray, "An Everyday Wilderness: Dwight Perkins and the Cook County Forest Preserve," *Future Anterior* 10 (Summer 2013): 1–19; and R. Stephen Sennott, *Dwight Heald Perkins*, February 2000, available at http://www.anb.org.flagship.luc.edu/articles/17/17-00671.html?a=1&g=m&f=%22Dwight%20Perkins%22&ia=-at&ib=-bib&d=10&ss=0&q=1 (accessed May 25, 2015); M. McCarthy, "Politics and the Parks"; Pamela Todd, "Perkins, Lucy Fitch," in *Women Building Chicago 1790–1990: A Biographical Dictionary*, ed. Rima Lunin Schultz and Adele Hast (Bloomington: Indiana University Press, 2001), 686–688; and Arthur Zilversmit, *School Architecture*, 2004, available at http://encyclopedia.chicagohistory.org/pages/1120.html (accessed May 25, 2015).

97. Condit, *The Chicago School of Architecture*, 203, for the quotation; and M. McCarthy, "Politics and the Parks."

98. Perkins, *Report of the Special Park Commission*, 75, for the quotation; ibid., 32–36, 74–79.

99. Jens Jensen, "Report of the Landscape Architect," in ibid., 80–106.

100. Grese, *Jens Jensen*, chaps. 1–2; and M. McCarthy, "Politics and the Parks." Also see William H. Tishler, ed., *Jens Jensen: Writings Inspired by Nature* (Madison: Wisconsin Historical Society Press, 2012).

101. William H. Tishler and Erik M. Ghenoiu, "Jens Jensen and the Friends of Our Native Landscape," *Wisconsin Magazine of History* 86 (Summer 2003): 6, 2–15.

102. Jensen, "Report of the Landscape Architect"; Jens Jensen, "Parks and Politics (1902)," in Tishler, *Jens Jensen*, 12–15; Jens Jensen, "Urban and Suburban Landscape Gardening (1903)," in ibid., 17–20; and Jens Jensen, "Street Trees for the City of Chicago," CT, April 28, 1904.

103. Jensen, "Report of the Landscape Architect." On the neighborhood unit, see Clarence Arthur Perry, *Housing for the Machine Age* (New York: Russell Sage Foundation, 1939). Also see his original formulation in Clarence Arthur Perry, "A Neighborhood Unit," in *Regional Plan of New York and Its Environs, vol. 7, Neighborhood and Community Planning* (New York: Committee for the Regional Survey of New York and Its Environs, 1929), 30–35; Dirk Schubert, "The Neighbourhood Paradigm: From Garden Cities to Gated Communities," in *Urban Planning in a Changing World: The Twentieth Century Experience*, ed. Robert Freestone (London: Spon, 2000), 118–138; Howard J. Gillette Jr., "The Evolution of Neighborhood Planning: From the Progressive Era to the 1949 Housing Act," *Journal of Urban History* 9 (August 1983): 421–444; and Robert Fishman, "The Metropolitan Tradition in American Planning," in *The American Planning Tradition: Culture and Policy*, ed. Robert Fishman (Washington, DC: Woodrow Wilson Center Press, 2000), 65–88.

104. Jacob A. Riis, *How the Other Half Lives: Studies among the Tenements of New York* (New York: Scribner, 1890); and Peter Bacon Hales, *Silver Cities—The Photography of American Urbanization 1839–1915* (Philadelphia: Temple University Press, 1983).

105. Perkins, *Report of the Special Park Commission*, 8–9, for a complete list of the fifty-nine illustrations. On the new communications media, see David Paul Nord, "Newspapers and New Politics: Midwestern Municipal Reform, 1890–1900" (Ph.D. diss., University of Wisconsin, Madison, 1981); and Laura E. Baker, "Civic Ideals, Mass Culture, and the Public: Reconsidering the 1909 Plan of Chicago," *Journal of Urban History* 36 (July 2010): 1–24.

106. Jensen, "Report of the Landscape Architect."

107. CT, August 21, 1927; July 26, 1932, for biographical information. Also see Bachin, *Building the South Side*, 145–155; and M. McCarthy, "Politics and the Parks."

108. CT, May 14, 2005, for the quoted phrase; ibid., for a description of the event. Also see Draper, "The Art and Science of Park Planning"; Mason, "Mary McDowell and Municipal Housekeeping"; M. McCarthy, "Politics and the Parks"; and Mary L. Gray, *A Guide to Chicago's Murals* (Chicago: University of Chicago Press, 2001), 260.

109. Henry G. Foreman, as quoted in CT, May 14, 2005; and ibid., for the presentation of the petitions.

110. CT, October 3, 25–26, 29, November 5–6, 9, 1905; February 2, 13, 18, 1908.

111. Ibid., April 27, 1909; September 10, October 20, December 7, 1910; September 22, 1914; May 8, 1915; February 18, 1916.

CHAPTER 2

1. Laura E. Baker, "Civic Ideals, Mass Culture, and the Public: Reconsidering the 1909 Plan of Chicago," *Journal of Urban History* 36 (July 2010): 1–24; and Carl S. Smith, *The Plan of Chicago: Daniel Burnham and the Remaking of the American City* (Chicago: University of Chicago Press, 2006). Also see Jens Jensen, "Regulating City Building," in *Jens Jensen: Writings Inspired by Nature*, ed. William H. Tishler (Madison: Wisconsin Historical Society Press, 2012), 37–40.

2. Harold L. Platt, "Jane Addams and the Ward Boss Revisited: Class, Politics, and Public Health in Chicago, 1890–1914," *Environmental History* 5 (April 2000): 194–222; and Harold L. Platt, "Chicago, the Great Lakes, and the Origins of Federal Urban Environmental Policy," *Journal of the Gilded Age and Progressive Era* 1 (April 2002): 122–153.

3. Joshua A. T. Salzmann, "The Creative Destruction of the Chicago River Harbor: Spatial and Environmental Dimensions of Industrial Capitalism, 1881–1909," *Enterprise and Society* 13 (June 2012): 235–275; James C. O'Connell, "Technology and Pollution: Chicago's Water Policy, 1833–1930" (Ph.D. diss., University of Chicago, 1980); and Carl Condit, *Chicago, 1910–1929: Building, Planning, and Urban Technology* (Chicago: University of Chicago Press, 1973). Also see the chapter on Frederic A. Delano in Patrick D. Reagan, *Designing a New America: The Origins of New Deal Planning, 1890–1943* (Amherst: University of Massachusetts Press, 1999), chap. 2. Delano was the original inspiration for a plan to modernize Chicago's rail and shipping facilities on a metropolitan scale. Reagan calls him "the father of New Deal Planning," in ibid., 28.

4. Kristen Schaffer, "Daniel H. Burnham: Urban Ideas and the 'Plan of Chicago,'" (Ph.D. diss., Cornell University, 1993); and Margaret Garb, "Race, Housing, and Burnham's Plan: Why Is There No Housing in the 1909 Plan of Chicago?" *Journal of Planning History* 10 (April 2011): 99–113.

5. Maureen A. Flanagan, *Seeing with Their Hearts: Chicago Women and the Vision of the Good City, 1871–1933* (Princeton, NJ: Princeton University Press, 2002), chap. 5.

6. *Chicago Daily Tribune* [hereafter cited as CT], August 13–18, 1908; April 19, 30, May 1, August 15, 1909.

7. Joseph P. Schwieterman, Dana M. Caspall, and Jane Heron, *The Politics of Place: A History of Zoning in Chicago* (Chicago: Lake Claremont Press, 2003); Robert Lewis, "Modern Industrial Policy and Zoning: Chicago, 1910–1930," *Urban History* 40 (February 2013): 97–113; and Wendy Plotkin, "Deed of Mistrust: Race, Housing and Restrictive Covenants in Chicago, 1900–1953" (Ph.D. diss., University of Illinois, Chicago, 1999).

8. Forest Preserve District of Cook County in the State of Illinois, *Forest Preserves of Cook County* (Chicago: Clohessy, [1923?]). Also see Jens Jensen, "A Greater West Park System," in *Jens Jensen*, 48–52.

9. Platt, "Chicago, the Great Lakes." Also see Harold L. Platt, *Shock Cities: The Environmental Transformation and Reform of Manchester and Chicago* (Chicago: University of Chicago Press, 2005); Kenneth Finegold, *Experts and Politicians—Reform Challenges to Machine Politics in New York, Cleveland, and Chicago* (Princeton, NJ: Princeton University Press, 1995), 119–168; Douglas Bukowski, *Big Bill Thompson, Chicago, and the Politics of Image* (Urbana: University of Illinois Press, 1998), 16–21; and the classic account, Lloyd Wendt and Herman Kogan, *Bosses in Lusty Chicago—The Story of Bathhouse John and Hinky Dink* (Bloomington: Indiana University Press, 1967).

10. Robin Einhorn, *Property Rules: Political Economy in Chicago, 1833–1872* (Chicago: University of Chicago Press, 1991).

11. Edward R. Kantowicz, "Carter H. Harrison II: The Politics of Balance," in *The Mayors: The Chicago Political Tradition*, ed. Paul M. Green and Melvin G. Holli (Carbondale: Southern Illinois University Press, 1987), 16–32; John D. Buenker, "Edward F. Dunne: The Limits of Municipal Reform," in ibid., 33–49; and Maureen A. Flanagan, "Fred A. Busse: A Silent Mayor in Turbulent Times," in ibid., 50–60.

12. Robin Einhorn, *Political Culture*, Newberry Library, 2004, available at http://www.encyclopedia.chicagohistory.org/pages/987.html (accessed August 29, 2015).

13. Roger Biles, *Machine Politics*, Newberry Library, 2004, available at http://www.encyclopedia.chicagohistory.org/pages/774.html (accessed August 29, 2015). Also see note 30.

14. Joel A. Tarr, *A Study in Boss Politics: William A. Lorimer of Chicago* (Urbana: University of Illinois Press, 1971); Maureen A. Flanagan, *Charter Reform in Chicago* (Carbondale: Southern Illinois University Press, 1987); and see note 23.

15. See Edwin Oakes Jordan, "Typhoid Fever and Water Supply in Chicago," *Journal of the American Medical Association* 39 (December 1902): 1561–1566; and Platt, "Jane Addams and the Ward Boss," for the epidemic in Jane Addams's Nineteenth Ward; CT, September 2, 1902, on water-supply safety and typhoid-fever outbreaks; ibid., February 27, 1903, on an epidemic in Evanston; and ibid., April 3, May 13, 17, 1910, on typhoid fever in the Calumet District. Also see Libby Hill, *The Chicago River: A Natural and Unnatural History* (Chicago: Lake Claremont Press, 2000).

16. Lyman E. Cooley, *The Lakes and Gulf Waterway as Related to the Chicago Sanitary Problem* (Chicago: Weston, 1890); Lyman E. Cooley, "Lake Levels Effects on Account of the Sanitary Canal of Chicago," a brief prepared for Harvey D. Goulder, Lake Carriers Association, November 27, 1894; Robert A. Waller, "Illinois Waterway from Conception to Completion, 1908–1933," *Journal of the Illinois State Historical Society* 65 (1972): 125–141; and Platt, "Chicago, the Great Lakes."

17. Salzmann, "The Creative Destruction of the Chicago River Harbor."

18. Paul M. Green, "Irish Chicago: The Multi-ethnic Road to Machine Success," in *Ethnic Chicago*, ed. Melvin G. Holli and Peter d'Alroy Jones (Grand Rapids, MI: Eerdmans, 1984), 412–459; Michael F. Funchion, "Irish Chicago: Church, Homeland, Politics, and Class—The Shaping of an Ethnic Group, 1870–1900," in ibid., 14–45; and Wendt and Kogan, *Bosses in Lusty Chicago*.

19. CT, October 5, 1895, on Smyth's nomination; ibid., April 16, 1927, on his biography and childhood memories; and Wendt and Kogan, *Bosses in Lusty Chicago*, 117–120, 220–221, on the Ogden Gas franchise.

20. CT, April 15, 1920; Alex Gottfried, *Boss Cermak of Chicago: A Study of Political Leadership* (Seattle: University of Washington Press, 1962), 122–124; "Roger Charles Sullivan," available at https://en.wikipedia.org/wiki/Roger_Charles_Sullivan (accessed August 29, 2015); and "Great Lakes Dock and Dredge Company," available at https://en.wikipedia.org/wiki/Great_Lakes_Dredge_and_Dock_Company (accessed September 9, 2015).

21. CT, August 30, 1892, for his selection as a ward delegate; ibid., October 7, 1943; and "Patrick Nash," available at https://en.wikipedia.org/wiki/Patrick_Nash, for his biography (accessed August 31, 2015). Also see CT, January 7, 1928, for the obituary of Thomas Nash, who lived to be 101 years old and one of Chicago's oldest citizens; "The Kelly-Nash Machine," *Fortune* 14 (1936): 47–52, 114–130, for the revelation that Nash's nephews, the Dowdle brothers, had also set up a construction company that had won at least $4 million in public contracts during the 1920s; and John T. Flynn, "These Are Our

Rulers," *Collier's* 105 (1940): 14–15, 40–43, who sets the value of the Dowdle brothers' business with the government at $13 million during the decade.

22. Horatio Alger, *Ragged Dick, or, Street Life in New York with the Boot-Blacks* (1868; repr., New York: Penguin Group, 1990).

23. Roger Biles, *Big City Boss in Depression and War: Mayor Edward J. Kelly of Chicago* (DeKalb: Northern Illinois University Press, 1984), chap. 1.

24. CT, August 9, September 23, 1899; and Kantowicz, "Carter H. Harrison II." Also see L. F. Hinds, "Builder of Chicago's Sewers; Hill an Expert in His Work," CT, May 22, 1910, for a detailed report on the work of the Sewer Department. Also see H. R. Abbott, "Intercepting Sewer Construction in the Northern Part of the Sanitary District of Chicago," *Journal of the Western Society of Engineers* 22 (June 1917): 373–405, on open cut methods; and CT, May 19, 1900, on tunnel methods of construction.

25. Commissioner of Public Works Lawrence E. McGann, as quoted in CT, August 9, 1899. Also see ibid., November 20, 1900.

26. Ibid., September 24, 1899, for the quotation.

27. Ibid., November 15, 18, December 17, 1899.

28. Ibid., November 15, 1899; January 4, February 21, 1900; and Finegold, *Experts and Politicians.*

29. Isham Randolph, as quoted in CT, August 10, 1900; and ibid., October 2, 1900, for the "assessment" meeting.

30. CT, December 6, 1905; January 11, April 27, 1906, for patronage within the SDC, including kickback "commissions" for the trustees; ibid., October 28, 1905, for Wenter's son's job; ibid., April 21, 1911, for Smyth's son's insurance business; and ibid., April 12, November 11, 1907, for exposure of the dredging and sewer contractors' cartels, respectively.

31. Thomas A. Smyth, "The Present and Future of the Drainage Canal," CT, January 11, 1903.

32. Ibid.; and CT, November 19, 1912, for an early plan for straightening the river. Also see Condit, *Chicago, 1910–1929*, 247; and Amy D. Finstein, "From 'Cesspool' to 'the Greatest Improvement of Its Kind': Wacker Drive and the Recasting of the Chicago Riverfront, 1909–1926," *Journal of Planning History* 14 (November 2015): 287–308.

33. See Chapter 1.

34. Amanda I. Seligman, *Cabrini-Green*, available at http://encyclopedia.chicagohistory.org/pages/199.html (accessed September 29, 2015). Also see the scientific study "Report on Sewage Disposal along the North Branch of the Chicago River and Its Tributaries," in *Illinois State Water Survey*, no. 11 (Urbana: Illinois Water Survey, 1913), 323–338; and the classic account, Harvey W. Zorbaugh, *The Gold Coast and the Slum: A Sociological Study of Chicago's Near East Side* (Chicago: University of Chicago Press, 1929).

35. Robert R. McCormick, as quoted in Richard Norton Smith, *The Colonel: The Life and Legend of Robert R. McCormick, 1880–1955* (Boston: Houghton Mifflin, 1997), 101; ibid., chap. 4; and CT, June 25, October 7, 19, 24–29, November 5, 9, 1905, for the 1905 election campaign. Ed Kelly also went down in defeat in a bid for a seat on the board of trustees.

36. CT, December 6–7, 1905; January 11, 1906.

37. Ibid., February 28, March 21, April 24–25, September 27, October 24, November 6, 11, 1908, for coverage of the deep-water campaign of both parties.

38. Ibid., November 6, 1910, for the first quotation; ibid., September 3, 1911, for

the second quotation; and Hill, *The Chicago River*, 138–150, for a history of the North Shore's Big Ditch. Also see CT, November 30, 1910, for a report on the opening ceremonies and celebration.

39. Platt, "Chicago, the Great Lakes"; and CT, August 4, 1908, for McCormick's announcement that construction of the canal would be on hold until the court case was settled.

40. CT, April 3, May 13–14, November 30, December 14, 1910.

41. Ibid., November 30, 1910.

42. W. A. Evans, as quoted in ibid., April 3, 1910. He also played a key role two years earlier in the formation of a four-state Lake Michigan Sanitary Association, which was dedicated to ending all sewage discharges into the lake. See ibid., April 12, 1908. Also see Platt, "Jane Addams and the Ward Boss Revisited," for additional information on Dr. Evans's leadership in public-health reform.

43. CT, May 13, 14, 17, 1910.

44. Ibid., July 25, 1907, on Randolph's decision to step down, join a private consulting firm, and anoint Wisner as his successor; ibid., April 3, 1910, on Wisner's remarks to the City Club; and ibid., October 27, 1911, on the McCormick commission report, which outlined a thirty-year, $12.6-million plan. Also see O'Connell, "Technology and Pollution," 138, for an assessment of Wisner's importance in the advancement of science and technology within the Sanitary District of Chicago; and CT, November 26, 1911, for a parallel article by Dr. Evans on the imminent death of the fish in the Illinois River as "trade wastes" depleted its oxygen levels.

45. Ibid., January 9, November 18, 1909. Also see Condit, *Chicago, 1910–1929*, 240–247.

46. CT, December 5, 1909, for the term "gray wolves"; and Charles Edward Merriam, *Chicago: A More Intimate View of Urban Politics* (New York: Macmillan, 1929), chap 2, for a firsthand account of "Big Fix," or machine politics in Chicago. Also see CT, December 9, 1909, for a profile of Merriam, including an illustration of him and members of his committee; Barry Dean Karl, *Charles E. Merriam and the Study of Politics* (Chicago: University of Chicago Press, 1974); and Reagan, *Designing a New America*, chap. 3.

47. CT, August 20, December 5, 1909.

48. Ibid., January 7, 28, 1910.

49. Ibid., February 1, 1910.

50. Ibid., January 18, 1910.

51. Ibid., December 1, 5, 1909. Also see Walter L. Fisher, "Legal Aspects of the Plan of Chicago," in *Plan of Chicago* (Chicago: Commercial Club, 1909), 127–156.

52. CT, December 1, 5, 31, 1909; January 7, 1910.

53. Ibid., January 24, February 1, March 14, 29, 1910.

54. Ibid., February 7, 10, May 27, July 7, 12, 1910.

55. Ibid., July 15–16, 1910; May 3, 1911; Karl, *Charles E. Merriam*, 61–72; and Kantowicz, "Carter H. Harrison II."

56. Robert R. McCormick, as quoted in CT, April 23, 1911; R. Smith, *The Colonel*, chap. 5; CT, April 29, 1910, for his work on the board to establish forest preserves on some of the district's land; ibid., August 20, November 4, 1908, for Sullivan's control of the party with P. A. Nash as one of his ward committeemen and Ed Kelly as a party insider; and ibid., September 20, November 4, 1910, for Sullivan's faction's victory in taking majority control of the sanitary district.

57. CT, September 3, 1911.

58. Thomas Smyth, as quoted in ibid., April 20, 1911, for the first quotation; ibid., June 10, 1911, for the second quotation; and ibid., September 3, 1911, for the official notifications.

59. Ibid., November 27, December 8, 27, 1911. However, Evanston would have to wait until 1919, when its interceptor sewer and pumping station finally went online.

60. Ibid., March 20, October 27, 1911; September 7, 1912, for the panel's report; and ibid., February 9, 1912, for its rejection. Also see Edward J. Kelly, "Progress of the Sewage Disposal Program at Chicago—I and II," *Engineering News-Record* 96, nos. 9, 10 (1911): 363–366, 394–400.

61. McGann, as quoted in CT, November 18, 1912; and ibid., November 20, 1912.

62. George M. Wisner, *Report on Sewage Disposal—The Sanitary District of Chicago* (Chicago: Klein, 1911); George M. Wisner, *Report on Industrial Wastes from the Stockyards and Packinghouses of Chicago* (Chicago: Barnard and Miller, 1914); G. B. Young, "The Result of the Experimental Employment of Hypochlorite Treatment to a Portion of the Chicago City Water Supply," *American Journal of Public Health* 4 (April 1914): 310–315; and Platt, *Shock Cities*, 432–441. Also see George A. Johnson and U.S. Department of the Interior, United States Geological Survey, "The Purification of Public Water Supplies," in *Water Supply Papers* (Washington, DC: Government Printing Office, 1913), for a national perspective.

63. Chicago Commission on Race Relations, *The Negro in Chicago: A Study of Race Relations and a Race Riot* (Chicago: University of Chicago Press, 1922). Also see Plotkin, "Deed of Mistrust"; Margaret Garb, *City of American Dreams: A History of Home Ownership and Housing Reform in Chicago, 1871–1919* (Chicago: University of Chicago Press, 2005), chap. 7; Margaret Garb, "Drawing the 'Color Line': Race and Real Estate in Early Twentieth-Century Chicago," *Journal of Urban History* 32 (July 2006): 773–778; Otis Dudley Duncan and Beverly Duncan, *The Negro Population of Chicago: A Study of Residential Succession* (Chicago: University of Chicago Press, 1957); Allan H. Spear, *Black Chicago: The Making of a Negro Ghetto, 1880–1920* (Chicago: University of Chicago Press, 1967); Thomas Lee Philpott, *The Slum and the Ghetto: Neighborhood Deterioration and Middle-Class Reform, Chicago, 1880–1930* (New York: Oxford University Press, 1978); William M. Tuttle, *Race Riot: Chicago in the Red Summer of 1919* (New York: Atheneum, 1970); Margaret Garb, *Freedom's Ballot: African American Political Struggles in Chicago from Abolition to the Great Migration* (Chicago: University of Chicago Press, 2014); and Michael McCoyer, "Darkness of a Different Color: Mexicans and Racial Formation in Greater Chicago, 1916–1960" (Ph.D. diss., Northwestern University, 2007).

64. Lewis, "Modern Industrial Policy and Zoning"; Philpott, *The Slum and the Ghetto*, chap. 7; and Colin Fisher, "African Americans, Outdoor Recreation, and the 1919 Chicago Race Riot," in *To Love the Wind and the Rain: African Americans and Environmental History*, ed. Dianne D. Glave and Mark Stoll (Pittsburgh: University of Pittsburgh Press, 2006), 63–76. Also see Joe Black, "A Theory of African-American Citizenship: Richard Westbrooks, the Great Migration, and the Chicago Defender's 'Legal Helps' Column," *Journal of Social History* 46 (Summer 2013): 896–915; and Colin Fisher, *Urban Green: Nature, Recreation, and the Working Class in Industrial Chicago* (Chapel Hill: University of North Carolina Press, 2015).

65. Clifford Edward Clark, *The American Family Home, 1800–1960* (Chapel Hill: University of North Carolina Press, 1986), 171, for the phrase "Bungalow Craze"; Homer Hoyt, *One Hundred Years of Land Values in Chicago—The Relationship of the Growth of Chicago to the Rise in Its Land Values, 1830–1933* (Chicago: University of Chicago Press, 1933), 196–232; and CT, May 22, 1910, on sewer construction.

66. Hoyt, *One Hundred Years of Land Values*, 196–232, 477, 483; Condit, *Chicago, 1910–1929*, 304; Carl Abbott, "'Necessary Adjuncts to Its Growth': The Railroad Suburbs of Chicago, 1854–1875," *Journal of the Illinois State Historical Society* 73 (Summer 1980): 117–131; and Jan Cigliano, "The Bungalow and the New American Woman," in *The Chicago Bungalow*, ed. Dominic A. Pacyga and Charles Shanabruch (Chicago: Arcadia, 2003), 97. Also see Joseph C. Bigott, *From Cottage to Bungalow: Houses and the Working Class in Metropolitan Chicago, 1869–1929* (Chicago: University of Chicago Press, 2001); and Joel A. Tarr and Gabriel DuPuy, eds., *Technology and the Rise of the Networked City in Europe and America* (Philadelphia: Temple University Press, 1988).

67. Duncan and Duncan, *The Negro Population of Chicago*, 484. Also see note 63.

68. Philpott, *The Slum and the Ghetto*, 102, 108, respectively, for the two quotations. For the reform campaign, see CT, January 8, October 8, 1910. It was funded by a $10,000 grant by the Russell Sage Foundation to a group of Chicago's settlement-house leaders, including Graham Taylor, Julia Lathrop, George Hooker, Walter Fisher, and Edith Abbott. Out of their research at the street level came classic studies of sociology, including Sophonisba P. Breckinridge and Edith Abbott, "Chicago Housing Conditions V: South Chicago at the Gates of the Steel Mills," *Journal of Sociology* 17 (September 1911): 145–176; and Edith Abbott, *The Tenements of Chicago, 1880–1935* (Chicago: University of Chicago Press, 1936). Also see Flanagan, *Seeing with Their Hearts*, 89–103.

69. Neal Samors, May Jo Doyle, Martin Lewin, and Michael Williams, *Chicago's Far North Side: An Illustrated History of Rogers Park and West Ridge* (Chicago: Rogers Park/West Ridge Historical Society, 2000), chaps. 1–2. Also see Barbara M. Posadas, "Suburb into Neighborhood," *Journal of the Illinois State Historical Society* 76 (Fall 1983): 162–176.

70. CT, September 24, 1911. Also see ibid., January 24, 1915, for a similar promotional article; and Samors et al., *Chicago's Far North Side*, chap. 2.

71. Samors et al., *Chicago's Far North Side*, chap. 2.

72. CT, April 28, 1912. Also see David Stradling, *Smokestacks and Progressives* (Baltimore, MD: Johns Hopkins University Press, 1999), 67–69, for the backstory of the company's move to the suburbs from the city center.

73. Ibid.; and *Summit, Illinois*, available at https://en.wikipedia.org/wiki/Summit,_Illinois (accessed November 1, 2015). Also see CT, July 16, 1912, for large land sales next to the corn-products factory; and ibid., June 30, 1912, for a summary of the growth of the western suburbs, including developments along the Des Plaines River.

74. CT, March 17, 1916.

75. Ibid., March 17–26, 1916.

76. Ibid., July 11, 21, 1919. Also see Dominic A. Pacyga, *Polish Immigrants and Industrial Chicago* (Columbus: Ohio State University Press, 1991); and Garb, *City of American Dreams*.

77. Garb, *Freedom's Ballot*, 149, chap 5. This important study includes an up-to-date bibliography of African American history in Chicago. Also see James R. Grossman, *Land of Hope: Chicago, Black Southerners, and the Great Migration* (Chicago: University of Chicago Press, 1989); and Richard C. Lindberg, *To Serve and Collect—Chicago Politics and Police Corruption from the Lager Beer Riot to the Summerdale Scandal* (New York: Praeger, 1991).

78. Bukowski, *Big Bill Thompson*; George Schottenhamel, "How Big Bill Thompson Won Control of Chicago," *Journal of the Illinois State Historical Society* 45 (Spring 1952): 30–49; and Lloyd Wendt and Herman Kogan, *Big Bill of Chicago* (Indianapolis: Bobbs-Merrill, 1953).

79. Bukowski, *Big Bill Thompson*; and Schottenhamel, "How Big Bill Thompson Won."

80. Garb, *Freedom's Ballot*, chap 6; C. Fisher, *Urban Green*, chap. 4; Grossman, *Land of Hope*, chap. 6; Adam Green, *Selling the Race: Culture, Community, and Black Chicago, 1940–1955* (Chicago: University of Chicago Press, 2007); Chicago Commission on Race Relations, *The Negro in Chicago*; and see note 64.

81. Bukowski, *Big Bill Thompson*; and C. Fisher, *Urban Green*. Also see Michael D. Innis-Jimenez, "Organizing for Fun: Recreation and Community Formation in the Mexican Community of South Chicago in the 1920s and 1930s," *Journal of the Illinois State Historical Society* 98 (Fall 2005): 144–161.

82. CT, March 20, 31, 1912; March 28, 1916; February 13, 1918; March 18, 1919, for reports of damaging floods in the western suburbs; and ibid., April 29, 1924, for a report on the virtual elimination of typhoid-fever outbreaks since the 1890s.

83. Wisner, *Report on Industrial Wastes*; CT, September 5, October 17, 1919, for the story of federal pressure on the SDC to force the meatpackers to treat their organic wastes, and the subsequent but unimplemented agreement for a 60-percent private and 40-percent public split of the costs. A start on a treatment plant came seven years later, after the U.S. Supreme Court took control of the agency. See ibid., November 30, 1919, for the report of the U.S. Army Corps of Engineers on Chicago's illegal diversion of 8,800 cubic feet per second and its "faulty" system of sewage disposal. Also see ibid., August 22, 1911, for Wisner's protest of Smyth's plan to build the Cal-Sag Canal at a snail's pace over four years, causing at least $1 million in waste: it actually took eleven years to complete.

84. Platt, "Chicago, the Great Lakes."

85. Finstein, "From 'Cesspool' to 'the Greatest Improvement.'"

86. CT, January 9, November 18, 1909; Harold L. Platt, "'Clever Microbes': Bacteriology and Sanitary Technology in Manchester and Chicago during the Progressive Era," in *Landscapes of Exposure: Knowledge and Illness in Modern Environments*, ed. Gregg Mitman, Michelle Murphy, and Christopher Sellers, *Osiris*, 2nd ser., vol. 19 (2004), 149–166; Carolyn G. Shapiro-Shapin, "'A Really Excellent Scientific Contribution': Scientific Creativity, Scientific Professionalism, and the Chicago Drainage Case," *Bulletin of the History of Medicine* 71 (Fall 1997): 385–411; and O'Connell, "Technology and Pollution."

87. Jane Addams, *Twenty Years at Hull-House* (New York: Macmillan, 1910), 295; and Philpott, *The Slum and the Ghetto*, chap 4.

88. Philpott, *The Slum and the Ghetto*, 107, for the quotation; ibid., 107–109; Maureen A. Flanagan, "Gender and Urban Political Reform: The City Club and the Women's City Club of Chicago in the Progressive Era," *American Historical Review* 95 (October 1990): 1032–1050; Kathryn Kish Sklar, *Florence Kelley and the Nation's Work* (New Haven, CT: Yale University Press, 1995); and Platt, "Jane Addams and the Ward Boss."

89. Henry M. Hyde, "Forest Preserve Held Up for Year," CT, May 8, 1915; Forest Preserve District of Cook County, ed., *The Forest Preserves of Cook County* (Chicago: Forest Preserve District of Cook County, 1918). The first preserve opened was Deer Grove, a 1,100-acre woodland in the far northwest township of the county, Palatine, about 26 miles from the city center. The Deer Grove site and others earmarked for purchase engendered real-estate speculators to buy adjacent properties. See CT, July 8, 1917. Also see Condit, *Chicago, 1910–1929*, for transportation improvements in the metropolitan region; and Flanagan, *Seeing with Their Hearts*, 103–109, on women's efforts on behalf of recreational open space.

90. CT, April 18, 1920.

91. Biles, *Big City Boss in Depression and War*, 8, for the quoted phrase; and "The Kelly-Nash Machine."

CHAPTER 3

1. *Sanitary District of Chicago v. U.S.*, 266 U.S. 405 (1925); *Wisconsin v. Illinois*, 278 U.S. 367 (1929); and Lt. Col. G. B. Pillsburg, "The Control of the Levels of the Great Lakes," *Journal of the Western Society of Engineers* 32 (1927): 250–262. Also see Harold L. Platt, "Chicago, the Great Lakes, and the Origins of Federal Urban Environmental Policy," *Journal of the Gilded Age and Progressive Era* 1 (April 2002): 122–153, for a legal history of the case; and Eric D. Craft, "The Value of Weather Information Services for Nineteenth-Century Great Lakes Shipping," *American Economic Review* 88 (December 1998): 1059–1076, for an economic analysis of the shipping industry.

2. Craft, "The Value of Weather Information Services"; and Douglas C. Ridgley, *The Geography of Illinois* (Chicago: University of Chicago Press, 1921), chap 5. Craft's essay on the relationship of the seasonal risks to shipping in terms of insurance premiums helps explain the annual rather than long-term perspectives of the experts. Ridgley's review of the state's climate and weather explicitly denies long-term climate change. He was a leading academic expert of his era. On the contrary, he believed that it was "unchangeable" (70). For a brief biography, see Guy H. Burnham, "Douglas C. Ridgley, 1868–1952," *Journal of Geography* 52 (1952?): 82. One study that did include long-term climate change might have been useful in deciding the case. See U.S. Department of Interior, U.S. Geological Survey, and Frank Leverett, "The Water Resources of Illinois," in *Annual Report* (Washington, DC: U.S. Geological Survey, 1896). Although its primary concern was the seasonal cycle of rainfall in relationship to agriculture, it noted that the period from 1875 to 1885 was an exceptionally wet period compared to the following ten years. "These observations," the report stated, "are certainly suggestive of periodic variations in rainfall, covering as they do an area of several States. It will be a matter of importance to note, as time goes on, whether the teaching of the weather records strains periodicity. If definite alternate wet and dry periods occur, the agriculture can be adjusted to the conditions and shortage of crops of certain kinds be foreseen" (28). For a contemporary viewpoint on climate change, see Linda D. Mortsch and Frank H. Quinn, "Climate Change Scenarios for Great Lakes Basin Ecosystem Studies," *Limnology and Oceanography* 43 (July 1996): 903–911; and Donald Wuebbles et al., "U.S. Global Change Research Program Climate Science Special Report (CSSR) [Fifth-Order Draft]," (Washington, DC: U.S. Global Change Research Program, June 28, 2017).

3. *Wisconsin v. Illinois*. But the chlorine added to the water supply reduced disease rates, not the divergence of lake water into the Sanitary and Ship Canal. The lake's water remained highly contaminated with disease-causing microbes.

4. *Missouri v. Illinois*, 200 U.S. 496 (1906); M. G. Barnes, "The Illinois Waterway," *Journal of the Western Society of Engineers* 26 (May 1921): 171–189, on the impracticality of making the Illinois River into a deep-water navigational channel as opposed to a 9-foot-deep barge canal; George W. Fuller, "The Sewage Disposal Problem of Chicago," *Journal of the Western Society of Engineers* 3 (February 1925): 113–125, for an introduction to the scale and scope of the problem at the time of the Supreme Court decision. For early studies of the effects of Chicago's sewage on the rivers, see James A. Egan, "Pollution of the Illinois River as Affected by the Drainage of Chicago and Other Cities," in *Report*

of the Sanitary Investigations of the Illinois River and Its Tributaries, ed. State Board of Health Illinois (Springfield, IL: Phillips Brothers, 1901), ix–xxxiv; John H. Long, "Chemical and Bacterial Examinations of the Waters of the Illinois River and Its Principal Tributaries," in ibid., 3–77; and F. Robert Zeit, "Identification of Bacteria Found in the Waters of the Illinois River and Its Principal Tributaries," in ibid., 78–93. Studies of pollution in rivers in New England began after the Civil War. See Donald J. Paisani, "Fish Culture and the Dawn of Concern over Water Pollution in the United States," *Environmental Review* 8 (Summer 1984): 117–131; and John T. Cumbler, *Reasonable Use: The People, the Environment, and the State—New England, 1790–1930* (New York: Oxford University Press, 2001). In contrast, serious scientific studies of the effects of pollution on the Great Lakes began relatively late in the 1920s. See Frank N. Egerton, "Pollution and Aquatic Life in Lake Erie: Early Scientific Studies," *Environmental Review* 3 (Fall 1987): 189–205; Margaret Beattie Bogue, "To Save the Fish: Canada, the United States, the Great Lakes, and the Joint Commission of 1892," *Journal of American History* 79 (March 1993): 1429–1454; and R. Peter Gillis, "Rivers of Sawdust: The Battle over Industrial Pollution in Canada, 1865–1903," *Journal of Canadian Studies* 21 (Spring 1986): 84–10.

5. On Stephen Forbes, see Daniel W. Schneider, "Enclosing the Floodplain: Resource Conflict on the Illinois River, 1880–1920," *Environmental History* 1 (April 1996): 70–96; Daniel W. Schneider, "Local Knowledge, Environmental Politics, and the Founding of Ecology in the United States: Stephen Forbes and 'the Lake as Microcosm' (1887)," *Isis* 91 (December 2000): 681–705; and Stephen P. Havera and Katie E. Roat, "Forbes Biological Station: The Past and the Promise," in *Illinois Natural History Survey Special Publication* 10 ([Urbana?]: Illinois Natural History Survey, [2003?]).

6. See Board of Trustees, Sanitary District of Chicago, *Memorandum Concerning the Drainage and Sewerage Conditions in Chicago and the Diversion of 10,000 C. F. S. From Lake Michigan at Chicago* (Chicago: Sanitary District of Chicago, December 1923), for the first inclusion of flood control in the discussion of the city's sanitation strategy. This document, however, came too late to be included in the 1925 decision by the court. It is in section 3 of this chapter.

7. See for example, CT, August 17, 1908; May 22, 1910; March 20, 1911; September 7, 1912; and notes 14–15.

8. William Dever, "Message on Water Waste [of June 27, 1923]," in City of Chicago, *City Council Proceedings*, April 16, 1923–April 9, 1924, 586–588, for the statistics on water use. I deal extensively with Chicago's corrupt and unfair system of water billing that generated excessive profits in Harold L. Platt, *Shock Cities: The Environmental Transformation and Reform of Manchester and Chicago* (Chicago: University of Chicago Press, 2005), chaps. 4, 11. In the 1880s, water managers realized that the installation of a universal system of meters was the only practical way of finding leaks in the underground network of mains and pipes. Meters also act as conservation devices, because ratepayers have to pay for the amount of their consumption. Without meters in Chicago, more than half of the water pumped leaked underground, creating a condition of chronic low pressure at the periphery of each service territory. Beside the extra jobs and coal contracts to maintain the system, the Water Department could justify building more and more pumping stations to boost the water pressure. More specific proposals for reform of the city's water-management policies follow below.

9. Richard Guy Wilson, Dianne H. Pilgrim, Dickran Tashjian, and Brooklyn Museum, eds., *The Machine Age in America, 1918–1941* (New York: Brooklyn Museum in association with Abrams, 1986); George E. Mowry and Blaine A. Brownell, *The Urban*

Nation, 1920–1980, rev. ed. (New York: Hill and Wang, 1981); Lizabeth Cohen, *Making a New Deal: Industrial Workers in Chicago, 1919–1939* (Cambridge, UK: Cambridge University Press, 1990); and Maureen A. Flanagan, *America Reformed: Progressives and Progressivisms, 1890s–1920s* (New York: Oxford University Press, 2007).

10. Homer Hoyt, *One Hundred Years of Land Values in Chicago—The Relationship of the Growth of Chicago to the Rise in Its Land Values, 1830–1933* (Chicago: University of Chicago Press, 1933), 239, 233–276.

11. U.S. Census Bureau (1920–1930), *Population*. Also see Edward V. Miller, "Industrialization on Chicago's Periphery: Examining Industrial Decentralization, 1893–1936," *Journal of Urban History* 43, no. 5 (2017): 720–743, for a discussion of how metropolitan, periphery, and fringe are defined by historians and geographers.

12. See CT, May 14, 1922, for a description of the proposed industrial zoning of the Calumet District; ibid., January 21, 1923, for a map that shows the existing and proposed industrial districts; Joseph P. Schwieterman, Dana M. Caspall, and Jane Heron, *The Politics of Place: A History of Zoning in Chicago* (Chicago: Lake Claremont Press, 2003), chap 3; and Robert Lewis, "Modern Industrial Policy and Zoning: Chicago, 1910–1930," *Urban History* 40 (February 2013): 97–113. On the SDC's obsession with a federally funded deep-water channel, see its attorney's statement, George F. Barrett, *The Waterway from the Great Lakes to the Gulf of Mexico* (Chicago: Sanitary District of Chicago, 1926).

13. CT, January 12, March 2, April 23, 1919. Also see ibid., October 26, 1919, for a critique from one of the world's leading planners, Thomas Adams, on Chicago's lack of comprehensive planning, resulting in suburban sprawl and mixed land uses.

14. Horace P. Ramey, as quoted in ibid., March 18, 1919. Heavy storms that sent Chicago River water into the lake, contaminating the water supply, continued to occur during the 1920s. See, for example, ibid., August 12, 1923; June 24, August 7, 1924.

15. Ibid., April 7, October 6, 1919. For other newsworthy, damaging storms during the 1920s, see ibid., June 15, 1920; May 26, 1922; April 9, 1926; April 22, June 14, 1929; Board of Trustees, Sanitary District of Chicago, *Memorandum Concerning the Drainage and Sewerage Conditions*, and note 14. The *Chicago Daily Tribune* was conspicuously silent on the great storm of August 23, 1923.

16. Charles Wacker, as quoted in CT, July 22, 1919; ibid., July 20, 21, 1919. Also see Carl Condit, *Chicago, 1910–1929: Building, Planning, and Urban Technology* (Chicago: University of Chicago Press, 1973), 22–24, 240–252.

17. Colin Fisher, "African Americans, Outdoor Recreation, and the 1919 Chicago Race Riot," in *To Love the Wind and the Rain: African Americans and Environmental History*, ed. Dianne D. Glave and Mark Stoll (Pittsburgh: University of Pittsburgh Press, 2006), 63–76; Michael D. Innis-Jimenez, "Organizing for Fun: Recreation and Community Formation in the Mexican Community of South Chicago in the 1920s and 1930s," *Journal of the Illinois State Historical Society* 98 (Fall 2005): 144–161; Gabriela F. Arredondo, "Navigating Ethno-racial Currents: Mexicans in Chicago, 1919–1939," *Journal of Urban History* 30 (March 2004): 399–427; and Joe Black, "A Theory of African-American Citizenship: Richard Westbrooks, the Great Migration, and the Chicago Defender's 'Legal Helps' Column," *Journal of Social History* 46 (Summer 2013): 896–915.

18. CT, October 10, 1924, on the opening of the stadium; ibid., July 7, 1924, and July 2, 1925, on openings of sections of Lake Shore Drive; ibid., May 22, 1922; February 24, 1923; March 20, 1924, on the politics of the park board; "The Kelly-Nash Machine," *Fortune* 14 (1936): 47–52, 114–130; and John T. Flynn, "These Are Our Rulers," *Collier's* 105 (1940): 14–15, 40–43, on the costs of stadium. Also see Roger Biles, *Big City Boss in Depression*

and War: Mayor Edward J. Kelly of Chicago (DeKalb: Northern Illinois University Press, 1984); and Elizabeth Halsey, *The Development of Public Recreation in Metropolitan Chicago* (Chicago: Chicago Recreation Commission, 1940), 88, for patronage figures. Also see Laura E. Baker, "Civic Ideals, Mass Culture and the Public: Reconsidering the 1909 Plan of Chicago," *Journal of Urban History* 36 (July 2010): 747–770.

19. Edward Kelly, as quoted in CT, October 10, 1924; and ibid., November 28, 1926. In 1928, Kelly was reelected without serious opposition as the president of the board of the South Park District. See ibid., March 3, 1928.

20. Charles Wacker, as quoted in Amy D. Finstein, "From 'Cesspool' to 'the Greatest Improvement of Its Kind': Wacker Drive and the Recasting of the Chicago Riverfront, 1909–1926," *Journal of Planning History* 14 (November 2015): 301, 287–308; Baker, "Civic Ideals, Mass Culture, and the Public"; and Douglas Bukowski, *Big Bill Thompson, Chicago, and the Politics of Image* (Urbana: University of Illinois Press, 1998).

21. Finstein, "From 'Cesspool' to 'the Greatest Improvement," 278.

22. Guy Debord, *The Society of the Spectacle* (1967; repr., Detroit: [Black and Red], 1970), for the quoted phrase; and Steven A. Riess, *Touching Base: Professional Baseball and American Culture in the Progressive Era* (Westport, CT: Greenwood Press, 1980).

23. South Park Board of Commissioners, *Annual Report* (1922): 97, as reported in Marian Lorena Osborn, "The Development of Recreation in the South Park System of Chicago" (master's thesis, University of Chicago, August 1928), 102, for statistics; L. Sue Greer, "The United States Forest Service and the Postwar Commodification of Outdoor Recreation," in *For Fun and Profit—The Transformation of Leisure into Consumption*, ed. Richard Butsch (Philadelphia: Temple University Press, 1990), 152–172; Peter J. Schmitt, *Back to Nature: The Arcadian Myth in Urban America, 1900–1930* (New York: Oxford University Press, 1969); and Galen Cranz, *The Politics of Park Design: A History of Urban Parks in America* (Cambridge: Massachusetts Institute of Technology Press, 1982).

24. Chicago Commission on Race Relations, *The Negro in Chicago: A Study of Race Relations and a Race Riot* (Chicago: University of Chicago Press, 1922); and Halsey, *The Development of Public Recreation*, passim, for a series of maps showing the location of the city's public parks.

25. James T. Farrell, *Studs Lonigan: A Trilogy Containing Young Lonigan, the Young Manhood of Studs Lonigan, Judgment Day* (New York: Modern Library, 1963); Charles Fanning and Ellen Skerrett, "James T. Farrell and Washington Park: The Novel as Social History," *Chicago History* 8 (Summer 1979): 80–91; Colin Fisher, *Urban Green: Nature, Recreation, and the Working Class in Industrial Chicago* (Chapel Hill: University of North Carolina Press, 2015); and John R. Schmidt, "William E. Dever: A Chicago Political Fable," in *The Mayors: The Chicago Political Tradition*, ed. Paul M. Green and Melvin G. Holli (Carbondale: Southern Illinois University Press, 1995), 82–98. Also see Meyer Levin, *The Old Bunch* (Payson, AZ: Rancho Lazarus, 1937), for a parallel, albeit more-upbeat, account by a Jewish Chicagoan coming of age in the 1920s.

26. Osborn, "The Development of Recreation," 67, 55–68.

27. Brian James McCammack, "Recovering Green in Bronzeville: An Environmental and Cultural History of the African American Great Migration to Chicago, 1915–1940" (Ph.D. diss., Harvard University, 2012); Fisher, *Urban Green*; and Thomas Lee Philpott, *The Slum and the Ghetto: Neighborhood Deterioration and Middle-Class Reform, Chicago, 1880–1930* (New York: Oxford University Press, 1978). As McCammack points out, working-class and elite African Americans had different relationships to urban and suburban nature. Like their more-affluent counterparts in the white community,

the black elite had greater access to Washington and Jackson Parks, the forest preserves, and recreational spaces in the countryside.

28. Innis-Jimenez, "Organizing for Fun"; and Arredondo, "Navigating Ethno-racial Currents." Also see Gabriela F. Arredondo, *Mexican Chicago: Race, Identity, and Nation, 1916–39* (Urbana: University of Illinois Press, 2008).

29. Charles H. Sergel, as quoted in CT, December 6, 1916. Sergel's wife, Annie Sergel, led the city's successful antismoke campaign to force the Illinois Central Railroad to electrify its lines on the South Side. See Harold L. Platt, "'Invisible Gases': Smoke, Gender, and the Redefinition of Environmental Policy in Chicago, 1880–1920," *Planning Perspectives* 10 (January 1995): 67–97.

30. For the reaction of the majority of five Democrats on the board, see CT, December 6, 8, 1916.

31. Wallace G. Clark, as quoted in CT, June 6, 1919.

32. Ibid., May 30, 1919, for the second quotation and account of the fight.

33. Ibid. Also see Joshua A. T. Salzmann, "Safe Harbor: Chicago's Waterfront and the Political Economy of the Built Environment, 1847–1918" (Ph.D. diss., University of Illinois, 2008).

34. CT, June 12, 20, December 12, 17, 1919; December 8, 1920.

35. Report of the Chief Engineers of the War Department, as reported in ibid., November 30, 1919, for the quotations; and ibid., September 5, October 17, 1919, on the stockyards.

36. Report of the Chief Engineers of the War Department, as reported in Board of Trustees, Sanitary District of Chicago, *Memorandum Concerning the Drainage*, 55, for the quotation; and CT, July 20, 1920, for the court ruling.

37. Christopher Hamlin, *A Science of Impurity—Water Analysis in Nineteenth Century Britain* (Bristol: Hilger, 1990); James C. O'Connell, "Technology and Pollution: Chicago's Water Policy, 1833–1930" (Ph.D. diss., University of Chicago, 1980); Christopher Hamlin, *Public Health and Social Justice in the Age of Chadwick—Britain, 1800–1854* (Cambridge, UK: Cambridge University Press, 1998); Michael Worboys, *Spreading Germs—Disease Theories and Medical Practice in Britain, 1865–1900* (Cambridge, UK: Cambridge University Press, 2000); Nancy Tomes, *The Gospel of Germs—Men, Women, and the Microbe in American Life* (Cambridge, MA: Harvard University Press, 1998); and Harold L. Platt, "'Clever Microbes': Bacteriology and Sanitary Technology in Manchester and Chicago during the Progressive Era," in *Landscapes of Exposure: Knowledge and Illness in Modern Environments*, ed. Gregg Mitman, Michelle Murphy, and Christopher Sellers, *Osiris*, 2nd ser., vol. 19 (2004): 149–166.

38. Egan, "Pollution of the Illinois River," xix, ix–xxxiv.

39. John W. Alvord and Charles B. Burdick, *Report of the Rivers and Lakes Commission on the Illinois River and Its Bottom Lands* (Springfield: Illinois State Journal Company, 1915), 64–83.

40. Ibid., 54–63; Schneider, "Enclosing the Floodplain"; and Schneider, "Local Knowledge, Environmental Politics." Also see Chester McArthur Destler, "Agricultural Readjustment and Agrarian Unrest in Illinois, 1880–1896," *Agricultural History* 21 (April 1947): 104–116; and Mary R. McCorvie and Christopher L. Lant, "Drainage District Formation and the Loss of Midwestern Wetlands, 1850–1930," *Agricultural History* 67 (Fall 1993): 13–39.

41. Alvord and Burdick, *Report of the Rivers and Lakes Commission*, 113, 113–135.

42. *Economy Light and Power Company v. U.S.*, 256 U.S. 113 (1921).

43. Ibid.; and Steven I. Apfelbaum, "The Role of Landscapes in Stormwater Management," paper presented at the 1EPA Seminar, 1995.

44. Apfelbaum, "The Role of Landscapes."

45. *Economy Light and Power Company v. U.S.*

46. Ibid.

47. Board of Trustees, Sanitary District of Chicago, *Memorandum Concerning the Drainage.*

48. Ibid., 29.

49. Ibid., 83, 83–92. Also see CT, June 24, 28, 1924, for another heavy rainfall, backflow of the Chicago River into the lake, denials of SDC of any risk to the public health, and refutation of these claims by the city's Health Department; and ibid., May 17, 1929, for an editorial reprinted from the Peoria Transcript making the same argument of more water as flood control.

50. *Sanitary District of Chicago v. U.S.*

51. CT, January 7, 1925. Also see ibid., January 6, 1925, for the call to action by the president of the SDC, Lawrence F. King.

52. Ibid., January 6, 11, 21, 1925.

53. U.S. Secretary of War John W. Weeks, "Permit [of March 3, 1925]," as reported in Sanitary District of Chicago, *Proceedings*, March 19, 1925, 201–202. Also see CT, January 14, 15, February 6, 1925, for coverage of the negotiations leading to the plan.

54. Sanitary District of Chicago, *Proceedings*, 202.

55. Sanitary District of Chicago, *Report of the Engineering Board of Review* (Chicago: Sanitary District of Chicago, 1925), pt. 2, 10.

56. Langdon Pearse, "The Sewage Treatment Program of the Sanitary District of Chicago," *Journal of the Western Society of Engineers* 31 (July 1926): 261–227; and Edward J. Kelly, "The Sanitary District of Chicago, Its Past, Present and Future," in ibid., 259–260. Also see ibid., 276–294, for several additional technical articles on the treatment plants. See CT, February 2, 1923, for Pearse's first announcement of plans for the world's largest treatment plant. For Wisner's resignation, see CT, June 15, 1920. For Sergel's removal as president of the board, see ibid., December 8, 1920. He lost the battle in court to reduce the term of president from a full six years to two years. For an obituary of Pearse, see ibid., July 21, 1956.

57. F. W. Mohlman, "Chemical and Biological Investigations of the Sanitary District," *Journal of the Western Society of Engineers* 31 (July 1926): 272, 267–276.

58. Ibid.; Edward Bartow, "Report on the Sewage Disposal along the North Branch of the Chicago River and Its Tributaries," in *Water Survey Series No. 11: Chemical and Biological Survey of the Waters of Illinois* (Urbana: University of Illinois, May 18, 1914), 323–338; Illinois State Water Survey, *Bulletin No. 20: Comparison of Chemical and Bacteriological Examinations Made on the Illinois River during a Season of Low and a Season of High Water, 1923–1924* (Urbana: State of Illinois, 1925); and Robert E Richardson, "Illinois River Bottom Fauna in 1925," *Bulletin of the Illinois State Natural History Survey* 15 (October 1925): 391–422, for the phrase "sludge worms." The Great Lakes were also under intense investigation by scientific experts. See CT, June 28, 1924; September 26, 1926; July 6, 1928.

59. Bartow, "Report on the Sewage Disposal," on the Northwest Sanitary group; and CT, May 12, 1921; August 10, 1926; July 16, 1927; October 2, 1924; April 4, June 7, 1928, on efforts to save the Des Plaines River. Also see the report by George Wisner and Langdon Pearse, Sanitary District of Chicago, *Report on Pollution of Des Plaines River and Remedies Therefor. Made to the Board of Trustees of the Sanitary District of Chicago, July, 1914* (Chicago: Sanitary District of Chicago, 1914).

60. William Hale Thompson, as quoted in CT, March 28, 1927; and ibid., September 1–3, 1925, for passage of the ordinance.

61. Ibid., March 20, 1927, April 16, May 6, 1927. Also see Schmidt, "William E. Dever."

62. See CT, February 14, April 11, 14, August 18, 1920; September 2, October 19, 22, 26, 1923; February 25, 1925; November 4, 7, 1928; March 26, 1929; and Flynn, "These Are Our Rulers," 14–15, 40–43, for the politics and policies of the SDC; and CT, December 25, 1920; July 21, 28, 1921; May 23, 1923; December 18, 1924; February 2, 1926, on its promotion of and spending on the Chicago River as a deep-water harbor. Also see Barrett, *The Waterway from the Great Lakes.*

63. CT, March 10–26, 1929, for initial use of the label. Also see Virgil W. Peterson, *Barbarians in Our Midst—A History of Chicago Crime and Politics* (Boston: Little Brown, 1952), 163–164, for a summary of the scandal.

64. CT, December 3, 1926; October 21, 1928.

65. William Hale Thompson and Timothy J. Crowe, as quoted in CT, November 24, 1927.

66. *Wisconsin v. Illinois*, 289 U.S. 395 (1933); and Walter Smith, compiler, *Stream Flow Data of Illinois* (Springfield: Illinois Division of Waterways, Department of Public Works and Buildings, 1937), 168.

67. See CT, November 7, 14, 21–24, 1928; and ibid., January 28, 1929, for the list of the patronage workers' names, addresses, salaries, and sponsors from Crowe's ward.

68. Ibid., February 15, 1929.

69. Ibid., March 14–15, May 4, 1929. Also see ibid., January 31, February 5, 8, 1929, for a failed attempt of reformers in the state legislature to conduct an investigation of the Sanitary District.

70. Ibid., March 15, May 30, 1930; June 6, December 4, 8–9, 12, 18, 1931; January 26–February 6, 1932; Flynn, "These Are Our Rulers"; and Alex Gottfried, *Boss Cermak of Chicago: A Study of Political Leadership* (Seattle: University of Washington Press, 1962), chap. 10, on Cermak's control of the prosecutor and the judge in the trial.

71. CT, October 24, 28, 1930; Flynn, "These Are Our Rulers"; and Peterson, *Barbarians in Our Midst*, 164.

72. *Wisconsin v. Illinois*, 278 U.S. 367 (1929).

73. Ibid., 281 U.S. 179 (1930).

74. Ibid., 289 U.S. 289 (1933).

75. CT, January 31, February 10, 21–22, 1929.

76. Jennifer S. Light, "The City as National Resource: New Deal Conservation and the Quest for Urban Improvement," *Journal of Urban History* 35 (May 2009): 531–560; Biles, *Big City Boss*; and Cohen, *Making a New Deal.*

77. John M. Allswang, *A House for All Peoples: Ethnic Politics in Chicago, 1890–1936* (Lexington: University Press of Kentucky, 1971). Also see Gottfried, *Boss Cermak.*

78. Gottfried, *Boss Cermak*, chaps. 1–8; CT, March 6–7, 1933; and Maureen A. Flanagan, "The Ethnic Entry into Chicago Politics: The United Societies for Local Self-Government and the Reform Charter of 1907," *Journal of the Illinois State Historical Society* 75 (Spring 1982): 2–14.

79. CT, July 1, 1922, on opening day. Also see ibid., May 13, 26, 1925, for a sample of reports on his team in action. The baseball park was located at 26th Street and Kostner Avenue. Also see Steven A. Riess, *City Games—The Evolution of American Urban Society and the Rise of Sports* (Urbana: University of Illinois Press, 1989).

80. CT, August 25, September 2, 1923; and Gottfried, *Boss Cermak*, chap. 8.

81. CT, July 2, 1925; May 29, 1926; August 27, 1927.

82. John M. Allswang, "The Negro Voter and the Democratic Consensus: A Case Study, 1918–1936," *Journal of the Illinois State Historical Society* 60 (Summer 1967): 145–175; John D. Buenker, "Dynamics of Chicago Ethnic Politics 1900–1930," *Journal of the Illinois State Historical Society* 67 (April 1974): 175–199; and Gottfried, *Boss Cermak*, chaps. 10–11. Also see Roger Biles, "Big Red in Bronzeville: Mayor Ed Kelly Reels in the Black Vote," *Chicago History* 10 (Summer 1981): 99–111; Christopher Robert Reed, "Black Chicago Political Realignment during the Great Depression and the New Deal," *Illinois Historical Journal* 78 (Winter 1985): 242–256; and Christopher Manning, *William L. Dawson and the Limits of Black Electoral Leadership* (DeKalb: Northern Illinois University Press, 2009).

83. Hoyt, *One Hundred Years*, 265–276.

84. City of Chicago, Chicago Recreation Commission, and Northwestern University, *Chicago Recreation Survey*, ed. Arthur J. Todd, William F. Byron, and Howard Vierow, vol. 5: *Recommendations of the Commission and Summary of Findings 1940*, 5 vols. (Chicago: Works Progress Administration, National Youth Administration, Illinois Emergency Relief Commission, 1937–1940), 9, for the quotation; ibid., vol. 1: *Public Recreation 1937*; and Halsey, *The Development of Public Recreation*, chap. 12, passim., for the statistics and a series of maps showing the location of the city's parks from 1900 to 1940.

85. City of Chicago, Chicago Recreation Commission, and Northwestern University, *Chicago Recreation Survey, 1937*, vol. 1, ed. Arthur J. Todd, William F. Byron, and Howard Vierow (Chicago: Works Progress Administration, National Youth Administration, Illinois Emergency Relief Commission, 1937), 24; Halsey, *The Development of Public Recreation*, chap. 12; and Cranz, *The Politics of Park Design*, chap. 3.

86. Cranz, *The Politics of Park Design*, 103, for the quotation; *Chicago Recreation Survey*, 1:162–164, for the statistics.

87. *Chicago Recreation Survey*, 1:164.

88. Jens Jensen, "Report of the Landscape Architect," in Dwight Heald Perkins, compiler, *Report of the Special Park Commission to the City Council of Chicago on the Subject of a Metropolitan Park System* (Chicago: City of Chicago, 1904), 67, for the quotation; ibid., 25, 37, 45, for photographs. Also see Libby Hill, *The Chicago River: A Natural and Unnatural History* (Chicago: Lake Claremont Press, 2000), 175–180.

89. CT, July 22, 1916, for the quotation; ibid., October 18, 1900; April 4, October 16, 1916.

90. Ibid., April 29, 1919; February 5, 19, March 31, 1920; January 29, 1922. Also see *Washburn v. Forest Preserve District of Cook County*, 327 Ill. 479 (1927); and CT, April 24, 1926, for the unsuccessful conclusion of the court battle with the real-estate developers after the FPDCC purchase by condemnation proceedings. The Illinois Supreme court ruled that the land was not a forest. In response, the state legislature would broaden the authority of county forest-preserve districts to include land other than forests.

91. Ransom Kennicott, as quoted in CT, January 29, 1922.

92. CT, January 20, September 28, 1926; July 16, 1927; May 11, 1929.

93. *The Chicago Defender*, August 6, 1934, for the quotation; CT, April 6, July 22, August 11, September 4, 1933; February 18, March 16, April 5, 1934. Also see Neil M. Maher, *Nature's New Deal: The Civilian Conservation Corps and the Roots of the American Environmental Movement* (Oxford, UK: Oxford University Press, 2008).

94. *The Chicago Defender*, March 16, 1935; April 2, 1938; August 3, 1940; CT, March 31, April 18, 25, 1935; September 4, 1939; January 28, 1940; December 9, 1942.

95. CT, April 18, 1935; September 4, 1939; January 28, 1940; December 9, 1942.

96. Gottfried, *Boss Cermak*, 179, for the quoted phrase; and CT, February 15–March 10, 1933.

97. Biles, *Big City Boss in Depression and War*, 13–21, passim; and Harold F. Gosnell, *Machine Politics, Chicago Model*, 2nd ed. (Chicago: University of Chicago, 1968).

98. Roberts Mann, *Landscape Engineering in the Forest Preserve District of Cook County* (Chicago: Forest Preserve District of Cook County, December 6, 1943), 1, available at http://collections.carli.illinois.edu/cdm/singleitem/collection/uic_ccfpdoc/id/30/rec/8 (accessed February 1, 2016).

99. Ibid. Also see Roberts Mann, *Outdoor Education in Metropolitan Areas—A Paper Presented before the Convention of the American Institute of Park Executives*, September 21, 1949, available at http://collections.carli.illinois.edu/cdm/singleitem/collection/uic_ccfpdoc/id/31/rec/9 (accessed February 3, 2016); and CT, January 28, 1940.

100. U.S. Census Bureau (1940), *Population*.

101. CT, November 15, 1936.

102. Ibid., July 17, 1929; July 6, 1930; July 25, 1937; Ralph E. Tarbett, "Present Status of Water Pollution in the Chicago-Cook County Area," in *The Chicago-Cook County Health Survey*, ed. U.S. Public Health Service (New York: Columbia University Press, 1949), 147–160; and Raymond I. Leland, "Municipal Sewer Systems in Cook County," in ibid., 135–146.

103. Tarbett, "Present Status of Water Pollution," 150, for the quotation; Leland, "Municipal Sewer Systems," in ibid., 135–146; and CT, July 27, 1919; June 3, 1922; September 21, December 30, 1930; September 30, 1937.

104. CT, March 17, September 10, 1935.

105. Horace P. Ramey, "Floods in the Chicago Area" (unpublished manuscript, February 25, 1958, Harold Washington Library Center, Chicago); and U.S. Comptroller General, *Metropolitan Chicago's Combined Water Cleanup and Flood Control Program: Status and Problems: Report to the Congress*, Washington, DC, May 24, 1978, 3.

106. Ramey, "Floods in the Chicago Area"; and CT, February 25, 1936; July 14, 1944. For other damaging floods and efforts to control them not recorded by Ramey during the period from 1930 to 1945, see ibid., June 6, 1932; October 26, November 2, 1941; February 7, August 8, 1942; April 11, July 17, 1943; and March 26, 1944.

CHAPTER 4

1. Andreas F. Prein, Roy M. Rasmussen, Kyoko Ikeda, Changhai Liu, Martyn P. Clark, and Greg J. Holland, "The Future Intensification of Hourly Precipitation Extremes," *Nature Climate Change*, December 5, 2016, available at http://www.nature.com.flagship.luc.edu/nclimate/index.html?cookies=accepted (accessed December 8, 2016), for the quotation; U.S. National Oceanic and Atmospheric Administration, National Climatic Data Center, *Record of Climatological Observations*, available at http://www.ncdc.noaa.gov, for rainfall data; and Elaine Tyler May, *Homeward Bound: American Families in the Cold War Era* (New York: Basic Books, 1988), for the quoted phrase. For an introduction to Chicago's ecological setting, see Joel Greenberg, *A Natural History of the Chicago Region* (Chicago: University of Chicago Press, 2002).

2. May, *Homeward Bound*, 10, 3–36; and Kenneth T. Jackson, *Crabgrass Frontier—The Suburbanization of the United States* (New York: Oxford University Press, 1985), for the quoted phrase.

3. City of Chicago and Chicago Plan Commission, *Master Plan of Residential Land Use of Chicago* (Chicago: Chicago Plan Commission, 1943); *Chicago Tribune* [hereafter

cited as CT], September 3, 1944; Dwight F. Metzler, "Housing," in *The Chicago-Cook County Health Survey*, ed. U.S. Public Health Service (New York: Columbia University Press, 1949), 389–454; and D. Bradford Hunt and Jon B. DeVries, *Planning Chicago* (Chicago: Planner's Press, 2013). The 1943 master plan became the foundation for subsequent plans, leading in 1958 to the official adoption of an updated version. Also see Roger Biles, *Big City Boss in Depression and War: Mayor Edward J. Kelly of Chicago* (DeKalb: Northern Illinois University Press, 1984); Howard J. Gillette Jr., "The Evolution of Neighborhood Planning: From the Progressive Era to the 1949 Housing Act," *Journal of Urban History* 9 (August 1983): 421–444; Robert Bruegmann, *Sprawl: A Compact History* (Chicago: University of Chicago Press, 2005), 33–41; Elaine Lewinnek, "Mapping Chicago, Imagining Metropolises: Reconsidering the Zonal Model of Urban Growth," *Journal of Urban History* 22 (December 2009): 1–29; Jennifer S. Light, *The Nature of Cities: Ecological Visions and the American Urban Professions, 1920–1960* (Baltimore, MD: Johns Hopkins University Press, 2009); and Laura McEnaney, "Nightmares on Elm Street: Demobilization in Chicago, 1945–1953," *Journal of American History* 92 (March 2006): 1265–1291.

4. Margaret Garb, *City of American Dreams: A History of Home Ownership and Housing Reform in Chicago, 1871–1919* (Chicago: University of Chicago Press, 2005); Joseph C. Bigott, *From Cottage to Bungalow: Houses and the Working Class in Metropolitan Chicago, 1869–1929* (Chicago: University of Chicago Press, 2001); and Dominic A. Pacyga and Charles Shanabruch, eds., *The Chicago Bungalow* (Chicago: Arcadia Publishing, 2003), 7, for the statistics.

5. Lizabeth Cohen, *A Consumers' Republic: The Politics of Mass Consumption in Postwar America* (New York: Knopf, 2003), 73, 115–165; Wendy Wall, *Inventing the "American Way": The Politics of Consensus from the New Deal to the Civil Rights Movement* (Oxford, UK: Oxford University Press, 2008); and Carl Condit, *Chicago, 1930–1970: Building, Planning, and Urban Technology* (Chicago: University of Chicago Press, 1974), 286–287, for housing statistics.

6. Kenneth Jackson, "Race, Ethnicity, and Real Estate Appraisal: The Home Owners' Loan Corporation and the Federal Housing Authority," *Journal of Urban History* 6 (August 1980): 419–452; and Jennifer S. Light, "Nationality and Neighborhood Risk at the Origins of FHA Underwriting," *Journal of Urban History* 20 (June 2010): 1–38.

7. Sarah Potter, "Family Ideals: The Diverse Meanings of Residential Space in Chicago during the Post–World War II Baby Boom," *Journal of Urban History* 39 (November 2012): 60, 59–78. Also see John T. McGreevy, *Parish Boundaries: The Catholic Encounter with Race in the Twentieth-Century Urban North* (Chicago: University of Chicago Press, 1996); Adam Green, *Selling the Race: Culture, Community, and Black Chicago, 1940–1955* (Chicago: University of Chicago Press, 2007); Brian James McCammack, "Recovering Green in Bronzeville: An Environmental and Cultural History of the African American Great Migration to Chicago, 1915–1940" (Ph.D. diss., Harvard University, 2012); and Michael D. Innis-Jimenez, "Organizing for Fun: Recreation and Community Formation in the Mexican Community of South Chicago in the 1920s and 1930s," *Journal of the Illinois State Historical Society* 98 (Fall 2005): 144–161.

8. U.S. Census Bureau (1940–2010), *Population*; Ann Durkin Keating, *Chicagoland—City and Suburbs in the Railroad Age* (Chicago: University of Chicago Press, 2005), 118, 114–139; William Cronon, *Nature's Metropolis—Chicago and the Great West* (New York: Norton, 1991). Also see Bruegmann, *Sprawl*, 1–20; Dolores Hayden, *Building Suburbia: Green Fields and Urban Growth, 1820–2000* (New York: Vintage, 2003); and

Christopher C. Sellers, *Crabgrass Crucible: Suburban Nature and the Rise of Environmentalism in Twentieth-Century America* (Chapel Hill: University of North Carolina Press, 2012).

9. Gregory C. Randall, *America's Original GI Town—Park Forest, Illinois* (Baltimore, MD: Johns Hopkins University Press, 2000).

10. Ibid.; and Joseph L. Arnold, *The New Deal in the Suburbs: A History of the Greenbelt Town Program, 1935–1954* (Columbus: Ohio State University Press, 1971).

11. Randall, *America's Original GI Town*; and William Hollingsworth Whyte, *The Organization Man* (New York: Simon and Schuster, 1956).

12. Clifford Edward Clark, *The American Family Home, 1800–1960* (Chapel Hill: University of North Carolina Press, 1986), 218, 216, respectively for the two quotations.

13. Ibid., chaps. 7–8; and Witold Rybczynski, "The Ranch House Anomaly," *Slate* (2007), available at http://www.slate.com/articles/arts/architecture/2007/04/the_ranch_house_anomaly.htm (accessed May 31, 2016).

14. Homer Hoyt, "The New Shopping Centers in the Chicago Region, September 1955," in *According to Hoyt: 53 Years of Essays, 1916–1969*, ed. Homer Hoyt ([Washington, DC?]: [Homer Hoyt Institute?], 1970), 631. Also see Robert Bruegmann, "Schaumburg, Oak Brook, Rosemont, and the Recentering of the Chicago Metropolitan Area," in *Chicago Architecture and Design, 1923–1993: Reconfiguration of an American Metropolis*, ed. John Zukowsky, Mark Jansen Bouman, and Art Institute of Chicago (Munich: Prestel and Art Institute of Chicago, 1993), 159–178; and Robert D. Lewis, *Chicago Made: Factory Networks in the Industrial Metropolis* (Chicago: University of Chicago Press, 2008).

15. CT, May 9, July 29, 1965. Appropriately enough, the reports were made by a new organization, the Opens Lands Project, an offshoot of the Welfare Council of Metropolitan Chicago. Also see Rutherford H. Platt, *Open Land in Urban Illinois: Roles of the Citizen Advocate* (Dekalb: Northern Illinois University Press, 1971). See Ralph E. Tarbett, "Sewerage in Chicago," in U.S. Public Health Service, *The Survey*, 103, 98–109, for CSD statistics.

16. Sellers, *Crabgrass Crucible*, 12.

17. City of Chicago, Chicago Recreation Commission, and Northwestern University, *Chicago Recreation Survey 1937*, ed. Arthur J. Todd, William F. Byron, and Howard Vierow (Chicago: Works Progress Administration, National Youth Administration, and Illinois Emergency Relief Commission, 1937), 1:98; CT, April 18, 1948. Also see Jane Addams, *The Spirit of Youth and the City Streets* (New York: Macmillan, 1909); and Robin Faith Bachin, *Building the South Side: Urban Space and Civic Culture in Chicago, 1890–1919* (Chicago: University of Chicago Press, 2004), 127–168.

18. CT, September 9, 1945; April 13, 1947.

19. Ibid., July 3, 1946; May 15, 1947; April 18, 1948.

20. Elaine Lewinnek, "Inventing Suburbia: Working-Class Suburbanization in Chicago, 1865–1919" (Ph.D. diss., Yale University, 2005)"; McCammack, "Recovering Green in Bronzeville"; Innis-Jimenez, "Organizing for Fun"; Marian Lorena Osborn, "The Development of Recreation in the South Park System of Chicago" (master's thesis, University of Chicago, August 1928); and Galen Cranz, *The Politics of Park Design: A History of Urban Parks in America* (Cambridge: Massachusetts Institute of Technology Press, 1982).

21. CT, April 26, May 3, 10–11, 28, 1908; May 24, 1946; June 29, 1950; and February 21, 1952. Also see ibid., April 12, 1958, for a history of the organization as it reached its 50th anniversary; Robert E. Grese, *Jens Jensen—Maker of Natural Parks and Gardens*

(Baltimore, MD: Johns Hopkins University Press, 1992); and William H. Tishler, and Erik M. Ghenoiu, "Jens Jensen and the Friends of Our Native Landscape," *Wisconsin Magazine of History* 86 (Summer 2003): 2–15.

22. Chicago, *Chicago Recreation Survey*, 1:164, passim; CT, July 21, 1947.

23. CT, July 1, 1952; Cranz, *The Politics of Park Design*, 110–123; and Sellers, *Crabgrass Crucible*, 1–35.

24. Donald L. Hey, "The Des Plaines River Wetlands Demonstration Project: Restoring an Urban Wetland," in *The Ecological City: Preserving and Restoring Urban Biodiversity*, ed. Rutherford H. Platt, Rowan A. Rowntree, and Pamela C. Muick (Amherst: University of Massachusetts Press, 1994), 83–92.

25. Harold L. Platt, *Shock Cities: The Environmental Transformation and Reform of Manchester and Chicago* (Chicago: University of Chicago Press, 2005); Cronon, *Nature's Metropolis*, for the quoted phrase; Craig E. Colten, "Illinois River Pollution Control, 1900–1970," in *The American Environment: Interpretations of Past Geographies*, ed. Lary M. Dilsaver and Craig E. Colten (Lanham, MD: Rowman and Littlefield, 1992), 193–214; Joshua A. T. Salzmann, "The Creative Destruction of the Chicago River Harbor: Spatial and Environmental Dimensions of Industrial Capitalism, 1881–1909," *Enterprise and Society* 13 (June 2012): 235–275; and Carl S. Smith, *City Water, City Life: Water and the Infrastructure of Ideas in Urbanizing Philadelphia, Boston, and Chicago* (Chicago: University of Chicago Press, 2013).

26. Harold L. Platt, "Chicago, the Great Lakes, and the Origins of Federal Urban Environmental Policy," *Journal of the Gilded Age and Progressive Era* 1 (April 2002): 122–153; and Horace P. Ramey, "Floods in the Chicago Area" (unpublished transcript, February 25, 1958, Harold Washington Library Center, Chicago). Ramey was chief engineer of the SDC.

27. Ramey, "Floods"; Raymond Leland, "Municipal Sewer Systems in Cook County," in *The Chicago-Cook County Health Survey*, ed. U.S. Public Health Service (New York: Columbia University Press, 1949), 135–146; and CT, August 18, September 29, 1946.

28. CT, January 17, 1945, for the quotations; ibid., January 18, June 3, 1945; Tarbett, "Sewerage in Chicago"; and Biles, *Big City Boss in Depression and War.*

29. CT, April 1, June 22, July 1, 7, 1945; October 6, 1946; May 11, 1947; February 12, 1950; March 5, October 29, 1954.

30. Ramey, "Floods"; CT, July 7, 1947; October 29, 1954.

31. Oscar Hewitt, as quoted in CT, July 6, 1947; and ibid., August 31, September 1, 1947; July 24, 1948.

32. Thomas D. Garry, as quoted in CT, July 11, 1946; ibid., July 29, 31, August 1–2, 13, October 3, 1946.

33. Ibid., September 29, 1946; February 28, March 13, 20, April 7–30, 1947. Macheijewski was a coal dealer, who served two terms in the U.S. Congress (1938–1942). He also served as the treasurer of Cicero before being elected to the board of trustees of the SDC. See his obituary, ibid., September 26, 1949.

34. Ibid., April 30, for the first quoted phrase; and ibid., April 10, 1947, for the following quoted phrases. Also see ibid., August 18, 1946; April 21, 1947; and February 3, 1952.

35. Arnold R. Hirsch, "Martin H. Kennelly—the Mugwump and the Machine," in *The Mayors: The Chicago Political Tradition*, ed. Melvin G. Holli and Paul M. Green (Carbondale: Southern Illinois University Press, 1987), 126–143; and CT, August 8, 1957.

36. See David Alexander Spatz, "Roads to Postwar Urbanism: Expressway Building and the Transformation of Metropolitan Chicago, 1930–1975" (Ph.D. diss., University

of Chicago, 2010), for a parallel and overlapping story of political reaction against Chicago's political culture of corruption and machine rule.

37. Ramey, "Floods," 14; CT, October 2–13, 19, November 1, 5, 1954.

38. U.S. Department of the Interior, Geological Survey, Warren S. Daniels, and Malcolm D. Hale, "Floods of October 1954 in the Chicago Area, Illinois, and Indiana," Washington, DC, 1955. For similar, local observations, see CT, July 7, 1947.

39. Anthony A. Olis, as quoted in CT, February 3, 1952; ibid., October 2–13, 29, November 1, 5, 1954; Tarbett, "Sewerage in Chicago," 103; and Frank E. Dalton and Raymond R. Rimkus, "The Chicago Area's Tunnel and Reservoir Plan," *Journal of the Water Pollution Control Federation* 57 (December 1985): 1114–1121, fig. 9, for the statistics on the capacity of the storm outlet sewers.

40. CT, October 15, 1954, for the quoted phrase; ibid., March 5, October 29, November 5, 1954; and U.S. Department of the Interior, "Floods of October 1954," 6–8, for the floodwater discharge statistics.

41. CT, October 21, 1954, for the Mazzolas' story. They lived at 9993 S. Yale Avenue, Chicago, which was about 2 miles/3.2 kilometers northwest of the marshy Lake Calumet. Also see ibid., July 14, 1957; *Chicago Defender*, October 23, 1954, for the story of Robbins; and U.S. Department of the Interior, "Floods of October 1954," 2–5, which also lists Palos Park, Worth, Oak Lawn, Alsip, Crestwood, Posen, Markham, Hazel Crest, Garden Homes, and Homewood as communities suffering flood damage. Homewood was also hit by a tornado during the peak of the storm.

42. U.S. Department of the Interior, "Floods of October 1954." This report's formulas cast the 1954 storms as a "hundred-year" event, but the authors believed the severity and frequency of rainfall events since the mid-1940s made the calculations unreliable as a forecast of future weather conditions.

43. Ibid., 5–6; CT, May 18, 1947; March 20, 1948; February 27, November 17, 1949.

44. Roy C. Blackwell, as quoted in CT, September 28, 1961.

45. David Nelson, as quoted in ibid., October 1, 1961.

46. Ibid., February 6, 1955.

47. Ibid., May 28, July 14–18, August 11, 1957; June 14, August 13, 1958; July 1–4, August 2–6, 1960; and September 13–15, 1961.

48. Roger Biles, *Richard J. Daley: Politics, Race, and the Governing of Chicago* (DeKalb: Northern Illinois University Press, 1995); Richard C. Lindberg, *To Serve and Collect—Chicago Politics and Police Corruption from the Lager Beer Riot to the Summerdale Scandal* (New York: Praeger, 1991); William J. Grimshaw, *Bitter Fruit—Black Politics and the Chicago Machine 1931–1991* (Chicago: University of Chicago Press, 1992); and Barbara Ferman, *Challenging the Growth Machine: Neighborhood Politics in Chicago and Pittsburgh* (Lawrence: University Press of Kansas, 1996). Also see Center for Neighborhood Technology, "The Prevalence and Cost of Urban Flooding: A Case Study of Cook County, Il," Chicago, May 2013, for a recent study of the emotional as well as property costs of damaging floods.

49. See Spatz, "Roads to Postwar Urbanism," for a parallel case and timing of a political revolt against city hall resulting in the state's preemption of building Chicago's metropolitan highway network. The state created a toll-road authority; also see note 48.

50. CT, October 15, 19, 1954; May 28, July 16–17, August 8, 30, September 13, 1957; March 31, August 13, 27, 1958; February 3, April 30, December 13, 1959; February 5, June 9, July 24, October 4, November 13, 1960; June 18, August 6, September 24, 28, October 1, 1961.

51. Ibid., April 11, May 10, August 13, 27, 29, 1958. Also see ibid., April 30, 1959, for the protest, including a federal lawsuit, by the mayor of Joliet against the SDC's and U.S.

Army Corps' plan to increase the Lockport Dam's maximum rate of release to 43,000 cubic feet/1,217.6 cubic meters per second.

52. Charles G. Sauers, as quoted in ibid., February 3, 1959; ibid., December 13, 1959; February 11, June 9, 1960; June 16, 1961; June 3, 1962.

53. Ibid., April 30, 1959; February 11, June 8, 21, July 21, October 4, 1960.

54. Ibid., May 13, 1961; and ibid., January 1, July 19, 1961; August 23, September 23, 30, October 25, November 21, 1962; April 7, June 2, 27, July 28, 1963, for CPD statistics.

55. Ibid., August 20, 1961; May 16, October 17, 1965; Rachel Carson, *Silent Spring* (Boston: Houghton Mifflin, 1962); and Linda J. Lear, *Rachel Carson: Witness for Nature* (New York: Henry Holt, 1997).

56. CT, October 4, 1964; May 9, June 6, July 29, 1965; April 13, 1969; May 14, July 26, 1970. Also see ibid., September 30, 1971; April 9, May 21, 1972.

57. Ibid., June 21, 1960; and ibid., September 21, 26, 28, October 1, 1961. The September rainfall totaled 13.6 inches (in.)/34.5 centimeters (cm), or 1.54 in./3.91 cm more than had fallen in October 1954. Also see Hunt and DeVries, *Planning Chicago*, 25–40.

58. CT, August 30–31, September 2, October 23, November 27, December 8–11, 19, 1962; March 19, April 26–28, May 10, 1963; and Biles, *Richard J. Daley*, chap. 3.

59. CT, August 28, September 21, October 8, 12, 14, November 1, December 13, 24, 1963; January 24, February 4, April 3, July 24, August 9–10, October 26, December 5, 1964; March 9, 12, 1965; May 28, June 26, August 23–25, September 8–9, October 5–21, November 5, 14, December 1–2, 9, 20, 1966; January 10, March 14, April 9, July 9, August 23, September 2, 1967; January 24, 27, March 9, May 2, 10–12, June 17, October 19, November 2, 1968; January 17–18, 22, April 4, October 27, November 22, 1969; January 22–23, March 7, 1970.

60. Vinton W. Bacon, as quoted in ibid., April 26–28, 1963; and Walter S. Baltis, as quoted in ibid., April 3, 1964, respectively; and see note 57.

61. CT, March 23, 1978; and Chicago Resource Coordination Policy Committee, "Our Community and Flooding: A Report on the Status of Floodwater Management in the Chicago Metropolitan Area, 1987," Chicago, April 1987.

62. CT, January 25, 1965; March 30, 1967. Also see Hazar Engineering Company and Metropolitan Sanitary District of Greater Chicago, "Flood and Pollution Control—A Deep Tunnel Plan for the Chicagoland Area" (Chicago: Metropolitan Sanitary District of Greater Chicago, May 1966); and Frank E. Dalton, Victor Koelzer, and William J. Bauer, "The Chicago Area Deep Tunnel Project: A Use of the Underground Storage Resource," *Journal of the Water Pollution Control Federation* 41 (April 1969): 515–534. For biographical information on Dalton, see his obituary in the *Chicago Sun-Times*, May 10, 2008.

63. See CT, October 6, 1946, for a description of the traditional methods of laying interceptor sewers underground; and Judith A. Martin and Sam Bass Warner Jr., "Local Initiative and Metropolitan Repetition: Chicago, 1972–1990," in *The American Planning Tradition: Culture and Policy*, ed. Robert Fishman (Washington, DC: Woodrow Wilson Centre Press and Johns Hopkins University Press, 2000), 263–296, for a critique of TARP.

64. Martin and Warner, "Local Initiative and Metropolitan Repetition," 363, 363–396. Also see U.S. Army Corps of Engineers, *The GLMRIS Report: Great Lakes and Mississippi River Interbasin Study*, U.S. Army Corps of Engineers, January 6, 2014, available at http://www.glmris.anl.gov/documents/docs/glmrisreport/GLMRIS_Report.pdf (accessed November 15, 2015). This definitive study also concludes that the size of TARP

needs to be at least twice as large to prevent flood damage and the pollution of Lake Michigan.

65. CT, June 6, 1965, for the quoted phrase; ibid., February 21, 1963; June 13, 1965.

CHAPTER 5

1. On the campaign, see *Chicago Tribune* [hereafter cited as CT], July 10, 1966; October 17, November 25, 26, 29, December 1, 1967. Also see Andrew Dribin, "Saving the Lake: Airports and Islands along Chicago's Lakefront, 1972," Paper presented at the annual meeting of the American Society for Environmental History, San Francisco, March 2014. On the biological process of excessive plant nutrients called eutrophication, see Terence Kehoe, "Merchants of Pollution? The Soap and Detergent Industry and the Fight to Restore Great Lakes Water Quality, 1965–1972," *Environmental History Review* 16 (Fall 1992): 21–46. Also see Rachel Carson, *Silent Spring* (Boston: Houghton Mifflin, 1962); Linda J. Lear, *Rachel Carson: Witness for Nature* (New York: Henry Holt, 1997); and note 4.

2. CT, June 11, August 11, 12, 20, September 17, 1966. On the pollution of the lake and waterways of the Chicago area, see Craig E. Colten, "Industrial Wastes in Southeast Chicago: Production and Disposal 1870–1970," *Environmental Review* 10 (Summer 1986): 93–105. On the ecology of Chicago, see Joel Greenberg, *A Natural History of the Chicago Region* (Chicago: University of Chicago Press, 2002); and Libby Hill, *The Chicago River: A Natural and Unnatural History* (Chicago: Lake Claremont Press, 2000).

3. CT, February 8, December 30, 1966; June 20, 27, July 8, 9, August 6, 15, 20, September 3, 21, November 26, 1967.

4. On the local crusade, see note 1. On the Clean Water Act, see Terence Kehoe, *Cleaning Up the Great Lakes: From Cooperation to Confrontation* (DeKalb: Northern Illinois University Press, 1997); Edmund P. Russell III, "Lost among the Parts per Billion: Ecological Protection at the United States Environmental Protection Agency, 1970–1993," *Environmental History* 2 (January 1997): 29–51; Paul Charles Milazzo, *Unlikely Environmentalists: Congress and Clean Water, 1945–1972* (Lawrence: University of Kansas Press, 2006); and David Stradling and Richard Stradling, "Perceptions of the Burning River: Deindustrialization and Cleveland's Cuyahoga River," *Environmental History* 13 (July 2008): 515–535. On earlier, Progressive Era reforms to save the Great Lakes, see Christine Meisner Rosen, "Businessmen against Pollution in Late Nineteenth Century Chicago," *Business History Review* 69 (Fall 1995): 351–397; Harold L. Platt, "Chicago, the Great Lakes, and the Origins of Federal Urban Environmental Policy," *Journal of the Gilded Age and Progressive Era* 1 (April 2002): 122–153; and Robin Faith Bachin, *Building the South Side: Urban Space and Civic Culture in Chicago, 1890–1919* (Chicago: University of Chicago Press, 2004).

5. Adam Rome, "'Give Earth a Chance': The Environmental Movement and the Sixties," *Journal of American History* 90 (September 2003): 525–554; Samuel P. Hays, *Beauty, Health, and Permanence: Environmental Politics in the United States, 1955–1985* (New York: Cambridge University Press, 1987); Andrew Hurley, *Environmental Inequalities—Class, Race, and Industrial Pollution in Gary, Indiana, 1945–1980* (Chapel Hill: University of North Carolina Press, 1995); Adam Rome, *The Bulldozer in the Countryside—Suburban Sprawl and the Rise of American Environmentalism* (New York: Cambridge University Press, 2001); and Dolores Hayden, *Building Suburbia: Green Fields and Urban Growth, 1820–2000* (New York: Vintage, 2003).

6. Kristin M. Szylvian, "Transforming Lake Michigan into the 'World's Greatest Fishing Hole': The Environmental Politics of Michigan's Great Lakes Sport Fishing, 1965–1985," *Environmental History* 9 (January 2004): 103–104, 102–127; and CT, September 26, 1967, for the second quotation.

7. Szylvian, "Transforming Lake Michigan."

8. Richard J. Daley, as quoted in CT, June 13, 1970. On "envirotechnical systems," see Sara B. Pritchard, *Confluence: The Nature of Technology and the Remaking of the Rhône* (Cambridge, MA: Harvard University Press, 2011). She defines them "as the historically and culturally specific configurations of intertwined 'ecological' and 'technological' systems, which may be composed of artifacts, practices, people, institutions, and ecologies" (19).

9. Tom McNamee, as quoted in the *Chicago Sun-Times*, May 9, 2005. McNamee was a "human-interest" reporter. Also see David Stradling and Richard Stradling, *Where the River Burned: Carl Stokes and the Struggle to Save Cleveland* (Ithaca, NY: Cornell University Press, 2015).

10. John Hall Fish, *Black Power/White Control: The Struggle of the Woodlawn Organization in Chicago* (Princeton, NJ: Princeton University Press, 1973); Bernard O. Brown, *Ideology and Community Action: The West Side Organization of Chicago, 1964–67* (Chicago: Center for the Scientific Study of Religion, 1978); Arnold R. Hirsch, *Making the Second Ghetto: Race and Housing in Chicago, 1940–1960* (New York: Cambridge University Press, 1983); William J. Grimshaw, *Bitter Fruit—Black Politics and the Chicago Machine 1931–1991* (Chicago: University of Chicago Press, 1992); Roger Biles, *Richard J. Daley: Politics, Race, and the Governing of Chicago* (DeKalb: Northern Illinois University Press, 1995); James R. Ralph, *Northern Protest: Martin Luther King, Jr., Chicago, and the Civil Rights Movement* (Cambridge, MA: Harvard University Press, 1993); Janet L. Abu-Lughod, *Race, Space, and Riots in Chicago, New York, and Los Angeles* (Oxford, UK: Oxford University Press, 2007); and Jeffrey Haas, *The Assassination of Fred Hampton: How the FBI and the Chicago Police Murdered a Black Panther* (Chicago: Lawrence Hill Books/Chicago Review Press, 2010).

11. Amanda I. Seligman, "'But Burn—No': The Rest of the Crowd in Three Civil Disorder in 1960s Chicago," *Journal of Urban History* 37 (March 2011): 230–255; Abu-Lughod, *Race, Space, and Riots*, chap 3; and Beryl Satter, *Family Properties: Race, Real Estate, and the Exploitation of Black Urban America* (New York: Metropolitan Books, 2009).

12. U.S. National Advisory Commission on Civil Disorders and Barbara Ritchie, *The Riot Report—A Shortened Version of the Report of the National Advisory Commission on Civil Disorders* (New York: Viking Press, 1969); Joel Rast, *Remaking Chicago: The Political Origins of Urban Industrial Change* (DeKalb: Northern Illinois University Press, 1999); and Amanda I. Seligman, *Block by Block: Neighborhoods and Public Policy on Chicago's West Side* (Chicago: University of Chicago Press, 2005). Also see Hurley, *Environmental Inequalities.*

13. Peter Marcuse, "The Ghetto of Exclusion and the Fortified Enclave: New Patterns in the United States," *American Behavioral Scientist* 41 (November/December 1997): 311–326; and Peter Marcuse, "Depoliticizing Globalization: From Neo-Marxist to Network Society of Manuel Castells," in *Understanding the City: Contemporary and Future Perspectives*, ed. John Eade and Christopher Mele (Oxford, UK: Blackwell, 2002), 131–158. Also see United Nations Human Settlements Programme, *Enhancing Urban Safety and Security* (London: Earthscan Publications, 2007); United Nations Human Settlements Programme, *The Challenge of Slums 2003* (London: Earthscan Publications,

2003); and Jon Bannister and Nick Fyfe, "Introduction: Fear and the City," *Urban Studies* 38 (May 2001): 807–813.

14. Satter, *Family Properties*, 93.

15. U.S. Census Bureau (1960–1980), *Population*.

16. Henry Binford, "Multicentered Chicago," in *The Encyclopedia of Chicago*, ed. James R. Grossman, Ann Durkin Keating, and Janice L. Reiff (Chicago: University of Chicago Press, 2004), 548–553.

17. Robert Bruegmann, "Schaumburg, Oak Brook, Rosemont, and the Recentering of the Chicago Metropolitan Area," in *Chicago Architecture and Design, 1923–1993 Reconfiguration of an American Metropolis*, ed. John Zukowsky, Mark Jansen Bouman, and Art Institute of Chicago (Munich: Prestel and Art Institute of Chicago, 1993), 159–178.

18. On the environmental engineering of Chicago, see Harold L. Platt, *Shock Cities: The Environmental Transformation and Reform of Manchester and Chicago* (Chicago: University of Chicago Press 2005). Also see John W. Larson, *Those Army Engineers: A History of the Chicago District, U.S. Army Corp of Engineers* (Washington, DC: Government Printing Office, 1980).

19. U.S. Comptroller General, *Metropolitan Chicago's Combined Water Cleanup and Flood Control Program: Status and Problems: Report to the Congress*, May 24, 1978, 3, available at http://catalog.hathitrust.org/Record/011420629 (accessed January 23, 2015); and Frank E. Dalton and Raymond R. Rimkus, "The Chicago Area's Tunnel and Reservoir Plan," *Journal of the Water Pollution Control Federation* 57 (December 1985): 1114–1121.

20. For an account by the original designers of TARP, see Frank E. Dalton, Victor Koelzer, and William J. Bauer, "The Chicago Area Deep Tunnel Project: A Use of the Underground Storage Resource," *Journal of the Water Pollution Control Federation* 41 (April 1969): 515–534; for the follow-up assessment at the time of the completion of the first phase of the project, see Dalton and Rimkus, "The Chicago Area's Tunnel and Reservoir Plan." Also see Timothy B. Neary, "Chicago-Style Environmental Politics: Origins of the Deep Tunnel Project," *Journal of Illinois History* 4 (Summer 2001): 83–102. For statistics on the storm-water overflow releases into Lake Michigan, see Metropolitan Water Reclamation District (MWRD), "Reversals to Lake Michigan (1985–present)," available at https://www.mwrd.org/irj/portal/anonymous/overview (accessed December 20, 2012).

21. David E. Nye, *American Technological Sublime* (Cambridge: Massachusetts Institute of Technology Press, 1994).

22. On the great floods of 1957–1958, see CT, July 14–17, September 13, 1957; June 14, August 29, 1958. On the MSD's follow-up studies of flood and pollution control, see ibid., April 10, 1963; January 25, 1965; March 27, 1966; March 30, October 4, 1967. On the culture of big technology, see Nye, *American Technological Sublime*; Maria Kaika, *City of Flows: Modernity, Nature, and the City* (New York: Routledge, 2005); and Erik Swyngedouw, *Social Power and the Urbanization of Water: Flows of Power* (Oxford, UK: Oxford University Press, 2004). On sanitation history in the United States, see Joel Tarr, *The Search for the Ultimate Sink: Urban Pollution in Historical Perspective* (Akron, OH: University of Akron Press, 1996); and Martin V. Melosi, *The Sanitary City—Urban Infrastructure in America from Colonial Times to the Present* (Baltimore, MD: Johns Hopkins University Press, 2000).

23. For an introduction, see Samuel P. Hays, *Conservation and the Gospel of Efficiency—The Progressive Conservation Movement 1890–1920* (Cambridge, MA: Harvard University Press, 1959); William H. Wilson, *The City Beautiful Movement* (Baltimore,

MD: Johns Hopkins University Press, 1989); Bachin, *Building the South Side*; Carl S. Smith, *The Plan of Chicago: Daniel Burnham and the Remaking of the American City* (Chicago: University of Chicago Press, 2006); and Margaret Garb, "Race, Housing, and Burnham's Plan: Why Is There No Housing in the 1909 Plan of Chicago?" *Journal of Planning History* 10 (April 2011): 99–113. For insight on contemporary urban planning, see Timothy J. Gilfoyle and Chicago History Museum, *Millennium Park: Creating a Chicago Landmark* (Chicago: University of Chicago Press, 2006).

24. CT, August 18, 1968.

25. See Russell, "Lost among the Parts per Billion"; and Milazzo, *Unlikely Environmentalists*. On federalism, see Sarah S. Elkind, *How Local Politics Shape Federal Policy: Business, Power, and the Environment in Twentieth-Century Los Angeles* (Chapel Hill: University of North Carolina Press, 2011).

26. For stories about TARP, see CT, January 25, 1965; March 27, 1966; March 30, 1967. Also see Dalton, Koelzer, and Bauer, "The Chicago Area Deep Tunnel Project." For stories of corruption within the MSD, see CT, March 18, 1963; August 20, 1976; May 8, 1978; and Neary, "Chicago-Style Environmental Politics." Also see Richard C. Lindberg, *To Serve and Collect—Chicago Politics and Police Corruption from the Lager Beer Riot to the Summerdale Scandal* (New York: Praeger, 1991); and Andrew J. Diamond, *Mean Streets: Chicago Youths and the Everyday Struggle for Empowerment in the Multiracial City, 1908–1969* (Berkeley: University of California Press, 2009). On the mayor, see Mike Royko, *Boss: Richard J. Daley of Chicago* (New York: Dutton, 1971); and Biles, *Richard J. Daley*.

27. CT, March 30, October 4, 1967; and Dalton and Rimkus, "The Chicago Area's Tunnel and Reservoir Plan."

28. CT, August 23, September 30, October 18, November 4, 1967; November 14, 1968. The court case was *Wisconsin v. Illinois* 388 U.S. 426 (1967). For a history of the case, see Platt, "Chicago, the Great Lakes."

29. CT, June 11, July 10, August 11, 12, 22, September 17, 1966.

30. John Egan, as quoted in CT, September 10, 1967.

31. CT, November 14, 1968, for the quotation. For the nonenforcement of the law by the MSD, see ibid., November 13–14, 1968; August 19, October 12, 1969; June 29, 1970; January 24, May 16, September 28, 1971; October 21, 1972; April 25, May 5, December 26, 1974; August 14, 1975; August 29, 1976; October 14, 1978; May 20, 1979; October 2, 1985.

32. Ibid., June 17, 1971, for the exemption of the MSD's rivers from the goal of a water-quality standard for swimming. Its waterways were downgraded to "restricted use." Also see ibid., October 21, 1972, for the MSD's lobby on behalf of a further lowering of these standards.

33. Ibid., March 21, 1968. Also see Dennis W. Dreher, "Stormwater Management Ordinance Approaches in Northeastern Illinois," in *Seminar Publication: National Conference on Urban Runoff Management*, ed. U.S. Environmental Protection Agency (Cincinnati, OH: U.S. Environmental Protection Agency, April 1995), 77–81.

34. CT, May 28, June 26, 1966.

35. Ibid., August 23–25, September 8, October 14, 17, 21, 1966.

36. George E. Mahin, as quoted in ibid., February 15, 1968; July 28, September 2, 1967; January 24, 27, March 9, 1968.

37. Ibid., May 2, 10–11, 1968; November 22, 1969. In the power struggle between the political boss and the efficiency expert, the reformer was handed a pyrrhic victory in a legislative contest that led to his final defeat. A state act of 1967 gave the superintendent

authority over jobs and contracts, but the trustees retained a veto power over his decisions. See ibid., July 28, September 2, 1967; January 27, June 17, October 19, 1968.

38. Abe Eiserman, as quoted in ibid., January 22, 1970; January 17–18, 22, 1969; January 21–23, 1970. Two months later, Bacon gained an appointment at the University of Wisconsin, Milwaukee. See ibid., March 7, 1970.

39. Ibid., February 20–21, 1978.

40. Richard J. Daley, as quoted in CT, June 13, 1970; and see note 18. For the MSD's refusal to prosecute the firms it had found in violation of local and state laws, see ibid., November 14, 1968. For its refusal to clean up the debris in the river, see ibid., August 19, October 12, 1969. And see note 31 for links between the MSD, its policy of nonenforcement, and political corruption.

41. Richard J. Daley, as quoted in CT, March 29, 1974; and Biles, *Richard J. Daley.*

42. For the deindustrialization of Chicago, see Bruegmann, "Schaumburg, Oak Brook, Rosemont," 159–178; Robert D. Lewis, *Chicago Made: Factory Networks in the Industrial Metropolis* (Chicago: University of Chicago Press, 2008); and Rast, *Remaking Chicago.* For the impacts of economic recession and housing policy on the city's residential neighborhoods, see Satter, *Family Properties.*

43. CT, February 20–21, 1978.

44. CT, February 21, 1978, for the quotation; ibid., February 21–22, 1978.

45. Frank Dalton, as quoted in ibid., February 23, 1978; ibid., February 24, 27, March 10, 1978. Also see ibid., February 28, 1978, for the Civic Federation's defense of the soaring cost of TARP due to inflation.

46. John Knighten, as quoted in ibid., August 18, 1968.

47. Ibid., August 18–19, 23, 1968.

48. Dalton, Koelzer, and Bauer, "The Chicago Area Deep Tunnel Project," 519, 515–534.

49. Mike Ripley, as quoted in CT, November 24, 1977.

50. For the flooding and TARP phase-two planning, see CT, August 18, 1968; February 10, 1972; November 5, 21, 1974; March 13, April 19, December 30, 1976; March 17, July 1, October 6, 1977; July 22, 1978. For funding by the state and federal governments, see ibid., June 20, 1975; July 3, August 29, October 11, 1976; March 10, 1977; September 20, 1979; July 23, 1981; May 24, 1985. For floods in the 1980s, see ibid., July 23, 1981; August 9, 1982; June 15, 1984; October 4–8, 1986; and MWRD, "Reversals to Lake Michigan."

51. *Chicago Sun-Times*, October 21, 1999; December 10, 2003. For additional information on the boring machine, see the TARP story as presented by its manufacturer, the Robbins Company, available at www.robbinstbm.com/case-study/tarp/ (accessed October 18, 2012). For some progress reports on the construction of the Deep Tunnel, see CT, November 5, 1974; August 29, December 30, 1976.

52. Dalton and Rimkus, "The Chicago Area's Tunnel and Reservoir Plan."

53. For statistics on the MWRD's pollution, see *Chicago Sun-Times*, April 22, 1990. For flooding, see CT, October 20, November 21, 1985; and MWRD, "Reversals to Lake Michigan."

54. U.S. Comptroller General, *Combined Sewer Flooding and Pollution: A National Problem: The Search for Solutions in Chicago*, May 15, 1979, 11, available at http://catalog.hathitrust.org/Record/011408779 (accessed January 23, 2015).

55. Ibid., ii, for the quotation; and U.S. Comptroller General, *Metropolitan Chicago's Combined Water Cleanup and Flood Control Program.*

56. U.S. Comptroller General, *Combined Sewer Flooding and Pollution*, 12, 13, for the quoted phrase and quotation, respectively; and Chicago Resource Coordination Pol-

icy Committee, "Our Community and Flooding: A Report on the Status of Floodwater Management in the Chicago Metropolitan Area, 1987," Chicago, April 1987.

57. U.S. Comptroller General, *Combined Sewer Flooding and Pollution*, chap 3.

58. Ibid., 42, chap. 4.

59. CT, August 14, 1975. Also see Rast, *Remaking Chicago.*

60. Albert E. Cowdrey, "Pioneering Environmental Law: The Army Corps of Engineers and the Refuse Act," *Pacific Historical Review* 44 (August 1975): 344, 331–349. For the case, see *U.S. v. Republic Steel,* 362 U.S. 482 (1960). The "Refuse Act" is actually section 13 of the 1899 Rivers and Harbors Act, a landmark of conservation during the Progressive Era. On the Calumet district of heavy industry, see Colten, "Industrial Wastes in Southeast Chicago."

61. CT, August 20, 1976. For the chamber's vision of the river as an industrial corridor and the lakefront as a recreational playground, see ibid., May 28, 1967.

62. Ibid., May 8, 1978, for the stamp-contest scandal. On MSD funding, see ibid., December 30, 1976; June 10, 1978. To complete its sandbagging of institutional corruption, the trustees also agreed to pay for the legal defense in the multi-million-dollar kickback case against Janicki, a second trustee, and the general superintendent. Janicki and the others were convicted in the kickback scheme.

63. Ibid., August 19, 1969.

64. James Fitzgerald, "Pollution Aplenty," CT, May 16, 1971.

65. Bill Harth, as quoted in ibid., April 14, 1973. Also see ibid., September 2, 1973, for a canoeist's description of the Des Plaines River as an open sewer; and ibid., May 2, 1974, for policy makers' demotion of the river to a low priority in the budget. For early government-sponsored efforts to clean up the river, see ibid., May 1, 1975; March 10, 1977; June 11, 1978.

66. Casey Bukro, "Old Chicago River: It Just Keeps Oozing Along," CT, April 25, 1974; and ibid., December 26, 1974, for scientific confirmation that Chicago's waterways were out of compliance with state standards.

67. David Kee, as quoted in CT, March 26, 1981; and ibid., November 5, 1974.

68. Ibid., November 4, 1983.

69. Ibid., October 17, 1965.

70. Ibid., December 27, 1973; April 27, 1975, for the River Rats Society; and ibid., June 28, 1984, for the wildflowers.

71. *Chicago*, October 1979; CT, August 6, 1979; and Lizabeth Cohen, *A Consumers' Republic: The Politics of Mass Consumption in Postwar America* (New York: Knopf, 2003).

72. Melvin G. Holli, "Jane M. Byrne: To Think the Unthinkable and Do the Undoable," in *The Mayors: The Chicago Political Tradition*, ed. Paul M. Green and Melvin G. Holli (Carbondale: Southern Illinois University Press, 1987), 172, 182, 172–182.

73. Mayor Jane Byrne, as quoted in CT, August 24, 1980.

74. CT, September 20, 1983; July 16, 1984, for the North Avenue Turning Basin Park. See ibid., May 20, 1979, for the MSD's no-plan plan of land use. For a partial, pre-FOTR record of what a contemporary freelance reporter called the city's "talk-but-don't-act approach" to planning the riverfront (Laura Green, as quoted in ibid., September 2, 1984), see ibid., July 28, 1963; June 28, 1965; August 20, 1967; February 15, March 7, May 30, 1968; July 21, October 6, 1977; April 2, 1978.

75. Ibid., July 18–19, 1977; September 2, 1984. Another advocate of the river as street front was the newspaper's architecture critic, Paul Gapp. See ibid., April 2, 1978.

76. Bernard Goldberg, as quoted in ibid., September 2, 1984. On the ordinance, see ibid., October 15, 1982.

77. On the membership of the FOTR, see ibid., March 19, 1983.

78. For the dedication of TARP, see ibid., May 24, 1985.

79. On the city's changing spatial form, see Binford, "Multicentered Chicago," 548–553; Bruegmann, "Schaumburg, Oak Brook, Rosemont," 159–178; and Ann Durkin Keating, "Chicagoland: More Than the Sum of Its Parts," *Journal of Urban History* 30 (January 2004): 213–230.

CHAPTER 6

1. Horace P. Ramey, "Floods in the Chicago Area," (unpublished manuscript. February 25, 1958, Harold Washington Library Center, Chicago), 14, for the quoted phrase.

2. For the flood, see *Chicago Tribune* [hereafter cited as CT], October 1–7, November 6, 9, 1986; Metropolitan Water Reclamation District [hereafter cited as MWRD], "Reversals to Lake Michigan," available at http://www.mwrd.org/irj/servlet/prt/portal/prtroot/pcd!3aportal_content!2fMWRD!2fMWRDInternet!2fRoles!2fServices_Facilities!2fCombinedSewerOverflows!2fOverview!2fOverview (accessed March 24, 2016); and Dennis W. Dreher, "Stormwater Management Ordinance Approaches in Northeastern Illinois," in *Seminar Publication: National Conference on Urban Runoff Management*, ed. U.S. Environmental Protection Agency (Cincinnati, OH: U.S. Environmental Protection Agency, April 1995), 77–81.

3. For policy formation, see CT, October 4, 7–8, November 9, 1986.

4. Ibid., November 9, 1986; and Rachel Carson, *Silent Spring* (Boston: Houghton Mifflin, 1962).

5. David Hunter, as quoted in CT, November 9, 1986. Also see the earlier, eco-critiques of Scott Bernstein of the Center for Neighborhood Technology in ibid., May 24, 1985; and Richard Semonin, chief of the Illinois State Water Survey, in ibid., November 9, 1986. For the "bibles" of eco-friendly design, see Ian L. McHarg and American Museum of Natural History, *Design with Nature* (Garden City, NY: Published for the American Museum of Natural History by the Natural History Press, 1969), in landscape architecture; Reyner Banham, *Los Angeles: The Architecture of Four Ecologies* (1971; repr., Berkeley: University of California Press, 2001), in urban planning; and Richard Joseph Neutra, *Building with Nature* (New York: Universe Books, 1971), in architecture.

6. Prairie Crossing, "History," available at http://prairiecrossing.com/ (accessed May 29, 2016).

7. Ibid. The development's early history and storm-water management practices can be traced in CT, September 21, 1992; January 10, 24, 1998; August 9, 2000; May 18, 2003; and the *Chicago Sun-Times* [hereafter cited as CST], September 3, 1995; June 16, 1997; September 18, 2000. Also see CST, August 27, 1999, for a list of five additional "conservation developments" in the Chicago area.

8. Joel Rast, "Creating a Unified Business Elite: The Origins of the Chicago Central Area Committee," *Journal of Urban History* 37 (June 2011): 583–605; and D. Bradford Hunt and Jon B. DeVries, *Planning Chicago* (Chicago: Planner's Press, 2013). For additional insight on the CCAC's vision for the river corridor, see CT, July 28, 1963; June 28, 1965; July 18, 1977; July 1, 1984.

9. For a list of these projects, see CST, August 17, 1990. The "Chinatown Square" project is located at Cermak and Archer Avenues. Altogether, the thirteen housing proj-

ects added about ten thousand residential units to the expanded, downtown area in addition to 16.3 million square feet (ft.)/1.5 million square meters (m) of office space and 6,250 hotel rooms. See ibid., March 13, April 22, August 17, 1990. The classic critique of the architecture of public housing in the wake of the "urban crisis" of the 1960s is Oscar Newman, *Defensible Space: Crime Prevention through Urban Design* (New York: Macmillan, 1972).

10. CST, June 8, 1997, for the quotation; ibid., April 26, 2011; and CT, June 1, 2010; May 12, 17, 2011. The basic chronology can be traced in Libby Hill, *The Chicago River: A Natural and Unnatural History* (Chicago: Lake Claremont Press, 2000).

11. CT, August 26, 1987.

12. French Wetmore, as quoted in ibid.; and ibid., August 26, 28, 1987, for the statistics.

13. Ibid., August 15, 1987, for the quotation; ibid., August 16, 1987, for weather conditions; and ibid., August 15–16, 1987, for stories on the airport.

14. Dorothy Halac, as quoted in ibid., August 16, 1987.

15. Ibid., August 20, 1987, for the quotation; ibid., August 25, 1987, for the Wood Dale story; and ibid., August 16–20, 1987, for other firsthand accounts.

16. James Thompson, as quoted in ibid., August 16, 1987.

17. Ibid., August 19, 1987.

18. Frank Dalton, as quoted in ibid., August 9, 1987.

19. Aurelia Pucinski, as quoted in ibid., August 16, 1987.

20. Robert Ginsburg, as quoted in ibid., August 16, 1987.

21. Richard M. Daley, as quoted in CST, March 13, 1990. For the funding of the river walk, see ibid., July 31, 1997. For the CCAC planner's characterization of city hall's river-protection policies, see CT, September 2, 1984. For the plan, see City of Chicago and Friends of the River, *Chicago River Urban Design Guidelines: Downtown Corridor* (n.p., [1990?]).

22. Joanne Alter, as quoted in CT, April 23, 1985. On Hey's wetland's project, also see CST, February 18, April 21, 1991. Also see Ruth Glass, *Clichés of Urban Doom and Other Essays* (Oxford, UK: Blackwell, 1989).

23. CT, April 25, December 26, 1974; November 4, 1983; CST, April 22, July 18, 1990; August 18, 1991; April 16, 1992; October 15, 2005. For the project on the Des Plaines to bring back beavers and eliminate carp, see CST, February 18, April 21, 1991. Also see notes 67 and 71.

24. Joyce Peralta, as quoted in CST, June 3, 1996.

25. Ibid., August 10, 16, September 17, 22, 1989; April 3, May 27, 1992. The MSD missed several deadlines to install equipment to monitor the types and sources of industrial effluents entering the sewer system. See CT, October 2, 1985.

26. CT, June 15, 19, 1984; January 23, 1985; CST, April 1, 1993. Also see CST, July 28, 1990, for a similar $32.5-million fine imposed on the steel company USX Corporation. The money financed the dredging of the rivers by the U.S. Army Corps to remove toxic sedimentation.

27. Jill Viehweg, as quoted in CST, June 24, 1994; ibid., April 16, 1997, on the crackdown on the five companies. In 2004, eight companies were assessed an even-more-costly $56 million in fines. See ibid., August 21, 2004. The EPA estimated that their toxic sediment was 20 ft./6 m deep in the Grand Calumet River.

28. Ibid., December 1, 1995, for the quotation; ibid., June 24, 1994, for the statistics.

29. On the floods of 1996, see ibid., July 18–21, 1996; June 1, 1997. On the amounts of storm water released into the lake, see MWRD, "Reversals to Lake Michigan."

30. Ken Kunkel, as quoted in CT, January 3, 1996. Also see ibid., July 19, 1996. Also see Chicago Wilderness, *Climate Considerations for Management of Natural Areas and Green Spaces in the City of Chicago*, 2012, available at https://adapt.nd.edu/resources/1107/download/Climate_Considerations_Chicago_FINAL.pdf (accessed May 26, 2016).

31. Dawn Psilos, as quoted in CT, June 18, 1996, for the U.S. Army Corps report; and ibid., May 21–22, 25–31, 1996, for flood-related stories.

32. Philip Bernstein, as quoted in CT, May 26, 1996.

33. Resource Coordination Policy Committee Chicago, "Our Community and Flooding: A Report on the Status of Floodwater Management in the Chicago Metropolitan Area, 1987," Chicago, April 1987; CT, July 19–21, 1996; and CST, July 18–21, 1996.

34. Earl Hunt, as quoted in CT, July 19, 1996; February 22, 1997; and CST, July 18, 1996.

35. CST, September 10, 1998.

36. Laurene Von Klan, Thomas Fuller, and Brook McDonald, as quoted respectively in CT, July 20, 1996.

37. Scott Bernstein, as quoted in ibid.

38. Ibid.; ibid., April 3, 1997, for an interview with ecologist Dennis Dreher; Dreher, "Stormwater Management"; and Resource Coordination Policy Committee Chicago, "Our Community and Flooding," for statistics on and maps of the flood-control reservoirs built in the metropolitan region.

39. CT, August 16, 1996. For reception of the plan, see ibid., December 26, 1996; January 14, 30, 1997.

40. Ibid., May 30, 1996. Also see Reid M. Helford, "Prairie Politics: Constructing Science, Nature, and Community in the Chicago Wilderness" (Ph.D. diss., Loyola University, August 2003), on the formation of Chicago's wilderness advocacy groups.

41. CT, February 21–25, 1997; CST, February 22, 1997; and MWRD, "Reversals to Lake Michigan."

42. For a survey of alternative approaches to flood and pollution control, see CT, July 20, 1996; CST, June 16, 1997; January 10, 2001; January 6, 2005.

43. David Dempsey, as quoted in CST, May 8, 1995. On the rebuilding of the locks, see ibid., December 10, 1997; April 26, 1999. On the settlement of the case, see ibid., October 10, 1996.

44. Frank Dalton, as quoted in ibid., May 8, 1995. For an appreciative obituary, see ibid., May 10, 2008.

45. See ibid., September 17, 1889; August 13, 1992.

46. Joel Greenberg, *A Natural History of the Chicago Region* (Chicago: University of Chicago Press, 2002), 184, for the quotation (my emphasis). Also see Theodore Steinberg, *Nature Incorporated: Industrialization and the Waters of New England* (Cambridge, UK: Cambridge University Press, 1991); and John T. Cumbler, *Reasonable Use: The People, the Environment, and the State—New England, 1790–1930* (New York: Oxford University Press, 2001).

47. For the invasive species count, see CST, July 7, 1996. For the upgrading of the fish population to include rainbow trout and largemouth bass, see ibid., June 8, October 18, 1997; March 4, 22–23, 1998; April 15, May 9, 1999; May 16, 2000; May 11, July 7, October 9, 2005; November 19, 2006. By 1999, the number of species of fish in the Chicago River had risen to sixty. See ibid., August 2, 1999. Also see ibid., November 19, 2006.

48. Greenberg, *A Natural History*, 140–176; CST, February 25, 1990, for an early warning about the zebra mussel; ibid., July 7, 1996, on the census of non-native species in Lake Michigan. Also see ibid., July 13, 1997, September 4, 2002.

49. On the electric fence and the goby invasion, see CST, April 22, 1998; October 23, 1999; October 27, 2000; December 13, 15, 2002. The fence was built near Romeoville. By 2002, the Asian carp had moved up the Mississippi River and into the Sanitary and Ship Canal. See ibid., September 4, December 13, 2002.

50. Ibid., June 8, 1997.

51. On bird sightings of eighteen different species, see ibid., March 4, 1998; January 10, March 30, June 24, 2001; March 26, 2004; December 15, 2005. The twenty-six-acre property was owned by a steel company. It was turned over to the local forest-preserve authority. For sightings of beavers and river otters, see ibid., August 13, 2003. For sightings of minks, muskrats, and raccoons, see ibid., July 7, 2005. These three animals are part of an ecosystem based on the return of mussels to the river. For beavers on the Main Branch, see ibid., October 15, 2005.

52. Mary Ellen Podmolik, as quoted in ibid., May 10, 1998. On real-estate development, also see ibid., July 30, 1999; March 12, 2001. On Cityfront Center, see CT, October 7, 1990; CST, March 24, 1998.

53. CST, July 27, 2001; May 18, 2003.

54. Ibid., April 15, 1998; February 14, 2003.

55. Joseph Zehnder and Laurene von Klan, as quoted respectively in CT, April 17, 1998. For the enthusiastic reception of the ordinance by the Metropolitan Planning Council, see CST, March 12, 2001. The downtown business group recognized the city with its Daniel Burnham Award.

56. Lee Bay, as quoted in CST, August 30, 2002, on the Lake Calumet marina. He was a deputy chief of staff in the mayor's office. Also see the "We think the river is clean enough," statement of the director of MWRD's monitoring and research department, in ibid., June 1, 2010. For the position of the FOTR, see ibid. March 12, 2001; November 17, 2003. Seizing the opportunity presented by the Asian-carp "crisis" to cast himself as the "environmental mayor," Daley hosted a summit of scientists and government agencies on the invasive species problem. See ibid., December 15, 2002. The meeting did not change his position in favor of a "working-river" solution.

57. Ibid., July 31, 1997; March 12, 2001.

58. Ibid., May 10, 1998; and Timothy J. Gilfoyle and Chicago History Museum, *Millennium Park: Creating a Chicago Landmark* (Chicago: University of Chicago Press, 2006).

59. Nicole Ott, as quoted in CST, August 4, 2000.

60. James O'Hara, as quoted in ibid., July 25, August 30, 2002; ibid., August 30, 2002.

61. Hugh McMillan, as quoted in ibid., August 4, 2002.

62. Jack Farnan, as quoted in ibid., August 30, 2002. For the progress of TARP, see ibid., December 10, 2003. At that point, TARP had 101 miles/162.5 kilometers (km) of tunnels, at a cost of $3.4 billion, plus sixteen workers who had died on the job. The McCook and Thornton reservoirs were expected to add 18.5 billion gallons (gal.)/70 billion liters (L) of storage capacity to the tunnels' 1.6 billion gal./6 billion L current capacity and the O'Hare reservoir (350 million gal./1.3 billion L capacity), which came online in 1999.

63. For the U.S. EPA orders, see ibid., July 7, 1996; May 23, 2000. For the battle over TARP funding, see ibid., February 3, June 24, July 25, November 11, 1998; May 19, October 21, 1999; June 14, 2001; December 10, 2003.

64. Ibid., August 2, 2001.

65. Richard M. Daley, as quoted by Andrew Herrmann, in ibid., March 14, 2007.

66. On the Eurasian ruffe, see CST, September 4, 2002; June 22, 2003. Also see INHS Report, "The Round Goby: An Example of the "Perfect" Invader?" available at

http://www.inhs.uiuc.edu/inhsreports/nov-dec98/goby.html (accessed December 20, 2012). On the flooding, see CST, August 3, September 1, October 9, 2001; August 23, 30, 2002.

67. CST, September 4, December 13, 15, 2002; June 22, August 13, 2003; October 9, 2005. A multitude of government agencies and private organizations have been studying the problem. For a useful list, see "Water Agencies and Organizations in the Great Lakes Region," available at http://www.great-lakes.net/links/envt/orgs_water.html (accessed December 20, 2012). On the Asian carp, see the following: National Invasive Species Information Center, available at http://www.invasivespeciesinfo.gov/index.shtml#.UN7nc4njlk4; Asian Carp Coordinating Committee, available at http://www.asiancarp.us/; U.S. EPA, Lake Michigan, available at http://www.epa.gov/glnpo/michigan.html; U.S. Army Corps of Engineers, Chicago District, Aquatic Nuisance Species Portal, available at http://www.lrc.usace.army.mil/Missions/CivilWorksProjects/ANSPortal.aspx; and Great Lakes Information Network, available at http://www.great-lakes.net/lakes/michigan.html (all accessed December 20, 2012).

68. Joel Brammeir, as quoted in CST, December 27, 2005; ibid., December 13, 2002; May 16, June 22, August 13, 2003; October 14, 2004; October 9, 2005.

69. Phil Moy, as quoted in ibid., May 16, 2003, for the scientist's justification; ibid., December 27, 2005, for the phrase "hydrologic separation"; Gary Wisby, as quoted in ibid., December 27, 2005, for the AGL report.

70. Ibid., August 13, 2003; May 9, 11, July 7, October 9, 2005; June 6, November 19, 2006. The "crisis" would reach a breaking point in June 2010. See ibid., June 3–10, 2010.

71. John Farman, as quoted in ibid., May 12, 2004. Farman was the general superintendent of the MWRD. On city planning and real-estate developments, see ibid., April 23, May 21, 2004; April 27, 2006. On stalled city projects, see ibid., October 11, 2004; April 27, 2006. On the kayakers, see ibid., August 27, 2004.

72. MWRD, "Reversals to Lake Michigan."

73. Ed Schwartz, as quoted in CST, July 15, 2007. For the flooding, see MWRD, "Reversals to Lake Michigan."

74. Unnamed report, as quoted in CST, July 22, 2007. Also see a similar, previous story in ibid., March 23, 1998. The disinfecting process used a three-step process of chlorination, dechlorination, and ultraviolet light.

75. Margaret Frisbie, as quoted in ibid., September 4, 2007. Also see follow-up editorials and reports in CT, September 17, October 12, 2007; February 3, July 25, 2008.

76. Ibid., February 3, July 13, 25, 2008.

77. Chris Parsons, as quoted in ibid., October 15, 2008. On the green candidates, see ibid., May 10, 2008.

78. Terrence J. O'Brien, as quoted in ibid., July 21, 2008. He was responding to an editorial titled, "A Green City Needs Clean Rivers. . . ," in ibid., July 13, 2008. The newspaper's response was its endorsement of the three "green" candidates. See ibid., October 15, 2008.

79. Richard M. Daley, as quoted in ibid., June 3, 2010.

80. Mark Brown, as quoted in ibid. For the federal order, see U.S. EPA, Letter to the Illinois EPA, May 16, 2011), available at http://www.epa.gov/region5/chicagoriver/pdfs/caws-letter-201216.pdf (accessed December 20, 2012); and U.S. EPA, "Basis for the EPA's Decision on Illinois' New and Revised Water Quality Standards for the Chicago Area Waterway System and Lower Des Plaines River," May 10, 2012, available at http://www.epa.gov/region5/chicagoriver/pdfs/caws-basis-for-decision-20120510.pdf (accessed December 20, 2012).

81. CST, January 6, May 10, June 25, July 25–28, 2010. Although dismissed by the courts, the states have kept appealing their case against Chicago based on new scientific evidence. In June 2017, a four-year-old, 8-pound/3.6-kilogram Asian carp was caught just 9 miles/14.5 km from the lake after evading three electric fences. The experts cannot explain how the fish got past them. See Dale Bowman, "Clash with the Carp SW of Chicago: Details of the Battle at the Front," CST, July 23, 2017, available at http://chicago.suntimes.com/sports/clash-with-the-carp-sw-of-chicago-details-of-the-battle-at-the-front/ (accessed July 27, 2017); and John Flesher, "Asian Carp Found in Little Calumet River Had Evaded Three Electric Fences," CT, August 18, 2017, available at http://www.chicagotribune.com/news/local/breaking/ct-asian-carp-near-lake-michigan-20170818-story.html (accessed August 18, 2017).

82. Henry Henderson, as quoted in National Resources Defense Fund, "Railing on the River: EPA Demands Chicago River Cleanup," *National Resources Defense Fund Press Release*, May 13, 2011. The failure to purify the city's sewage per the promised agreement continues. See Michael Hawthorn, "Chicago River Still Teems with Bacteria Flushed from Sewers after Storms," CT, June 23, 2017.

83. On Emanuel's advocacy in support of saving the Great Lakes while he was a congressional representative, see his editorial, CT, July 18, 2004. On the change of policy at the MWRD, see ibid., June 3–13, 2010; May 31, 2011. For the denunciation of the MWRD's "junk science," see Ann Alexander, "Alexander's Blog," August 9, 2010, available at witchboard.nrdc.org/blogs/aalexander/Chicago_river_disinfection_hea.html (accessed December 22, 2012).

84. U.S. EPA, Letter to the Illinois EPA, November 3, 2011, available at http://www.epa.gov/region5/chicagoriver/pdfs/chicagoriver-letter-20111103.pdf (accessed December 20, 2012). Also see CT, May 12, 17, June 1, November 6, 2011; CST, June 8, 2011; and John T. Slania, "Troubled Waters?" *Time Out Chicago*, no. 396 (September 27–October 2, 2012): 12–16.

85. Rahm Emanuel, as quoted in CST, September 20, 2011; also see ibid., November 7, 2011; and CT, October 8, 2012. On Jeanne Gang Studio, including projects for the four beach houses, Northerly Island, and damming the interbasin connections, see "Inside Studio Gang," an exhibit at the Art Institute of Chicago, 2011–2012; and Whet Moser, "Jeanne Gang and Robert Cassidy: Two Generations of Urban Planners on the Chicago River," *Chicago Magazine*, December 2011, available at http://www.chicagomag.com/Chicago-Magazine/The-312/December-2011/Jeanne-Gang-and-Robert-Cassidy-Two-Generations-of-Urban-Planners-on-the-Chicago-River/ (accessed November 27, 2012). The Gang Studio website has information on its Northerly Island and exhibit projects, available, respectively, at http://www.studiogang.net/work/2007/northerlyisland and http://www.studiogang.net/work/2012/building (accessed January 5, 2013).

86. CST, April 26, 2011.

87. Elizabeth Kocs, "Finding Nature in the City: A Case Study of Ecological Restoration in an Urban Park" (Ph.D., City University of New York, 2013).

88. For an interim report, see U.S. Army Corps of Engineers, GLMRIS Natural Resources Team, "Non-native Species of Concern and Dispersal Risk for the Great Lakes and Mississippi River Interbasin Study," available at http://glmris.anl.gov/documents/docs/Non-Native_Species.pdf (accessed December 20, 2012). For the full report, see U.S. Army Corps of Engineers, *The GLMRIS Report: Great Lakes and Mississippi River Interbasin Study*, January 6, 2014, available at http://www.glmris.anl.gov/documents/docs/glmrisreport/GLMRIS_Report.pdf (accessed November 15, 2015).

89. U.S. Army Corps of Engineers, "The Electric Dispersal Barriers," Chicago, November 2011, available at http://www.lrc.usace.army.mil/Portals/36/docs/projects/ans/docs/ElectricBarrierBrochure.pdf (accessed December 20, 2012); also see note 88.

CONCLUSION

1. *Chicago Sun-Times* [hereafter cited as CST], July 24, 2011. This rainfall superseded the previous record by 0.2 inches/0.5 centimeters. Also see D. Bradford Hunt and Jon B. DeVries, *Planning Chicago* (Chicago: Planner's Press, 2013), for a more-comprehensive account of planning in Chicago.

2. CST, July 24, 2011; and Metropolitan Water Reclamation District [hereafter cited as MWRD], "Reversals to Lake Michigan," available at http://www.mwrd.org/irj/servlet/prt/portal/prtroot/pcd!3aportal_content!2fMWRD!2fMWRDInternet!2fRoles!2fServices_Facilities!2fCombinedSewerOverflows!2fOverview!2fOverview (accessed March 24, 2016).

3. Andreas F. Prein, Roy M. Rasmussen, Kyoko Ikeda, Changhai Liu, Martyn P. Clark, and Greg J. Holland, *The Future Intensification of Hourly Precipitation Extremes Nature Climate Change*, available at http://www.nature.com.flagship.luc.edu/nclimate/index.html?cookies=accepted (accessed December 8, 2016).

4. Ibid.

5. Center for Neighborhood Technology, "The Prevalence and Cost of Urban Flooding: A Case Study of Cook County, Ill.," Chicago, May 2013.

6. Ibid.

7. Ibid., on the flooding; and CST, December 10, 2003; May 10, 2008, for the completion of phase two of TARP and the plans for a final, third phase, respectively. Also see Friends of the River and Openlands, "Our Liquid Asset: The Economic Benefits of a Clean Chicago River" (Chicago: Friends of the River and Openlands, May 2013); Whet Moser, "Chicago: 150 Years of Flooding and Excrement," *Chicago Magazine*, April 18, 2013, available at http://www.chicagomag.com/Chicago-Magazine/The-312/April-2013/Chicago-150-Years-of-Flooding-and-Excrement/ (accessed February 6, 2014), for flood-damage statistics; and Daniel Cusick, "Can Chicago Handle the Coming Rains?" *Scientific American*, April 4, 2012, available at http://www.scientificamerican.com/article/can-chicago-handle-the-coming-rains/ (accessed May 12, 2014), for perspectives on the impacts of climate change.

8. Guy Stuart, *Integration or Resegregation: Metropolitan Chicago at the Turn of the New Century* (Cambridge, MA: Civil Rights Project at Harvard University, May 1, 2002), available at https://civilrightsproject.ucla.edu/research/metro-and-regional-inequalities/integration-or-resegregation-metropolitan-chicago-at-the-turn-of-the-new-century/stuart-integration-resegregation-chicago-2002.pdf (accessed September 1, 2002).

9. Center for Neighborhood Technology, "The Prevalence and Cost of Urban Flooding."

10. *Chicago Tribune* [hereafter cited as CT], April 15, 2016.

11. Ibid.

12. See Friends of the River and Openlands, "Our Liquid Asset," for a useful summary of progress on the reclamation of the river and current plans for its future. Taxpayer support for the expansion of the forest-preserve districts in the collar counties can be traced in CST, May 25, August 4, 1998; March 16, 1999; April 16, 2007; and CT, November 3, 2008. Also see the websites of the six forest-preserve districts for mission statements and restoration plans.

13. Jacqueline Peterson, "'Wild Chicago': The Formation and Destruction of a Multiracial Community on the Midwestern Frontier, 1816–1837," in *The Ethnic Frontier*, ed. Melvin Holli and Peter d'A Jones (Grand Rapids, MI: Eerdmans, 1977), 25–71; and Harold M. Mayer, "The Launching of Chicago: The Situation and the Site," *Chicago History* 9 (Summer 1980): 68–79.

14. *Crain's Chicago Business* [hereafter cited as CCB], May 30–31, 2015; Ben Joravsky, "A Development in Sheep's Clothing," *Chicago Reader*, December 6, 2012, 12–15; and John T. Siania, "Troubled Waters?" *Time Out Chicago*, no. 396 (September 27–October 3, 2012): 12–16.

15. Jim Walch, as quoted in CCB, April 18, 2016. Also see the optimistic retrospective by Robert Cassidy, "River, Front and Center," *Planning*, January 2013, 25–33. In 1979, he helped launch the formation of the Friends of the River with his article, Robert Cassidy, "Our Friendless Chicago River," *Chicago Magazine*, August 1979, available at http://www.chicagomag.com/Chicago-Magazine/August-1979/Our-Friendless-Chicago-River/ (accessed October 12, 2014); and a progress report on the Wolf Point development, Alby Gallun, "Wolf Point Developers Land $200 Mil Loan for 60-Story Tower," *CCB*, July 3, 2017, 13

16. CCB, November 12, 2015; October 18–20, 2016; Siania, "Troubled Waters?"; and Friends of the River and Openlands, "Our Liquid Asset."

17. CT, March 14, April 27, 2007; March 29, 2013; August 16, September 21, October 21, 2016; Friends of the River and Openlands, "Our Liquid Asset"; and Hunt and DeVries, *Planning Chicago*, chap. 19. Recent zoning changes have opened up the North Branch near Goose Island for intense residential development. See Alby Gallun, "Massive New Development Plans on Tap for Lakeshore East," CCB, July 3, 2017, 12–13; and Mark Brown, "Chicago's Skyline Heads Up North," CST, August 5, 2017. available at chicago.suntimes. 13.2 billion liter section of the McCook Reservoir will be completed. See //chicago.suntimes.com/sports/clash-with ?the baic pcom.chicago-politics/brown-chicagos-skyline-heads-up-river/ (accessed August 8, 2017).

18. U.S. EPA, Letter to the Illinois EPA, November 3, 2011, available at http://www.epa.gov/region5/chicagoriver/pdfs/chicagoriver-letter-20111103.pdf (accessed December 20, 2012). Also see CT, May 12, 17, June 1, November 6, 2011; CST, June 8, 2011; and Siania, "Troubled Waters?"

19. City of Chicago, Department of Housing and Economic Development, *Chicago Sustainable Industries, Phase One: A Manufacturing Work Plan for the 21st Century* (Chicago: City of Chicago, 2012). Also see Hunt and DeVries, *Planning Chicago*, chap. 17; and Derek S. Hyra, "Conceptualizing the New Urban Renewal: Comparing the Past to the Present," *Urban Affairs Review* 48 (July 2012): 498–527.

20. City of Chicago, Department of Housing and Economic Development, *Chicago Sustainable Industries.*

21. Little Village Quality of Life, as quoted in Delta Institute, *Little Village Vacant Property and Brownfield Redevelopment Strategy* (Delta Institute, March 2016), available at http://deltainstitute.org/delta/wpcontent/uploads/Little-Village-VPB-Redevelopment Strategy.compressed.pdf (accessed April 20, 2016).

22. City of Chicago, Department of Housing and Economic Development, *Chicago Sustainable Industries*; and CT, November 30, 2014; March 28, September 21, 2016. On the protest movement, see Robert R. Gioielli, *Environmental Activism and the Urban Crisis: Baltimore, St. Louis, Chicago* (Philadelphia: Temple University Press, 2014), chap 4. For a completely different, community-oriented plan for this stretch of the Chicago

River, see Jeanne Gang, Reuben P. Keller, Kari Lydersen, Studio Gang Architects (Firm), Natural Resources Defense Council, and Harvard University Graduate School of Design, *Reverse Effect: Renewing Chicago's Waterways* (Chicago: Studio Gang Architects, 2011); and Geoffrey Johnson, "Jeanne Gang: Reversing the River Could Transform Chicago," *Chicago Magazine*, December 6, 2011, available at http://www.chicagomag.com/Chicago-Magazine/December-2011/Jeanne-Gang-Reversing-the-River-Could-Transform-Chicago/ (accessed May 6, 2016).

23. CT, November 30, 2014; March 28, 2016. Also see Richard C. Lindberg, *To Serve and Collect—Chicago Politics and Police Corruption from the Lager Beer Riot to the Summerdale Scandal* (New York: Praeger, 1991).

24. CT, December 21, 2016, for the quoted phrase; ibid., for current census reports. Also see CCB, February 15, 2011; March 23, 2016; CT, March 24, 2016; Dahleen Glanton, "Are Quality-of-Life Issues Spurring Chicagoans to Move Out?" CT, March 28, 2016; and Derek S. Hyra, "Racial Uplift? Intra-racial Class Conflict and the Economic Revitalization of Harlem and Bronzeville," *City and Community* 5 (March 2006): 71–92. In 2016, births counterbalanced deaths and people moving out to result in a net loss in Illinois of almost 38,000 residents.

25. William J. Grimshaw, *Bitter Fruit—Black Politics and the Chicago Machine 1931–1991* (Chicago: University of Chicago Press, 1992); and Gary Rivlin, *Fire on the Prairie: Chicago's Harold Washington and the Politics of Race* (New York: Henry Holt, 1992).

26. Thomas J. Gradel and Dick Simpson, *Corrupt Illinois: Patronage, Cronyism, and Criminality* (Urbana: University of Illinois Press, 2015), chap. 6, for an overview of Cook County government and insight on its corruption by machine politics.

27. Steve Neal, "Stroger's Obstinance Keeps Forest Preserves in Squalid State," CST, August 29, 2003, for the first quoted phrase; John Sheerin [president of the Friends of the Forest Preserves], "Forest Preserves Need Action Now," ibid., July 22, 2003, for the second quoted phrase; ibid., January 16, 2003, for the third quoted phrase.

28. CST, January 19, 2008; and Wikipedia, "John Stroger," 2016, available at https://en.wikipedia.org/wiki/John_Stoger (accessed April 22, 2016). Also see Grimshaw, *Bitter Fruit*.

29. Report of the FoP and the FoFP, as quoted in CST, October 10, 2002.

30. Ibid., January 19, 2008, for the label; and ibid., December 7, 2003, for the data on the FPDCC.

31. Michael Quigley, "Lost Mission," as quoted in CT, January 7, 2002.

32. Ibid., August 12, 2002, for the editorial indictment of "shabby neglect"; CST, October 10, 2002; July 22, December 7, 2003, for the subsequent three quoted phrases, respectively.

33. CST, May 25, August 4, 1998; March 16, May 28, June 23, 1999; CT, February 20, 25, 2001; January 7, 2002. For Stroger's defense, see his letter to CST, October 15, 2004. Also see Reid M. Helford, "Prairie Politics: Constructing Science, Nature, and Community in the Chicago Wilderness" (Ph.D. diss., Loyola University, August 2003), on the frustrated efforts of environmentalists to restore patches of the prairie.

34. Charles Coleman, as quoted in CST, August 2, 2002; ibid., October 10, 2002.

35. Ibid., August 2, October 10, 2002; January 16, 21, 23, July 6, 2003.

36. Chuck Lueder, as quoted in ibid., July 6, 2003; and ibid., for the first quotation.

37. Ibid., January 16, December 7, 2003.

38. Tony Peraica, as quoted in ibid., September 3, 2003; ibid., July 9, 2003; November 16, 2007.

39. Ibid., August 28–29, September 8, October 10, November 19, December 7, 10–11, 2003.

40. CT, July 2, 24, 26, 2006; and Wikipedia, "Todd Stroger," 2016, available at https://en.wikipedia.org/wiki/Todd_Stroger (accessed April 22, 2016).

41. CST, February 14, 21, May 3, 2007; April 2, 2008; ibid., November 16, 2007; January 26, March 17, 24, April 2, June 30, 2008, on the political struggles over the county government budget; and ibid., March 2, 2006; March 24, 2008; April 18–24, June 15, 2009; February 19, April 17, 2010, for the Dunnings scandal and rehiring.

42. Ibid., April 16, 2007; November 3, 2008.

43. Ibid., January 7, 10–13, 22, 25, 31, February 3, November 3, 2010.

44. CT, November 30, 2014; Forest Preserve District of Cook County, *Natural and Cultural Resources Master Plan—Managing Our Ecosystems and Heritage for the Next Century*, March 9, 2015, available at http://fpdcc.com/downloads/plans/FPCC-Natural-Cultural-Resources-Master-Plan_3-9-15_WEB.pdf (accessed April 29, 2016); Forest Preserve District of Cook County, *5 Year Implementation Strategy for the Next Century Conservation Plan*, October 28, 2015, available at http://fpdcc.com/downloads/FPCC-NCCP-5-year-implementation-strategy-10-28-15.pdf (accessed April 29, 2016); and Forest Preserve District of Cook County, *Gateway Master Plan*, November 2015, available at http://fpdcc.com/downloads/plans/FPCC-Gateway-Master-Plan-FINAL-1-16.pdf (accessed April 29, 2016).

45. Yegiao Wang and Debra K. Moskovits, "Tracking Fragmentation of Natural Communities and Changes in Land Cover: Applications of Landsat Data for Conservation in Urban Landscape (Chicago Wilderness)," *Conservation Biology* 15 (August 2001): 835–843. The authors conclude, "Analyses with geographic information system models reveal rapid acceleration of urban and suburban sprawl over the past 12 years" (835).

46. Chicago Metropolitan Agency for Planning, *2040—Comprehensive Regional Plan* (Chicago Metropolitan Agency for Planning, October 2010), available at http://www.cmap.illinois.gov/documents/10180/17842/long_plan_FINAL_100610_web.pdf/1e1ff482-7013-4f5f-90d5-90d395087a53 (accessed May 5, 2016). Also see CT, August 28, 2011.

47. Chicago Metropolitan Agency for Planning, *Water 2050—Northeastern Illinois Regional Water Supply/Demand Plan* (Chicago Metropolitan Agency for Planning, March 2010), 55, available at http://www.cmap.illinois.gov/documents/10180/14452/NE+IL+Regional+Water+Supply+Demand+Plan.df/26911cec-866e-4253-8d99-ef39c5653757 (accessed May 5, 2016). The plan lays out five different climate scenarios of temperature and rainfall with a 5.0-inch/12.7-centimeter (cm) range of variation above and below the historic mean of 34.5 in./87.6 cm each year. Also see Chicago Wilderness, *Chicago Wilderness Climate Action Plan for Nature*, [2011?], available at http://www.openlands.org/filebin/images/plans_reports/Chicago_Wilderness_Climate_Action_Plan_for Nature.pdf (accessed May 26, 2016).

48. U.S. Army Corps of Engineers, *The GLMRIS Report: Great Lakes and Mississippi River Interbasin Study*, January 6, 2014, available at http://www.glmris.anl.gov/documents/docs/glmrisreport/GLMRIS_Report.pdf (accessed November 15, 2015).

49. Ibid.; and Judith A. Martin and Sam Bass Warner Jr., "Local Initiative and Metropolitan Repetition: Chicago, 1972–1990," in *The American Planning Tradition: Culture and Policy*, ed. Robert Fishman (Washington, DC: Woodrow Wilson Centre Press and Johns Hopkins University Press, 2000), 263–296, for a critical analysis of TARP's original calculations of storm-water runoff. They were limited to a half-block area and 0.5 in./1.3 cm of water in the first hour of rainfall.

50. Friends of the River and Openlands, "Our Liquid Asset." In 2017, a 3.5-billion-gallon/13.2-billion-liter section of the McCook Reservoir will be completed. See Neil Steinberg, "Soon You Can Look Here for the Water That Used to Be in Your Basement," CST, July 9, 2017, available at http://chicago.suntimes.com/columnists/soon-you-can-look-here-for-the-water-that-used-to-be-in-your-basement/ (accessed July 23, 2017).

51. Steinberg, "Soon You Can Look Here"; and CT, June 4, 8, August 9, 2010, on the disinfection controversy. Also see ibid., March 26, 2016, for a confirmation that O'Brien's obfuscation of the issue was "junk science," as the U.S. EPA charged.

52. Friends of the River and Openlands, "Our Liquid Asset"; Chicago Metropolitan Agency for Planning, *Water 2050*; and David E. Nye, *American Technological Sublime* (Cambridge: Massachusetts Institute of Technology Press, 1994).

53. Martin Felsen and UrbanLab, *Growing Water*, 2007, available at http://www.urbanlab.com/h2o/ (accessed May 10, 2016).

54. Ibid. Also see Donald L. Hey, Jill A. Kostel, and Inc. Wetlands Research, *Des Plaines River Wetlands Demonstration Project Water Level Fluctuation Study Final Report* (Wadsworth, IL: U.S. Army Corps of Engineers, March 2012).

55. Henry Nash Smith, *Virgin Land: The American West as Symbol and Myth* (Cambridge, MA: Harvard University Press, 1950); Barbara Novak, *Nature and Culture: American Landscape and Painting, 1825–1875* (New York: Oxford University Press, 1980); Leo Marx, *The Machine in the Garden: Technology and the Pastoral Ideal in America* (New York: Oxford University Press, 1964); and Peter J. Schmitt, *Back to Nature: The Arcadian Myth in Urban America, 1900–1930* (New York: Oxford University Press, 1969).

56. William H. Wilson, *The City Beautiful Movement* (Baltimore, MD: Johns Hopkins University Press, 1989); Robert Bruegmann, *Sprawl: A Compact History* (Chicago: University of Chicago Press, 2005); and Christopher C. Sellers, *Crabgrass Crucible: Suburban Nature and the Rise of Environmentalism in Twentieth-Century America* (Chapel Hill: University of North Carolina Press, 2012).

57. Chicago Wilderness, *The State of Our Chicago Wilderness: A Report Card on the Ecological Health of the Region*, 2006, available at https://www.csu.edu/cerc/documents/StateOfOurChicagoWilderness.pdf (accessed May 26, 2016).

58. Chicago Metropolitan Agency for Planning, *2040*, 91 (fig. 17), 92 (the one healthy riverine ecosystem is a stretch of the Upper Fox River in the northwest section of McHenry County); CT, April 17–20, 2013; CBS, "Stormy Weather: Wettest June on Record for Illinois: Second Wettest Month Ever," June 30, 2015, available at http://chicago.cbslocal.com/2015/06/30/stormy-weather-wettest-june-on-record-for-illinois-second-wettest-month-ever/ (accessed April 14, 2016); and U.S. Department of Interior, Geological Survey, Elizabeth A. Murphy, and Jennifer B. Sharpe, "Flood-Inundation Maps for the DuPage River from Plainfield to Shorewood, Illinois, 2013," Reston, Virginia, 2013.

59. Nelson Algren, *Chicago, City on the Make* (Garden City, NY: Doubleday, 1951).

60. Sam Bass Warner, *The Private City: Philadelphia in Three Periods of Its Growth* (Philadelphia: University of Pennsylvania Press, 1968). Also see Robin Einhorn, *Property Rules: Political Economy in Chicago, 1833–1872* (Chicago: University of Chicago Press, 1991).

61. Robert Cassidy, "Our Friendless Chicago River," *Chicago Magazine*, August 1979, available at http://www.chicagomag.com/Chicago-Magazine/August-1979/Our-Friendless-Chicago-River/ (accessed November 27, 2012).

62. John B. Jentz and Richard Schneirov, *Chicago in the Age of Capital: Class, Politics, and Democracy during the Civil War and Reconstruction* (Urbana: University of Illinois Press, 2013); and Gradel and Simpson, *Corrupt Illinois*, chaps. 1, 9.

63. Martin V. Melosi, *The Sanitary City—Urban Infrastructure in America from Colonial Times to the Present* (Baltimore, MD: Johns Hopkins University Press, 2000).

64. Paddy Woodworth, *Our Once and Future Planet: Restoring the World in the Climate Change Century* (Chicago: University of Chicago Press, 2013), chap 5; Helford, "Prairie Politics"; Gioielli, *Environmental Activism and the Urban Crisis*, chap. 4; and Gang et al., *Reverse Effect*.

Index

Harold L. Platt is Professor of History Emeritus at Loyola University Chicago. He is the author or editor of several books, including *The Electric City: Energy and the Growth of the Chicago Area, 1880–1930*; *Shock Cities: The Environmental Transformation and Reform of Manchester and Chicago*; and *Building the Urban Environment: Visions of the Organic City in the United States, Europe, and Latin America* (Temple). He has twice won the book-of-the-year award from the American Public Works Association.

Printed in the USA
CPSIA information can be obtained
at www.ICGtesting.com
LVHW052237250823
756269LV00011B/90